Overexploitation and Contamination of Shared Groundwater Resources

NATO Science for Peace and Security Series

This Series presents the results of scientific meetings supported under the NATO Programme: Science for Peace and Security (SPS).

The NATO SPS Programme supports meetings in the following Key Priority areas: (1) Defence Against Terrorism; (2) Countering other Threats to Security and (3) NATO, Partner and Mediterranean Dialogue Country Priorities. The types of meeting supported are generally "Advanced Study Institutes" and "Advanced Research Workshops". The NATO SPS Series collects together the results of these meetings. The meetings are co-organized by scientists from NATO countries and scientists from NATO's "Partner" or "Mediterranean Dialogue" countries. The observations and recommendations made at the meetings, as well as the contents of the volumes in the Series, reflect those of participants and contributors only; they should not necessarily be regarded as reflecting NATO views or policy.

Advanced Study Institutes (ASI) are high-level tutorial courses intended to convey the latest developments in a subject to an advanced-level audience

Advanced Research Workshops (ARW) are expert meetings where an intense but informal exchange of views at the frontiers of a subject aims at identifying directions for future action

Following a transformation of the programme in 2006 the Series has been re-named and re-organised. Recent volumes on topics not related to security, which result from meetings supported under the programme earlier, may be found in the NATO Science Series.

The Series is published by IOS Press, Amsterdam, and Springer, Dordrecht, in conjunction with the NATO Public Diplomacy Division.

Sub-Series

A.	Chemistry and Biology	Springer
B.	Physics and Biophysics	Springer
C.	Environmental Security	Springer
D.	Information and Communication Security	IOS Press
E.	Human and Societal Dynamics	IOS Press

http://www.nato.int/science
http://www.springer.com
http://www.iospress.nl

Series C: Environmental Security

Overexploitation and Contamination of Shared Groundwater Resources

Management, (Bio)Technological, and Political Approaches to Avoid Conflicts

edited by

Christophe J.G. Darnault
University of Illinois at Chicago,
Chicago, IL, U.S.A.

Springer

Published in cooperation with NATO Public Diplomacy Division

Proceedings of the NATO Advanced Study Institute on
Overexploitation and Contamination of Shared Groundwater Resources:
Management, (Bio)Technological, and Political Approaches to Avoid Conflicts
Varna, Bulgaria
1–12 October 2006

A C.I.P. Catalogue record for this book is available from the Library of Congress.

ISBN 978-1-4020-6984-0 (PB)
ISBN 978-1-4020-6983-3 (HB)
ISBN 978-1-4020-6985-7 (e-book)

Published by Springer,
P.O. Box 17, 3300 AA Dordrecht, The Netherlands.

www.springer.com

Printed on acid-free paper

CONTENTS

PREFACE

Environmental security is based on complex interactions between society and the environment. Forty percent of the world's population depends upon increasingly scarce and shared water resources. This situation is critical at both international and national levels not only for socio-economic development, but also for regional stability. From this perspective, water resources scarcity, quality and distribution are the factors that would most likely lead to political and socio-economic conflicts, setting human security at risks. The intensive use of groundwater for irrigation and water supply is adding pressures to scarce water resources and the environment through overexploitation and contamination which result in situations that may lead to conflicts.

In a world where water crises are approaching, the use of sustainable and integrated management principles is most valuable in order to avoid conflicts. Successful management of shared groundwater resources should be based on multi-disciplinary approaches and effective and equitable water sharing for all water users through cooperation and within a compassionate ecological framework along with the need for sustainable resources development and the quest for environmental and human security.

In this context, the NATO ASI on "Overexploitation and Contamination of Shared Groundwater Resources: Management, (Bio)Technological, and Political Approaches to Avoid Conflicts" was organized in Varna, Bulgaria from October 1 to 12, 2006. A total of 74 participants including 13 invited lecturers and 61 selected participants from 19 different countries and diverse backgrounds have attended the NATO ASI and shared their knowledge and experiences. The program of the institute included key lectures from invited speakers, oral and poster presentations from participants and two roundtable discussions. The participants had the chance to visit the city of Varna and a field trip was also organized to explore the key hydrogeological features of the Dobrudja area, including the Durankulak historical archaeological site and the Durankulak lake.

This book is dedicated to the memory of Professor Georgi Gergov, Head of the River Sediment Transport and Fluvial Processes Division at the National Institute of Meteorology and Hydrology from the Bulgarian Academy of Sciences, Sofia, Bulgaria, who was the Partner-Country Director of the NATO ASI. It is with great sadness that we heard of his untimely death after the NATO ASI.

The book titled "Overexploitation and Contamination of Shared Groundwater Resources: Management, (Bio)Technological, and Political Approaches to Avoid Conflicts" from the NATO Science for Peace and Security Series C: Environmental Security is composed of 21 chapters reflecting the lectures from the ASI and is divided into five parts. The first part aims to give an overview of the depletion and contamination of shared groundwater resources across the world. The techniques to assess the depletion and contamination of shared groundwater resources are presented in the second part. The third part covers the tools to predict/forecast the future depletion and contamination of shared groundwater resources. Sustainable measures to palliate the overexploitation of groundwater and decontamination of aquifers compose the fourth part. The fifth and last part encompasses the approaches to the management of shared groundwater resources for conflict prevention and resolution.

First and foremost, I wish to thank the NATO Science Committee for their support in realizing an interesting idea into a deeply engaging project and for awarding me a grant that funded this NATO ASI. I express my sincere acknowledgements to Dr. Deniz Beten, Programme Director of Environmental Security of the NATO Public Diplomacy Division, for her unwavering support. I am indebted to Lynne Campbell-Nolan, Administrative Assistant of the NATO Public Diplomacy Division, for her valuable assistance throughout the project. I extend my thanks to Annelies Kersbergen, Springer Publishing Editor, for her continuous help. I am thankful to Professor Philippe Baveye from University of Abertay Dundee, Scotland (formerly Cornell University, U.S.A.) for his guidance in the concept development and realization of this NATO ASI. Finally, I am tremendously grateful to my wife, Dr. Sezim Sezer Darnault for her unfailing encouragement and support.

Professor Christophe J. G. Darnault, Ph.D.
NATO-Country Director of NATO ASI
University of Illinois at Chicago
Chicago, Illinois, U.S.A.
October 7, 2007.

CONTRIBUTORS

Prof. Rachid Ababou
Institut de Mécanique des Fluides de Toulouse, Allée du Professeur Camille Soula, 31400 Toulouse, France

Dr. Ahmad Al-Bitar
Institut de Mécanique des Fluides de Toulouse, Allée du Professeur Camille Soula, 31400 Toulouse, France

Dr. Alice Aureli
UNESCO, Division of Water Sciences, International Hydrological Programme (IHP), Natural Sciences Sector, 1, rue Miollis, 75732 Paris Cedex 15, France

Baptiste Dafflon
University of Lausanne, Institute of Geophysics, CH-1015 Lausanne, Switzerland

Prof. Christophe J. G. Darnault
University of Illinois at Chicago, Department of Civil and Materials Engineering, 842 W. Taylor St., Chicago, IL 60607, U.S.A.

África de la Hera
Geological Survey of Spain, Ríos Rosas 23. 28003 Madrid, Spain

Dr. Jean Fried
UNESCO - International Hydrological Programme (IHP), and University of California at Irvine, Urban Water Research Center, Engineering Tower 844C, Irvine, CA 92697, U.S.A.

Prof. Jacques Ganoulis
UNESCO Chair and Network INWEB, International Network of Water/Environment Centres for the Balkans, Aristotle University of Thessaloniki, Department of Civil Engineering, 54124 Thessaloniki, Greece

Itzel G. Godinez
University of Illinois at Chicago, Department of Civil and Materials Engineering, 842 W. Taylor St., Chicago, IL 60607, U.S.A.

Prof. Klaus Holliger
University of Lausanne, Institute of Geophysics, CH-1015 Lausanne, Switzerland

Dr. W. Todd Jarvis
Oregon State University, Institute for Water and Watersheds, 210 Strand Hall, Corvallis, Oregon, 97331-5506, U.S.A.

Dr. Ewa Kmiecik
University of Science and Technology, Department of Hydrogeology and Water Protection, 30-059 Krakow, Poland

Prof. Manuel R. Llamas Madurga
Royal Academy of Sciences, Valverde 22. 28034 Madrid, Spain

Dr. Marta Machkova
Bulgarian Academy of Sciences, National Institute of Meteorology and Hydrology, 66 Tsarigradsko Shosse Blvd, 1784 Sofia, Bulgaria

Jose Luis Martin-Bordes
UNESCO - International Hydrological Programme (IHP), 1, rue Miollis, 75732 Paris Cedex 15, France

Prof. Pedro Martínez-Santos
Universidad Complutense de Madrid, Jose Antonio Novais s/n. 28040 Madrid, Spain

Dr. Hendrik Paasche
University of Potsdam, Institute for Geosciences, 14415 Potsdam, Germany

Dr. Shammy Puri
UNESCO - International Hydrological Programme (IHP), and UNESCO - International Shared Aquifer Resource Management (ISARM) Initiative, 1, rue Miollis, 75732 Paris Cedex 15, France

Prof. Krishna R. Reddy
University of Illinois at Chicago, Department of Civil and Materials Engineering, 842 West Taylor Street, Chicago, Illinois 60607, U.S.A.

Dr. Sebastian Stefaniak
Polish Academy of Sciences, Institute of Environmental Engineering, 41-819 Zabrze, Poland

Raya M. Stephan
UNESCO - International Hydrological Programme (IHP), and UNESCO - International Shared Aquifer Resource Management (ISARM) Initiative, 1, rue Miollis, 75732 Paris Cedex 15, France

Prof. Jadwiga Szczepańska
University of Science and Technology, Department of Hydrogeology and Water Protection, 30-059 Krakow, Poland

Dr. Jens Tronicke
University of Potsdam, Institute for Geosciences, 14415 Potsdam, Germany

Dr. Irena Twardowską
Polish Academy of Sciences, Institute of Environmental Engineering, 41-819 Zabrze, Poland

Prof. Ülkü Yetiş
Middle East Technical University, Department of Environmental Engineering, 06531 Ankara, Turkey

Part I
Depletion and Contamination of Shared Groundwater Resources: An Overview

Chapter 1
Shared Groundwater Resources: Global Significance for Social and Environmental Sustainability

Shammy Puri*, Alice Aureli, Raya M. Stephan

Abstract It is now widely accepted that groundwater, though not as visible as surface water, is ubiquitous in the global landmass, is contained in the pore spaces of rock formations (aquifers), and its science, hydrogeology, has rapidly developed over the last 35 years. This science has contributed to the well-being and development of the human population in all parts of the globe (Burke and Moench, 2000). While aquifer systems, due to their partial isolation from surface impacts, on the whole contain excellent quality water, there has been poor resource management because "economic externalities" have been ignored. In many countries, aquifers have been fully evaluated and extensively used for municipal and oher demands, as a common good. Aquifer resources represent a substantial hidden global capital, which if abused, leads to future economic costs and environmental-social conflicts. As a result, the beneficial use of groundwater should be more particularly subjected to socio-economic, institutional, legal, cultural, ethical and policy considerations than surface water. Its national sustainable management seems to be hampered by weak social and institutional capacity, and poor legal and policy frameworks. In a transboundary context, this can be even further amplified because of contrasting levels of knowledge, capacities and institutional frameworks on either side of many international boundaries. UNESCO and its partners have established a global programme to address the long-term sustainability for groundwater resources, through a series of actions, including ISARM and PCCP. In this chapter, which is an overview designed to provide the NATO ASI with the necessary background, these actions, and the progress achieved over the past decade will be described so that

Shammy Puri, Raya M. Stephan
UNESCO - International Hydrological Programme (IHP) and UNESCO - International Shared Aquifer Resource Management (ISARM) Initiative, 1, rue Miollis, 75732 Paris Cedex 15, France

Alice Aureli
UNESCO, Division of Water Sciences - International Hydrological Programme (IHP), Natural Sciences Sector 1, rue Miollis, 75732 Paris Cedex 15, France

* To whom correspondence should be addressed. E-mail: shammypuri@aol.com

C.J.G. Darnault (ed.), *Overexploitation and Contamination of Shared Groundwater Resources.*

participants can gain the necessary background to take the agenda forward. Some lessons learned from the inventory process of the Americas are presented, followed by a review of the linkage between transboundary aquifers and globalization, and a presentation of groundwater dependent ecosystems and governance issues. The chapter concludes with an outlook of the UNESCO programme on transboundary water resources encompassed in two very successful initiatives, the PCCP and ISARM.

Keywords Shared groundwater resources, global distribution and assessment, transboundary aquifer governance, groundwater dependent ecosystems

1.1 Introduction

This chapter provides a summary of the notable developments in groundwater resources management that have been achieved over the past decade through actions carried out by many international agencies, including UNESCO and its partners. The scope of the chapter relates to the key issues that were discussed in the NATO Advance Study Institute (ASI) (October, 2006), i.e. the concerns about overexploitation and contamination of groundwater resources that are shared. The concept of "shared" needs to be clarified, as it can mean one of several things—shared across sectors of water demand (irrigation vs. municipal demands), shared among communities which may have competing needs (industrial development vs. agricultural production), shared with the needs of the environment (requisite baseflow in streams and wetlands vs. human needs) and, finally, shared across international boundaries. The latter is termed as transboundary water resources. This chapter will mainly consider shared aquifer resources in their international transboundary context, though references will be made to the national context, as appropriate, to provide the participants of the ASI with adequate background for a discussion and taking forward an agenda of sustainable management of transboundary aquifers in the Eastern European and Central Asian regions.

1.1.1 Purpose of the ASI

As stated in the institute documents, an Advanced Study Institute (ASI) is a high-level teaching activity where a carefully defined subject is treated in depth by lecturers of international standing, and new advances in a subject, not taught elsewhere, are reported in tutorial form. While the following chapter (Chapter 2) aims to fulfill these requirements, it will of necessity be an abbreviated resources base. The full knowledge base of resources can be found in a series of UNESCO publications and the Web sites of the programmes supported by UNESCO (e.g. http://www.unesco.org/water). References are provided where appropriate so that documents, reports and guidelines can be sought out.

1.1.2 Anticipated Outcomes of the ASI Workshop

At the outset, the anticipated outcome of the ASI was the strengthening of the capacity of agencies that nominate their representatives to the ASI. As ASI participants should normally come from NATO, Partner and Mediterranean Dialogue countries, information is focused on the transboundary aquifer resources in the region of interest. Previous work in the region has included an inventory of transboundary aquifers in the United Nations Economic Commission for Europe (UNECE) region (UNECE, 1999), an inventory of transboundary aquifers in the southern coastal region of the Mediterranean (Puri, 2002a) and the development of the inventory of the Balkan Region (ISARM-INWEB, 2004). Based on these science driven efforts, and the UNECE Convention that deals with transboundary water resources (UNECE, 1999)—the Helsinki Convention—it is also anticipated that the national governments in the region, and experts, will continue to foster greater cooperation in the sound use and management of transboundary aquifers. Since the participants were asked by the organizers to provide some background data on their own national transboundary aquifers, this national background data and information would, in essence, be similar to the inventory that has been conducted by the UNESCO-ISARM Programme in the Americas and Africa (Aureli, 2004). Some points of interest, as lessons learnt from the Americas as they relate to availability of data, follow in Table 1.1.

Table 1.1 Comments on the scope of a questionnaire that may assist in the "inventory" processes

National Information to Be Provided	Comments from the Lessons Learnt and Experience Gained in the ISARM Programme
Overview/historical background on transboundary water issues in your country	The initial assessment provided by the national experts groups may not identify all of the shared aquifers—a second round is often needed.
Geographic area affected	Definition of the extent of aquifers and the relevant geological formations is often inaccurate and not easy to correlate across boundaries.
Nature of issues (e.g. diminished water quantity, degraded water quality, industry vs. agriculture vs. other consumptive uses, etc.)	Generally aquifers in the border regions do not have this type of data, but it can be obtained
Brief history (temporal, major milestones)	As above.
Regional importance of addressing issues (impacts: economic, public health, political)	Groundwater resource data and socio-economic data are difficult to match, since information is collected by diverse agencies, infrequently coordinated.

(continued)

Table 1.1 (continued)

National Information to Be Provided	Comments from the Lessons Learnt and Experience Gained in the ISARM Programme
Status of transboundary issues	
Actively being addressed (e.g. legislatively, judicial, mediation, regional accord, other arrangement)	In the European region, there are several bilateral agreements that generally refer to surface water.
Passively being addressed (formation of regional commission, informal discussions)	At the local administrative level there is often dialogue.
Détente	
Lessons learned from historical successes and failures in addressing these transboundary issues	Aquifer resources are generally a background issue to others e.g. transboundary national parks.
Obstacles adversely affecting these issues	
Shortage or low quality of data and information	National databases may not be as helpful as those held by the local administrations e.g. municipal water supply agencies.
Jurisdictional/ethnic issues	Fewer constraints in the European Union (EU) accession countries, many constraints in others.
Political problems	As above.
Impediments to public participation	As above.
Lack of sense of urgency	As above.
Other	
Recommended actions needed to address/resolve issue	Previous experience shows that capacity strengthening is needed at both national and local levels.
Potential of unresolved disagreement leading to social/military conflict	"Water" is not generally an issue that directly leads to military conflicts, through it can be an irritant that may affect social conflicts.

1.2 Aquifers, Groundwater Resources and Advances in Hydrogeology

1.2.1 Basic Definitions

An aquifer is a geological formation capable of storing and transmitting water in significant quantities, such that the water can be extracted by manual or mechanical means. Its dimensions can vary greatly, from a few hectares in surface area to thousands of square kilometers, while the depth can range from several meters to hundreds or thousands of meters. The term "aquifer" denotes the geological rock formation and the water within the rock formation.

1.2.2 Groundwater Resources

Groundwater comprises all the water below the land surface, and specifically, the water lying beneath the phreatic level, completely filling all the pores and fissures in the ground. This water flows naturally out to the surface through springs, seepage areas and water courses, or directly into the sea. It can also be channeled artificially into wells, galleries or other types of catchments. A significant proportion of what we call surface water originates from groundwater, as base flow in rivers. Groundwater is derived from recharge and, after passing through the aquifers, it may flow into rivers or out to the land surface through seepage areas, springs and diffuse discharge areas.

Groundwater is constantly renewed through natural recharge. This recharge is mainly derived from precipitation, but it can also be caused by surface runoff or resulting from surface water courses (particularly in arid climates), from nearby aquifers or from use-returns (notably the irrigation).

Groundwater passes through aquifers very slowly. Its normal velocity ranges from less than a meter to a few hundred meters per year; only in the case of karstic aquifers and severely fractured rocks can water travel through at speeds similar to those of surface currents. Thus, a drop of water that falls onto a watershed located 200 km from the coast and that is incorporated into a river flow would take a few days to enter the sea. If the similar drop, however, traveled underground (through a sandy aquifer), it would take tens or even hundreds of years to reach the sea. This slow movement by water through the unsaturated and saturated zones provides opportunities for us to manage, exploit and protect groundwater. In the latter case, it means that we can act before a contaminating agent has time to spread throughout the whole aquifer.

1.2.3 Groundwater Extraction

Groundwater is accessible, in principle, to all the inhabitants of the area overlying an aquifer, thus many more than those living close to streams. We refer to those who have the means and the right to exploit this source and who find it advantageous because it is the source of supply which is often the least costly, the most practical and the best suited to use by individuals and their family. In fact, the handling of groundwater is not, or hardly dependent, on community-owned equipment, contrary to surface water, which requires diversion works and often regulation, as well as transport which is usually beyond the means of private individuals. Groundwater is a local resource par excellence.

1.2.4 Advances in Hydrogeology and Increase in Exploitation

Groundwater extraction has evolved since the beginning of the twentieth century. The extraction rates increased generally during 1950 to 1980. The portion of groundwater extraction compared to the total extraction of freshwater also increased evidently (Puri, 2003).

In the developing countries, the increase in identified extractions was generally strong starting in the 1970s and 1980s, especially in those countries with strong demographic and economic growth and great expansion in irrigation, such as India and China, the biggest exploiters. The strongest growth took place in arid or semi-arid countries, in countries whose petroleum revenues encouraged the development of deep groundwater exploitation for irrigation. The growth in the level of exploitation in European countries was also evident: + 54% in 25 years (1950–1975) in the United Kingdom, + 70% in 7 years (1970–1977) in Denmark, + 15% in 10 years in Spain and in the ex-USSR, + 12% in 5 years (1971–1976) in The Netherlands. Extraction rate growth in Australia almost tripled in 30 years (1970–2000). Of all the raw materials extracted from underground, the greatest tonnage by far, is that of groundwater. However, these extractions are unequally distributed around the world (Table 1.2).

Table 1.2 Groundwater extraction on continents (From Margat, 2006)

Groundwater Extraction	km^3/year
Africa	35
North and Central America	150
South America	25
Asia	500
Australasia and Oceania	10
Europe	80

Figures rounded off

1.2.5 Groundwater Uses and Social Benefits

In most countries, the groundwater extraction exceeds 1/5 and often 1/3 of the total water extraction for all uses (including the cooling of thermoelectric plants). Groundwater exploitation covers about 1/5 of the world's current water needs, for all uses (estimated at 2000–3700 km^3/year) (Shiklomanov, 2003). Irrigation and community potable water supplies are the predominant uses.

The contribution of groundwater is naturally predominant in the majority of the countries in the arid zone that have no water streams originating in neighboring countries, notably when it concerns non-renewable resources (Saudi Arabia, Libya and Algeria), or when the local or even foreign surface water resources (e.g. in Bangladesh, Botswana and Congo) are almost uncontrollable, or even when it derives from an arbitrarily established priority, such as in Denmark, Georgia, and Croatia.

On a worldwide scale, the exploitation of groundwater covers approximately 50% of the potable water demand (community supply: households, public services, supplied activities, etc.), 20% of the irrigation water demands and 15% of the demand for those industries not directly connected to a water supply. Currently, at least half of the world's population must be supplied with potable water from underground sources. This applies not only to the rural populations, for whom groundwater is often the most appropriate local supply source, but also to a significant part of urban populations. A number of the world's large cities receive their water supplies exclusively or mostly from groundwater sources. In particular, groundwater is by far the most utilized source to cover the domestic needs of the dispersed or isolated non-serviced dwellings. In the United States, for example, 98% of domestic needs (which amounted to a total of 4.95 km^3/year in 2000, i.e. 8% of the national potable water demand) are satisfied by individual extractions of groundwater, which adds up to 22 km^3/year of groundwater extracted for supplies to communities.

Globally, according to the information collected, groundwater uses are

- 2/3 of the groundwater extracted in the world—at least 500 km^3/year (up to 800 km^3/year as estimated by International Water Management Institute (IWMI) (2005)—is used for irrigation;
- 1/4, about 200 km^3/year, contributes to potable water supplies;
- 1/10, roughly 100 km^3/year, is used by industries which are not supplied or is extracted by mining.

However, these proportions vary more or less in accordance with the region:

- In Europe, more than 50% of the extracted groundwater serves as potable water supplies, but only 20–25% in North America and in Africa and less than 20% in Asia;
- In Africa, more than 75% of extracted groundwater is used for irrigation and almost as much in Asia and in North America (70%). This distribution by country amongst the major use sectors is even more diverse.

There is a contribution of groundwater in the energy sector for cooling, and for geothermal energy (which only uses water that is not suitable for other purposes). Its utilization as a cold source for heat pumps is beginning to be developed in few countries for individual or especially collective heating.

In most developed countries and in many developing ones, groundwater is the main supply source of potable water and quasi-exclusive sometimes in arid zones. For example, this part amounts to nearly 100% in Austria and Denmark, more than 90% in Italy and Hungary, 84% in Switzerland, 79% in Russia, 74% in Germany, 66% in Poland, and more than 60% in Belgium, France and the Netherlands. It nearly reaches 70% on average for the ensemble of the European Union. It amounts to 100% in Pakistan, 64% in India and is significant in many African countries (Figure 1.1).

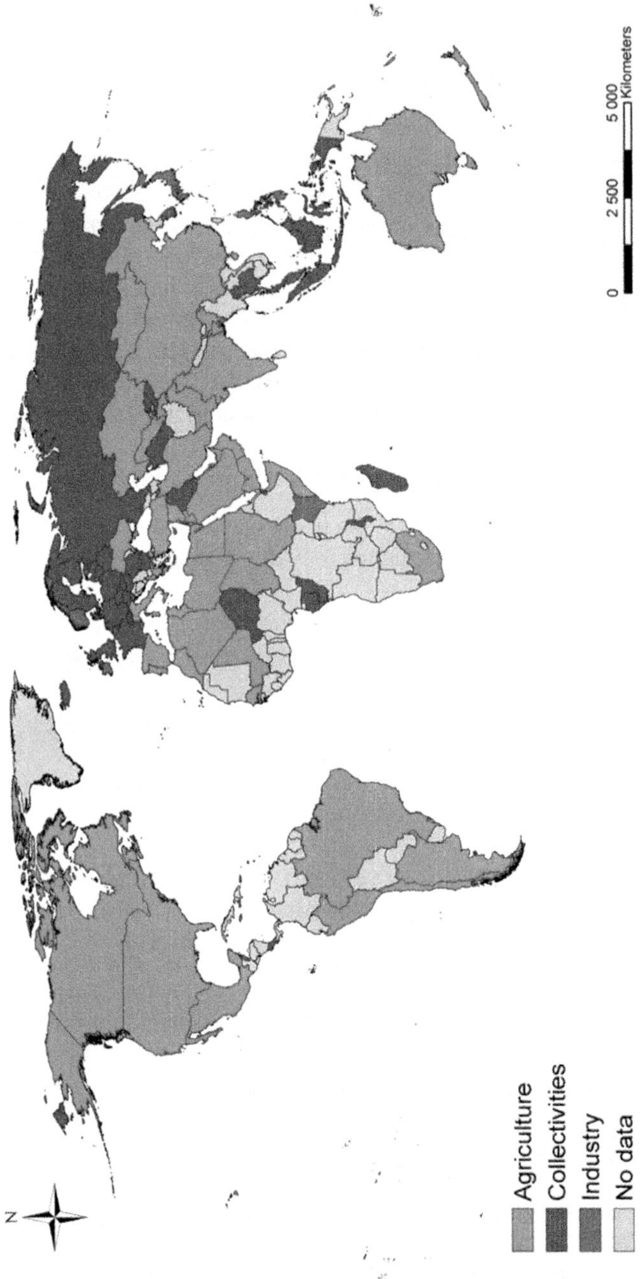

Fig. 1.1 Predominant use of groundwater in the World

The agricultural sector is by far the principal consumer and user of groundwater—almost as much so in the developed countries, where agricultural irrigation is the predominant water use sector:

- 87% in Greece, 71% in Spain, 70% in United States, 57% in Italy and in Australia;
- 90% in Saudi Arabia and in Libya, 89% in India, 86% in Bangladesh;
- 85% in Tunisia, 84% in South Africa, 79% in Argentina, 60% in Turkey;
- 64% in Mexico, and 54% in China.

The utilization of groundwater for agriculture is very concentrated in a few countries, as is illustrated by the geography of areas irrigated by groundwater in the world (Figure 1.1). According to Food and Agriculture Organization (FAO) (AQUASTAT, 2003), five countries, of which four are in Asia, account for 4/5 of the territories irrigated by groundwater:

- India: 38.6% of the world total;
- United States: 15.8% of the world total;
- China: 12.3% of the world total;
- Pakistan: 7.1% of the world total;
- Iran: 5.3% of the world total.

1.2.6 Specific Economic Aspects

The variety of groundwater uses result in noticeable differences in usage values while the diversity in proportions of groundwater used as a supply source differentiates the relative importance of the economic agents concerned.

- The potable water supply sector corresponds to uses of great economical and social values for public and commercial services, as well as for intermediary production–distribution agents that are either public (state, municipalities, unions) or private (delegated management). The frequent preponderance of this sector in the exploitation and utilization of groundwater globally increases the portion and commercial value of groundwater;
- In those industrial sectors not directly connected to a water supply, the groundwater exploiters are private or public enterprises that are direct users, and the water is not a "commercial" goods. Only its extraction can be subject to tax;
- In the groundwater irrigation sector, there is a very large predominance of individual exploiters who are the direct users (e.g. irrigating farmers) covering only the direct exploitation costs (more often subsidized rather than taxed) and who operate generally by virtue of water rights attached to the ownership of land. However, extraction can also be subject to licensing.

Consequently, the use of groundwater for irrigation is generally more economical and value-adding than the usage of surface water distributed by the community networks. As it was shown by R. Llamas in the example of Spain (Chapter 20), groundwater irrigation is five times more productive—in terms of economic return: production value per m^3 of water utilized—than surface water irrigation, both due to the very superior average efficiency which permits the reduction of water demands per hectare (an average of 5,000 m^3/year with groundwater as opposed to 10,000 m^3/year with surface water) and due to the choices of agricultural speculations with strong value added. Groundwater irrigation supplies 30–40% of the total production of agricultural irrigation whereas it uses only 20% of the total irrigation water volume (Llamas and Custodio, 2003). According to the International Water Management Institute (IWMI) 2005, the production value of worldwide groundwater irrigated agriculture is on the order of 150–170 billion US dollars, which would correspond to an average productivity of 0.2 US dollars per m^3 of water utilized. Thus, in general, the use of groundwater, which use is greater prevalent in the private sector than in the public water economy sector, has an economic value clearly superior to the mere proportion of water volumes at stake.

One can conclude that water extracted from underground is generally better utilized than surface water and that groundwater generally constitutes the portion of water resources that has the greatest economic value. Therefore, maintaining its production and preserving its quality should be a priority objective in water resources management.

1.3 Transboundary Aquifers: Global Distribution and Assessment

While debates about the management of transboundary rivers basins have been taking place for many years (e.g., UNESCO, 1993), the same cannot be stated about transboundary aquifers. In the same way that there are internationally shared river basins, there are also internationally shared, or transboundary, groundwater resources hidden below the ground surface, in all parts of the world. Some transboundary aquifers contain large fresh water resources, enough to provide safe and good quality drinking water for the needs of all humanity for tens of years.

Competition for visible transboundary surface waters, based on available international law and hydraulic engineering is evident in all continents. However the hidden nature of transboundary groundwater and lack of legal frameworks invites misunderstandings by many public policy makers. Not surprisingly therefore, transboundary aquifer management is still in its infancy, since its evaluation is difficult, suffering from a lack of institutional will and finance to collect the necessary information.

Although there are fairly reliable estimates of the resources of rivers shared by two or more countries, no such estimates exist for transboundary aquifers (World Bank, 1999). In order to respond to this gap, the UNESCO-ISARM Programme, in conjunction with From Potential Conflict to Co-operation Potential (PCCP) and others, has been developing its activities since 2000, following the conference of experts in Tripoli (UNESCO, 2001). One of the drivers of the ISARM Programme is to support cooperation among countries to develop their scientific knowledge and to eliminate potential for conflict, particularly where conceptual differences might create tensions. ISARM aims to contribute to the improvement of the scientific knowledge on internationally shared aquifers and their risk based management, providing a framework for regional and scientific cooperation. It aims to educate, inform and provide inputs for policies and decision-making, based on good technical and scientific understanding (Puri, 2001).

1.3.1 Global Distribution

In order to gain a global assessment to study and better manage aquifer resources, a number of agencies joined together in a major venture, the World-wide Hydrogeological Mapping and Assessment Programme (WHYMAP). WHYMAP was created in 1999 at the UNESCO General Conference of the Commission for the Geological Map of the World (CGMW). The programme aims at collecting, collating and visualizing hydrogeological information at the global scale, to convey groundwater related information in an appropriate way for global discussion on water issues and to give recognition to the invisible underground water resources within the World Heritage Programme. WHYMAP also brings together the colossal efforts in hydrogeological mapping, at regional, national and continental levels. The principle focus of the WHYMAP program is the establishment of a modern digital Geographic Information System (GIS) in which all data relevant to groundwater are stored together with its geographic reference. In its final form the WHYMAP-GIS will contain a significant number of thematic layers, local and shallow aquifers, transboundary aquifer systems, depth/thickness of aquifers, groundwater vulnerability, geothermalism, stress situations of large groundwater bodies and “at risk” areas. The first of such thematic maps based on the WHYMAP, the “transboundary” distribution map (scale 1:50,000,000) has been published (Struckmeir et al., 2006).

The WHYMAP-Transboundary confirms that regional aquifers sometimes extend over large areas and the flow paths of groundwater in them, crossing national boundaries, can extend over tens or hundreds of square kilometers. The area of the largest systems known on our planet can even reach over 2 million sq km, and be shared by several countries. With thick saturated sediments, of up to 1,000 m, they form vast underground water storages. Although there could be massive groundwater resources in stock, in arid regions, with little contemporary renewal from rainfall, aquifers can be particularly vulnerable to overexploitation. The preliminary summary-inventory of the global distribution of transboundary aquifers shows that there are at least 517 such aquifers in the global continental land mass.

1.3.2 Global Assessment: Inventories Under the UNECE and ISARM Programme

1.3.2.1 The UNECE Inventory and Some Findings

The earliest of the regional inventories was conducted by the UNECE, which resulted in a comprehensive knowledge base of the transboundary aquifers in Europe. It was one of the activities of the Groundwater Programme 1996–1999 under the UNECE "Water Convention". The Convention on the Protection and Use of Transboundary Watercourses and International Lakes was drawn up under the auspices of the Economic Commission for Europe (ECE) and was adopted in Helsinki in March 1992. The inventory was based on 25 country replies to a comprehensive questionnaire; as a result of which 89 transboundary aquifers were reported by one or more of the countries. During the ECE inventory process, actions were taken to communicate the results between the riparian countries in order to harmonize the differences in data and information. The conclusion of this pioneering inventory was that the information and knowledge of the transboundary aquifers was rather poor and the results of the inventory were considered as an assessment of the knowledge at that time. However, several important valuable lessons were learnt:

- Main types of utilization and human interference.
  Like national aquifers, transboundary aquifers are equally important for utilization purposes. In a number of cases, the inventory found that the lowering of water tables was caused by human interference—overexploitation—in the neighboring country. In fewer cases, human interference was mentioned as a reason for changing the groundwater quality, but these data where qualitative rather than quantitative.
- Possible sources and types of pollution.
  Industry, households, agricultural, dumping sites/landfill, mining municipalities, military sites and others have been indicated as possible sources of pollution.
- Monitoring quantity and quality levels.
  From the questionnaire, it was proved that in more than 85% of the transboundary aquifers, there is some groundwater quantity monitoring. The density of monitoring sites varies both in space and in time. Groundwater quality monitoring has been reported in 80% of the investigated transboundary aquifers.

1.3.2.2 The Launch of the TARM/ISARM Programme

In March 2000, UNESCO, International Association of Hydrogeologists (IAH) and FAO launched a global activity on shared aquifers and drafted a framework document for an international initiative on transboundary aquifer resource management (Puri, 2001), that was designated as ISARM. In the course of the period to 2006, the ISARM Programme has finalized the full inventory of transboundary aquifers in the Americas, the preliminary inventory of the transboundary aquifers in Africa, the detailed inventory of the transboundary aquifers in the Balkans as well as in the southern rim of the Mediterranean Sea.

1.4 Significant Issues Being Addressed

1.4.1 Sustainable Management: The Lessons from the Inventory of the Americas

Since the most advanced of the ISARM inventories is from the Americas, several useful lessons and a preliminary regional level assessment can be made. The Americas inventory is expected to be finalized in three steps: (1) scientific hydrogeological assessment, (2) review and assessment of the legal and institutional frameworks in use, and (3) provide guidance for a regional strategy. The results of the first step are to be shortly published (UNESCO-ISARM Americas, 2007).

The transboundary aquifers that have been identified in the inventory of the Americas, provide a good basis for consideration of some of the significant issues that affect sustainable environmental management. The distribution of groundwater resources in the Americas and the Caribbean is depicted on an extract of the WHYMAP, shown in Figure 1.2, indicating major groundwater basins, areas with complex hydrogeological structures and areas with local and shallow aquifers. Clearly groundwater can occur in all locations but is uneconomic for development where the unit cost of its extraction for the user is greater than the unit cost of using alternate sources, such as surface water i.e. when the well yields are low, as in local and shallow aquifers. High yielding aquifers, in major basins, provide a very competitive resource for economic development; in many such regions of the Americas, aquifers have been extensively developed, providing enormous economic benefits to users, such as the High Plains Aquifers, which is a national aquifer system, though shared by several States of The United States. Even in such a national aquifer, the sustainable and joint management of the aquifer has been beset by inter-jurisdictional issues. As the demand increases and the competition for scarce resources grows, more production can be expected from the resources of aquifers. While national aquifers are within the jurisdictions of nations, transboundary aquifers, much as transboundary rivers, require international collaboration, as it has been demonstrated in the course of the development of the inventory of the Americas.

In a preliminary spatial analysis of the aquifers of the Americas (Figure 1.2), 61% of the North American region is underlain by local and shallow aquifers, dominated by the Canadian Shield. The high precipitation in this region and ease of access to surface water means there is relatively low demand on these aquifers, and thus only local management strategies will suffice. In the South American region, 44% of the region is underlain by local and shallow aquifers. Where these aquifers occur in the humid zones (Amazon Basin), again the resources of these aquifers will only require local management strategies. However, in the southern part of the Americas, along the Pacific Coastal lands, these aquifers are important sources of water for local communities and even for groundwater dependent ecosystems and wetlands, as shown in studies conducted in North central Chile, in the Romeral Basin (Squeo et al., 2006).

Thus, for the sound management of such aquifers in the transboundary context, early joint collaboration is needed for the following reasons: the aquifers are small and recharge is very vulnerable to changing rainfall patterns, the ecosystems that

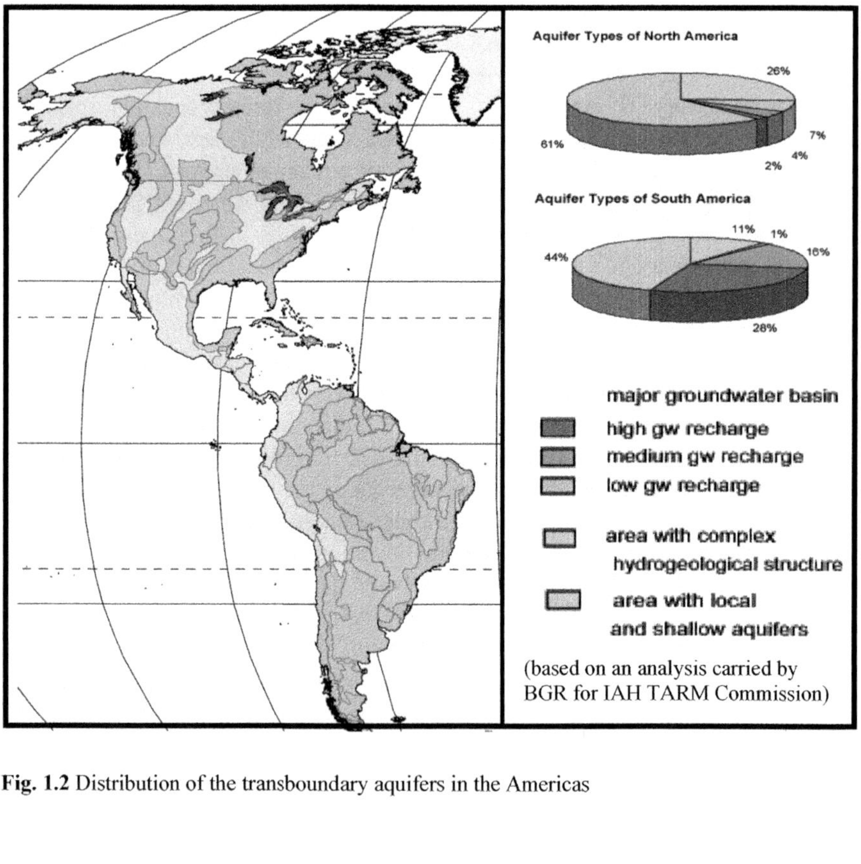

Fig. 1.2 Distribution of the transboundary aquifers in the Americas

rely on their discharge are fragile, and the coastal wetlands are reliant on aquifer discharge. Sharing countries will then have to adopt strategies that will require them to jointly take measures such as sustainable land use that incorporate means to enhance rainfall recharge.

Hydrogeologicaly complex structures in aquifers are found in 26% of North America and 11% of South America. Therefore, high investment in exploration and preliminary analysis of the aquifer systems are required before any major schemes for their development can be undertaken. Countries sharing such aquifer systems will need to cooperate to a great degree of details to well define the aquifer system, the resources that can be exploited, and the ecosystems that rely on the resources of the aquifers. The ecosystems that rely on aquifer discharge can be wide, ranging from wetlands to coastal swamps and inland lakes. For example, 34% of the water entering Lake Michigan comes from transboundary aquifers (SLMWSA, 2006; Cherkauer and Hensel, 1986).

Aquifers located in major basins have been categorized according to regions ranging from low to high annual recharge. In South America, 28% of the area is underlain by major aquifers that receive high recharge—here aquifer development could start in advance of major investigations, generating the equity needed to put into place the joint sustainable management plans. In North America, this category of aquifers appears to be quite restricted and consists of only in 2% of the area, while low recharge of aquifers is observed in 7% of the area. In general, the lower recharge of the aquifer, the higher the need to put in place good management strategies early in the development, since the users will draw on aquifer storage. Poor management practices will mean that aquifer storage is depleted before sustainable land use (and thus enhanced recharge) can be introduced, as it can be illustrated by the High Plains Aquifers. It has been estimated that the volume of storage that has been depleted from this aquifer is equivalent to 0.025 mm sea level rise (Konikov and Kendy, 2005).

In conducting the inventory of transboundary aquifers on the background of the regional assessment of aquifer resources, the regions where there is an urgent need for sharing countries to come together and to collaborate before embarking on a major development program have been highlighted. The inventory also enables the countries to prioritize the degree of attention and the level of investment that should be put into place, so that they can achieve environmental sustainability through the use of good resource management.

1.4.2 Transboundary Aquifers and the Impacts of "Globalization"

The ASI program also includes sessions devoted to the principles of depletion and contamination of shared groundwater resources. While in the national context these principles are well observed and understood, though arguably not well managed, in the transboundary context these same issues take on an order of magnitude greater in complexity. What does this mean? From a pure natural science perspective, no impact other than that initiated from, and transmitted through, the aquifer system is feasible. However, there is increasing evidence among the socio-ecological community that globalization is a central feature of the coupled human-environmental systems, or as termed by them, socio-ecological systems. Given that approximately 40% of the global population lives in one or another transboundary river or aquifer system, this is an intuitive observation in terms of the impact on the more obvious human ecologies, in the mass media, in consumer products and in preferences of food, fiber and entertainment, the impacts are less tangible, but no less significant on natural resources—transboundary aquifer systems—and interconnected with it. This is discussed below.

In considering the drivers that affect quality and quantity in transboundary aquifers, most researchers accept that apart from the impacts of climate change, they are also subject to easily discernible local changes such as land use change, urbanization and associated with these changes a bewildering input of substances that enter the groundwater flow systems. They also accept that such change is global i.e. worldwide, and it is proceeding at an unprecedented pace, and in geographic scope, affecting river basins and aquifers in term of quantity and quality. This is confirmed in various assessments such as the Millennium Ecosystem Assessment that has shown that global ecosystem degradation involving loss of ecological capital is intense—in 2002, humanity's global ecological footprint was exceeded by 23%. If this trend continues, the globalized human economy (comprising of demand on natural resources, including water from aquifers) is in ecological overshoot, from the impacts of global interdependence on goods and services.

In translating the impact of this change into aquifers, it is widely reported that aquifers are being overdrawn—i.e. the groundwater resource is being pumped beyond the rate at which it is recharged—consequently, in the majority of such aquifers, water is being drawn from storage. As yet there are no definitively agreed figures of the global total of withdrawal from storage. Nevertheless, if one were to simply adopt the figure provided for one globally significant aquifer system, the High Plains Ogallala Aquifer in the United States, this alone is stated by Konikov and Kendy (2005) to be equivalent to a sea level rise of 0.025 mm, as noted above. If, to this is also added the volume drawn from storage in the North China Plains Aquifers, the Indo-Gangetic Plains Aquifers, the Gujeral Aquifers and the Mexico Aquifers, for which somewhat unreliable figures are available, though the order of magnitude is well known, then the global "loss of transboundary aquifer storage" become an issue beyond intellectual interest. When combined with the risks to aquifer functions and to aquifer dependent ecosystems, the issue needs urgent quantification.

Why does loss of global aquifer storage require urgent quantification? Because economic losses, translated through environmental and livelihood losses will be difficult to reverse. The decline in resilience of ecosystems that are linked closely with aquifers and groundwater in the lower income countries, may reach a "tipping point" beyond which they cannot be revived, i.e. the ecological overshoot mentioned in the global interdependence above.

Unfortunately, scientists and economists are at present lacking agreed and sound methodologies for taking these observations from science to policy, and so far the IWRM paradigm may have failed to address this, specifically for transboundary aquifers and their dependent ecosystems; despite measures such as the Ramsar Convention. So would an "EcoSystem Services" approach may be more effective? This issue is considered in the following section.

1.4.3 Transboundary Natural Resources: Groundwater Dependent Ecosystems

The ASI proposes to consider techniques to assess the depletion and contamination of shared groundwater resources, and tools to predict/forecast the future depletion and contamination of shared groundwater resources. Again, it is noted that in the national context these issues may be more tractable than in the transboundary context. To gain a better insight into this, it would seem appropriate to adopt an "out of the box" perspective. As long as aquifers are thought of as a single dimensional resource, i.e. one that holds water singly for human use, the sustainable management of aquifers will remain threatened in the same way as any other common good, such as natural transboundary forests, transboundary biodiversity, etc. To overcome this one-dimensionality, aquifer resources should be considered as an undeniable component of "ecosystems", thus invoking the so called "ecosystems approach". Once this is accepted, then the framework for the Millennium Ecosystem Assessment can be used to address both national and transboundary aquifers, thus precisely putting a value on the ecosystems services that aquifers provide.

Looked at from another perspective and as noted in a recent report by the International Union for the Conservation of Nature and Natural Resources (IUCN) (Emerton and Bos 2006), and quoting extensively from it, there is a growing—although by no means universal—recognition that the environment demands water, whether from groundwater or from surface resources. For example, both the 1993 Dublin Statement on Water and Sustainable Development and the World Summit on Sustainable Development (WSSD) Plan of Implementation highlight the need to maintain freshwater flows (often provided from aquifer baseflows) for the environment. This relationship however, remains implicit and is not translated into useable tools. Inversely, the role of transboundary ecosystems and the supply of transboundary water have received far less attention, consequently the link between water (particularly groundwater) and the environment has rarely been perceived by national institutions beyond pollution and water quality concerns. Leaving ecosystems out of the water rhetoric and practice may ultimately undermine the very sustainable development and poverty alleviation goals that the international community is working hard, and investing heavily, to achieve: i.e. cost-effective, equitable and sustainable access to water resources and services for all. Recognizing transboundary ecosystem values will help increase the sustainability of the efforts. But, there is an added bonus: ecosystem values may also offer a pathway to increase investment and human well-being. If these values are made visible, they can also be integrated into existing economic arrangements and lead to a new field of incentives, investments and value chains that support the Millennium Development Goals. As a positive response to this, pilot actions and schemes of payment for environmental services (PES) are underway that may lead to the emergence of a new economic sector in the management of transboundary resources.

Ecosystems, through their demand for (ground)water, provide a wide range of goods and services for human production and consumption—for example fish, timber, fuel, food, medicines, crops and pasture. On the supply-side of the equation, natural ecosystems such as forests and wetlands generate important economic services which maintain the quantity and quality of water supplies. Furthermore, they help to mitigate or avert water-related disasters such as flooding and drought. Often ecosystems provide a far more effective, cost-efficient, equitable and affordable means of providing these goods and services than artificial alternatives. Yet, typically, groundwater dependent ecosystems are not allocated sufficient water or funding when water related decisions are made and water investments are planned.

1.4.4 Governance of Transboundary Aquifers: Technical and Financial Issues

Groundwater, and especially transboundary groundwater, is often referred to as an "undervalued" resource, where an ecosystem approach underwritten by economic valuation can help in qualifying this significant, but vague characteristic, to quantify the degree of under-valuation in social and economic terms. The core of the issue is, paradoxically that groundwater presents many opportunities and advantages for national economic development and environmental sustainability. In many regions of the world, groundwater represents a reliable and flexible source of freshwater, upon which people have become increasingly dependent. Groundwater development has significantly enhanced local productivity and makes the resource accessible to a wide range of individual users. Mechanized boreholes have allowed the access to be "on demand" and "just-in-time". However, as a consequence of this apparent success the social, economic and environmental systems that depend on groundwater are under threat from groundwater overexploitation and pollution.

The evolution is dynamic and it is important to recognize the change of attitudes over time to these recent and rapidly expanding problems. Cheap and reliable borehole pumps were only introduced in the post-Second World War period and the scale and intensity of the abstraction and pollution have only been apparent in the last quarter of the twentieth century. Prior to this, groundwater was seen as a ubiquitous and reliable source of high quality water. As groundwater exploitation from increasingly deep transboundary aquifers grow, often with incentives of inappropriate national (or transnational) energy and agricultural pricing subsidies, drawdown externalities are extending beyond national boundaries. However, the impacts of increases in energy prices are slowly being felt and as some countries experience dwindling tax revenues, they are forced to phase out national energy subsidy schemes. If the phase outs are unharmonized across national boundaries in transboundary aquifers, the consequences can be significant. Access to affordable energy is emerging as a priority groundwater issue and, as the economic fundamentals are imposed, there are also adverse social impacts as groundwater falls out of the economic reaches of the individual farmers as the consequence of limited economic access to energy and capital to pump increasingly deep groundwater. However, as groundwater depletion is being increasingly recognized as a major environmental issue, international funding institutions can also be expected to restrict financing groundwater resources development.

1.4.5 Governance of Transboundary Aquifers: Legal and Institutional Issues

The scientific principles involved in the sound management of transboundary aquifers are well known and understood by groundwater specialists. These include an appreciation of the full system, i.e. from sources of recharge, to the regions of discharge, as well as the quantity and quality issues along the flow path. Usually the system is well described by the use of conceptual models through which groundwater specialists from across national boundaries can communicate well. Unfortunately sustainable management of transboundary aquifers goes well beyond developing consistent conceptual models. It needs in addition, harmonization of legislation (L), equivalence in institutional structures (IS) and consistency in socio-economics (SE) drivers and also a coherent application of the environmental protection (EP) criteria, abbreviated to LISSEEP. Developing cooperation for sound management therefore requires an equal attention to these other drivers, which must follow upon the hydrogeological conceptual consistency. One of the key issues in developing cooperation is strengthening institutions such as basin Commissions or Joint Bodies, so that these aspects can be addressed. There exists extensive literature and substantial experience in developing cooperation for the sound management of transboundary river basins. While many of the principles from this experience can be applied to aquifers, there are issues peculiar to aquifer behavior that should be defined in the LISSEEP so that cooperation can be made effective (Puri, 2002b).

The scope of tasks to be entrusted to a commission should not preclude alternative options that could be adopted (Table 1.3). Since the existing agencies may well have river basin management responsibilities, then consideration should be given to strengthening the existing basin management agency. in their complete absence a new authority, or a specialized management institution e.g. irrigation

Table 1.3 Scope of responsibility for aquifer commissions

Scope	Responsibilities
Technical	Establishing a sound conceptual model of the whole aquifer and interaction with surface water. Formulation of a sustainable basin development plan and coordination, including prioritization, water quality and pollution prevention plans. Control of beneficial uses – allocations for municipal demands, agricultural demands, industrial demands. Establishing other aquifer uses e.g. thermal energy, balneological needs, natural discharges to wetlands, etc.
Economic and financial	Internal financing, including cost sharing and sharing criteria. Financing specific projects, management of international funds, compensation criteria, sharing benefits, payment of interest and repayment of debts. Assessment of collection of revenues, setting of tariffs. External financing.
Legal and administrative	Administration of the right to use water at the national level and coordination with national agencies and institutions, establishing water users associations. Prevention and settlement of disputes between water users. Drafting and implementing required legislation—international agreements, ministerial resolutions, harmonization of legislation. Other legal advice.
Public participation	Ensuring full involvement of the stakeholders. Empowering water user associations and defining property rights. Implementing the full scope of sustainability in resource use.

agency, could be developed and entrusted with acting as the "apex" body for water resources, including groundwater. other aspects such as duration, constitution of the commission, procedures for decision-making and the lisseep factions should also be taken into account.

1.4.6 Governance of Transboundary Aquifers: International Legislation

An initiative that has been underway for several years in cooperation with the UN International Law Commission, has now in 2006, reached a fruitful stage of the preparation of legal articles on the use of transboundary aquifers. The UNESCO-IHP cooperation with the UNILC was based on providing a technical expert group that has assisted a Special Rapporteur that was appointed by the Commission. Working in close cooperation with the Rapporteur, all major issues that affect transboundary aquifers have been covered in draft Articles.

More information on this topic is provided in Chapter 3 that covers this issue in more detail.

References

AQUASTAT, 2003, International water related database held at the UN Food & Agriculture Organisation, FAO, (September 9, 2007); http://www.fao.org.

Aureli, A., 2004, *International Shared Aquifers Resources Management Programme A UNESCO Initiative*, presented at the Assemblée Mondiale du Réseau et des Organismes de Bassins Transfrontaliers, Dakar.

Burke, J. J., and Moench, M. H., 2000, *Groundwater & Society, Resources. Tensions and Opportunities*, UN DESA, New York.

Cherkauer, D. S., and Hensel B. R., 1986, Groundwater flow into Lake Michigan from Wisconsin, *Journal of Hydrology* **84**(3–4):261–271.

Emerton, L., and Bos, E., 2006, *Value. Counting Ecosystems as an Economic Part of Water Infrastructure*, IUCN, Gland, Switzerland and Cambridge, UK. p. 88.

ISARM-INWEB, Balkans, 2004, *Inventory of Transboundary Aquifers in the Balkan Region*, J. Ganoulis, ed., Aristotle University of Thessaloniki, Greece.

IWMI, 2005, *The Global Groundwater Situation: Overview of Opportunities and Challenges*, 2nd World Water Forum, The Hague.

Konikov, L., and Kendy, E., 2005, Groundwater depletion, a global problem, *Hydrogeology Journal* **13**:317–320.

Llamas, R., and Custodio, E., eds., 2003, *Intensive Use of Groundwater*, eds., Balkema Publishers, Lisse, The Netherlands, pp. 1–479.

Margat, J., 2006, *Water Resources of the World*, UNESCO (in press).

Puri, S., 2001, *Internationally Shared Transboundary Aquifer Resources Management – A Framework Document*, UNESCO. IHP-VI Serial Documents in Hydrology.

Puri, S., 2002a, *Inventory of Transboundary Aquifers in the South Mediterranean Region*, Internal Report, UN ESCWA, Beirut.

Puri, S., 2002b, Issues in developing cooperation for the sustainable management of transboundary aquifers, in: *From Conflict to Cooperation; Challenges & Opportunities in Water Resources Management*, Proceedings of UNESCO International Conference, 21–22 November 2002, Delft.

Puri, S., 2003, Intensive use of groundwater in transboundary aquifers, in: *Intensive Use of Groundwater*, R. Llamas and E. Custodio, eds., Balkema Publishers, pp. 415–441.

Shiklomanov, I. A., 2003, *World Water Resources*, UNESCO Publication in Hydrology.

SLMWSA, 2006, *Southern Lake Michigan Water Supply Association*, (July 30, 2007); http://www.nipc.org/environment/slmrwsc/Overview.htm.

Squeo, F. A., Aravena, R., Aguirre E., Pollastric, A., Jorquera, C. B., and Ehleringer, J. R., 2006, Groundwater dynamics in a coastal aquifer in north-central Chile: Implications for groundwater recharge in an arid ecosystem, *Journal of Arid Environments* **67**(2):240–254.

Struckmeir, W., Richts, A., Acworth, I., Arduino, G., Bocanegra, E., Cunningham, W., Droubi, A., de Franca, N., Gilbrich, W., Girman, J., van der Gun, J., Klingbeil, R., Margat, J., Poitrinal, D., Puri, S., Rivera, A., Safar-Zitoun, M., Tujchneider, O., Vasak, S., Vrba, J., Winter, P., Zaepke, M., Zaisheng, H., and Zekster, I., 2006, *WHYMAP and the World Map of Transboundary Aquifer Systems*, Special Edition from World-wide Hydrogeological Mapping and Assessment Programme (WHYMAP) for the 4th World Water Forum., Scale 1:50,000,000, BGR/UNESCO, Paris, France and Hanover, Germany.

UNECE, 1999, Inventory of transboundary Groundwaters in Europe. *Task Force on Monitoring & Evaluation*, Vol I, E. Almassy and Z Buzas, eds., Published Riza Institute for Water Management, Lelystad, the Netherlands.

UNESCO, 1993, *Transboundary River Basin Management & Sustainable Development*, 2 Volumes Technical Documents in Hydrology.

UNESCO, 2001, International Conference on Regional Aquifer Systems in the Arid Zones—Managing no Renewable Resources, Tripoli, Technical Documents in Hydrology, No. 42.

UNESCO-ISARM Americas, 2007, *Atlas of Transboundary Aquifers of the Americas*, Published by UNESCO in press (UNESCO-IHP Office in Montevideo).

World Bank, 1999, *Groundwater, Legal & Policy Perspectives*, Proceedings of a World Bank Seminar, Technical Paper 456.

Chapter 2
Education and Training for Transboundary Groundwater Management as an Instrument of Dialogue and Communication

Alice Aureli*, Jean Fried, Jose Luis Martin-Bordes, Shammy Puri, Raya M. Stephan

Abstract The management of groundwater is complex, and when complicated by political boundaries, it can become a source of conflicts. Education and training are major instruments in teaching people to jointly manage their common water resources, according to sustainable development principles, especially emphasizing partnerships and the knowledge of each other's historical and human dimensions. Therefore, UNESCO's International Hydrological Programme (IHP), through its Internationally Shared Aquifer Resources Management (ISARM) project has started to design training programmes, which after presenting the complete picture of the hydrological cycle, focus on transboundary groundwater in a pluridisciplinary way. The preliminary phase was discussed during an experts' workshop held at UNESCO Headquarters in November 2006. During this meeting, the invited experts identified the challenges and contents of the courses, both for university graduates and professional training, as well as the pedagogical methods. They also adopted the proposals of the preliminary phase and decided to start the next phase, such as an experimental course to test the first drafts and progress toward a more elaborate training programme.

Keywords Education, transboundary water resources, groundwater management

---

Alice Aureli
UNESCO, Division of Water Sciences - International Hydrological Programme (IHP), Natural Sciences Sector 1, rue Miollis, 75732 Paris Cedex 15, France

Jean Fried
UNESCO - International Hydrological Programme (IHP) and University of California at Irvine, Urban Water Research Center, Engineering Tower 844C, Irvine, CA 92697, U.S.A.

Jose Luis Martin-Bordes, Shammy Puri, Raya M. Stephen
UNESCO - International Hydrological Programme (IHP) 1, rue Miollis, 75732 Paris Cedex 15, France

* To whom correspondence should be addressed. E-mail: a.aureli@unesco.org

C.J.G. Darnault (ed.), *Overexploitation and Contamination of Shared Groundwater Resources.*

2.1 Introduction

UNESCO launched at the 14th Session of the Intergovernmental Council of the International Hydrological Programme of UNESCO in June 2002 the worldwide initiative on Internationally Shared Aquifer Resources Management (ISARM), as a multi-agency effort aimed at improving the understanding of scientific, socio-economic, legal, institutional and environmental issues related to the management of transboundary aquifers. Since then—in the framework of ISARM—UNESCO and its partners have prepared inventories of the transboundary aquifers from Africa, the Balkans and the Americas and have initiated the inventory of the Asian component of ISARM. The activities mentioned above have been generating information necessary to address the management of transboundary aquifers in a multidisciplinary manner.

UNESCO, based on the achievements of the ISARM programme and recognizing the importance of education and training in the creation of local and regional capacity to manage shared groundwater resources, is proposing a specific training program in transboundary groundwater management to complete the existing curricula on this issue. The training program will also be an important element in support of the 2nd UN World Water Development Report: "Water, a shared responsibility" published in March of 2006.

As the report states, sustainable management of groundwater is essential since groundwater represents "96 percent of the Earth's unfrozen freshwater," used by most people around the world. The rising demand for water resulting from the constant increase in population worldwide combined with the rising scarcity of water resources in many countries, including traditionally humid regions, is leading to greater need and use of groundwater for human consumption and usage. Especially, a significant majority of drinking water in both Europe and the United States comes from groundwater.

2.2 Education and Training for Transboundary Groundwater Management: Why?

The management of groundwater is a complex task and further complicated by the presence of political and/or administrative boundaries. Issues related to use, ownership, access and development as well as protection can become a source of conflict and mismanagement. Regional and national security is at stake as are issues of economic development impacting jobs, health and investment. The science of the hydrologic system, as well as the laws, rights, agreements and treaties that govern this critical resource now require a deeper understanding and leadership capacity; education and training become major instruments in building such capacity and expertise.

Although groundwater management has been included in higher education curricula in engineering and other related disciplines for decades, the subject of transboundary issues related to how this resource is managed across political boundaries has not emerged as an area of study to be covered in a comprehensive and global manner. The work of the UNESCO-IHP projects and centers such as ISARM, World-wide Hydrogeological Mapping and Assessment Programme (WHYMAP) or International Groundwater Resources Assessment Center (IGRAC), identify the many internationally shared aquifers in addition to those within a Nation State that may also have regional impacts but are not necessarily managed under international law or treaties (Puri, 2001; ISARM, 2007; WHYMAP, 2007; IGRAC, 2007). One assumption made by this program is that practitioners and policy makers have not benefited from education and training specifically in the area of transboundary groundwater and its many complex factors. Such education and training would provide the added benefit of identifying best practices while also focusing on methodology and management tools essential to their work.

The management of a transboundary resource calls for legal rules giving guidance on the common use and protection of the resource. In the case of transboundary groundwater, the rules of international law are still in a stage of development. However, there is a new trend acknowledging the importance of the legal tool for reaching cooperation over a common groundwater resource. At a very high level, the UN International Law Commission (ILC) has just adopted at first reading a full set of draft articles on the law of transboundary aquifers. These draft articles were prepared and developed with the scientific assistance of UNESCO-IHP, which has through its ISARM project established a multi-disciplinary and international group of groundwater experts to provide guidance to the ILC members on the science of hydrogeology. These draft articles are intended to give riparian States a framework for cooperating over a transboundary aquifer.

At the level of States, cooperation can be completely developed as to reach the level of joint management and establish a joint management mechanism (authority or commission), or can limit itself to the first steps of cooperation such as the regular exchange of data, or the establishment of a common database. If for a long period, States have concluded agreements only on their transboundary surface water resources, there are some moves in the direction of transboundary groundwater resources. The arrangement on the protection, utilization and recharge of the Franco-Swiss Genevese aquifer (June 9, 1977) is the only example of a treaty dealing exclusively with the management of a transboundary aquifer, but there are other examples of cases and agreements showing first steps of cooperation over transboundary aquifers. Projects on transboundary aquifers are now integrating a legal component, in order to develop the cooperation between the concerned States. This is the case for example in Africa for the Iullemeden aquifer system, or the coastal aquifers of the gulf of Guinea. In the Americas, the ISARM national coordinators have acknowledged the importance of the legal tool and have decided for their project to enter into a second phase (the preliminary phase concerned the assessment of the transboundary aquifers in the American continent) consisting of the diagnostic of the existing institutional and legal framework in the participating countries.

The development of the legal component of a course on the management of transboundary aquifers can use these cases (and others) as success stories and examples enhancing cooperation towards joint management. However, if the legal tool is certainly an important one in the management of transboundary groundwater resources, its development cannot be done separately without a basic knowledge of the hydrogeological characteristics of the aquifer in concern.

Why a specific groundwater course and not a more general course concerning the hydrological cycle which groundwater is part of ? The logical integrated water management approach would be to address all aspects of the cycle, but the very specific characteristics of groundwater, such as its vulnerability to pollution and especially diffuse pollution on large time-scales, combined with an almost impossible rehabilitation due to its geological conditions, the uncertainties over its physical, biochemical and flow dynamic properties due to the difficulties of its monitoring, do encourage us to propose specific groundwater courses.

2.3 Objectives of the Training Programme

Taking into account the cultural differences between neighboring countries, our program has the following general objectives:

- Build the capacities necessary to participate in the implementation of the UN sustainable development principles and the Millennium Development Goals (MGDs), especially the careful use of shared groundwater resources, the integration of shared groundwater resources management in the sector of economic policies of the concerned countries, and the establishment of efficient partnerships, including public participation, at international level for a joint management of the shared groundwater resources; the MDG 7 and 8 are particularly concerned: "Ensure environmental sustainability" and "Develop a global partnership for development";
- Develop communication capacities between policy, science and technology at international level, such as the mutual understanding of scientific and legal languages and methods; and an interest in the cultural and historical dimensions on the other side of the border, keeping in mind the specificities of groundwater management;
- Develop the sensitivity to conflict resolution and harmonization of languages, and build the capacities necessary to answer crises where transboundary groundwater plays a role;

Our program has the more specific following objectives:

- Review the economic principles, methods and instruments concerning water resources management, and their adaptation to groundwater in an international context;
- Review the existing financial instruments, especially international, of use for transboundary groundwater management;
- Provide information on UN activities and the international community in general and in the domain of natural resources;
- Show the significance of groundwater in general and transboundary groundwater in particular, in terms of related political and technical problems, such as allocation of water, preservation of recharge in quantity and quality, protection of recharge areas, and land-use policies;
- Identify the various types of transboundary groundwater and the related specificities, questions and difficulties, including climate change and global warming aspects;
- Give a general introduction to transboundary groundwater management practices;
- Give a general introduction to Integrated Water Resources Management, showing the role of groundwater in the hydrological cycle, stressing the transboundary specificities and difficulties, and giving a broad vision of water management policies, methods and techniques.

We have concluded the preliminary phase of conceiving and drafting such a program by an expert workshop which was held in November 2006 at the UNESCO headquarters in Paris. The challenges of the program have been identified, a first draft of the contents of two types of courses, for university graduates on the one hand and for professional training on the other hand, has been produced and it has been decided to organize an experimental course to test the contents and the teaching methods, before producing a more complete program.

2.4 General Organization of the Training Programme

We identified two levels of courses, the policy level and the practitioners' level:

- The policy level aims at creating the capacity to have a broad vision of transboundary groundwater issues, including language and other cultural barriers, in order to enhance negotiations, political and managerial decisions, and to create institutions and instruments necessary for transboundary management, especially multinational institutions and partnerships;
- The practitioners' level aims at training those who, in the field, will work in political, institutional and managerial organizations that ultimately implement decisions based on science, law, economics and politics.

While the policy level should be the domain of the social sciences and humanities with knowledge of water resources and management, the practitioners' level is generally the domain of environmental scientists, engineers, hydrologists and those trained in the physical understanding of resource management. It is important to create linkages between policy and science so that there is common understanding and ability to communicate across regional and international boundaries. For example, language, history, cultural practices and traditions are important elements for negotiating and mutual decision-making.

Therefore, although all courses follow a pluridisciplinary approach and put the emphasis on the communication dimensions, they differ by the type of competence they should build: for the political and more general level, they will be university graduate courses, while for the practitioner's level, they will be professional training courses, addressing already well-formed specialists.

2.5 Target Groups

- Group 1, the political level, includes future decision-makers and planners, intergovernmental negotiators, such as graduate students specializing in political science, public policy, human science focusing on politics, business and management with an interest in resources economics and policies. They are concerned by university graduate courses;
- Group 2, the practitioners' level, is comprised of confirmed practitioners encountering the transboundary groundwater management questions listed in the objectives mentioned above (e.g. civil engineers, environmental scientists, geographers, political scientists, human scientists, lawyers, economists). They are concerned by professional training courses;
- Group 3 reflects the present needs of sustainable development or new communication by considering the involvement of the population in participating in its development decision-making and implementation, according to the UN sustainable development principles. Elementary and high school teachers, media and information decision-makers, as well as military officers are important representative of this group. This group is concerned with both university graduate courses and professional training courses.

2.6 Conclusion

It was decided that a course should be comprised of two stages:

- A leveling stage, to harmonize the communication and languages between the participants, for instance, the hydrological cycle and basic groundwater flow and characteristics would be taught to policy students;

- An integration level where the participants would follow the same teaching, emphasizing the specific aspects of transboundary aquifers.

Another conclusion concerned the follow-up of the preliminary phase, it was decided that the contents of a practitioners' course would be identified and related to a specific geographical area; and that an experimental course would be organized in 2007 to test these contents and also the two-stage method. The elaboration of the program would then proceed from the results of this experiment.

References

Puri, S., 2001, *Internationally Shared Aquifer Resources Management, Their Significance and Sustainable Management, A Framework Document*, S. Puri, ed., UNESCO-IHP, Series on Groundwater No1; (July 30, 2007); http://unesdoc.unesco.org/images/0012/001243/124386e.pdf.

IGRAC (July 30, 2007), http://www.igrac.nl.

ISARM (July 30, 2007), http://www.isarm.net.

WHYMAP (July 30, 2007), http://www.bgr.bund.de/EN/Themen/Wasser/Projekte/Berat_Info/whymap/whymap_projektbeschr.html.

Chapter 3
Transboundary Aquifers in International Law: Towards an Evolution

Raya M. Stephan*

Abstract This chapter presents the development of international law in transboundary aquifers. While few years ago, State practice over transboundary aquifers was very scarce and almost non-existent, the situation is now showing signs of change. States sharing an aquifer are seeking to establish cooperation mechanisms over their common resource, and are introducing the consideration of a legal component in projects on transboundary aquifers. In June 2006, draft articles on the law of transboundary aquifers have been adopted at first reading by the UN International Law Commission.

Keywords UN International Law Commission, draft articles on law of transboundary aquifers, EU Water Framework Directive

3.1 Introduction

Groundwater represents as much as 97% of the Earth's freshwater fraction in liquid form (excluding water locked in polar ice caps) (Foster, 1999). In arid and semi-arid regions, it is often the only source of water. Most of this groundwater is found in transboundary aquifers. Despite an increased dependency on groundwater resources leading to overexploitation, depletion and pollution (World Water Assessment Program, 2003, Yamada, 2003b), groundwater, and more precisely transboundary aquifers, have received less attention at the international level than surface water. International law has typically considered groundwater only as a subsidiary to surface water (Stephan, 2006). This is the case in the UN Convention on the Law of Non-Navigational Uses of International Watercourses of 1997, and of the wide State practice over transboundary waters. In this chapter, I will argue

Raya M. Stephan
UNESCO - International Hydrological Programme (IHP) 1, rue Miollis, 75732 Paris Cedex 15, France
* To whom correspondence should be addressed. E-mail: r.stephan@unesco.org

C.J.G. Darnault (ed.), *Overexploitation and Contamination of Shared Groundwater Resources.*

that a change seems to be occurring. The UN International Law Commission, after considering the topic of transboundary aquifers since 2003, has adopted *at first reading* in June 2006, a set of draft articles on the law of transboundary aquifers. Also, the legal element is emerging more and more as a central component in projects over transboundary aquifers.

3.2 Groundwater as a Subsidiary to Surface Water

Until recently, international law has given little attention to groundwater resources and transboundary aquifer systems, following a global general trend (Campana, 2005). In the only global convention on the use of water resources, and in most interstate treaties and agreements on transboundary waters, groundwater is considered only when it is related to a surface water body. There are though few exceptions.

3.2.1 The UN Convention on the Law of Non-Navigational Uses of International Watercourses (May 21, 1997)

On May 21, 1997, the UN General Assembly adopted with a large majority of votes (106 affirmative votes, 26 abstentions and 3 negative votes) the Convention on the Law of Non-navigational Uses of International Watercourses, based on the draft articles submitted by the International Law Commission (ILC) which is the UN body in charge of the progressive development and the codification of international law (UN Doc. A/RES/51/229, available at http://www.un.org/law/cod/watere.htm, hereinafter UN Watercourse Convention).

3.2.1.1 The Definition of a "Watercourse"

The Convention represents the latest authority in international water law. It has however not yet formally entered into force, as the number of 35 required ratifications (article 36) has not been reached yet. The Convention includes groundwater in its coverage but in a very limited way. In article 2 on the "Use of terms", the Convention defines a watercourse as "a system of surface waters and groundwaters constituting by virtue of their physical relationship a unitary whole and normally flowing into a common terminus" (article 2 paragraph a). An international watercourse is defined as "a watercourse, parts of which are situated in different States" (article 2 paragraph b). This leaves out important aquifer systems which are unrelated to surface water such as the North Western Sahara Aquifer System (Algeria, Libya, Tunisia). However, groundwater and surface water, even when they are related, do not necessarily "share" a common terminus. In reality, surface water and groundwater rarely flow to a common terminus.

Therefore, regarding groundwater, the UN Watercourse Convention appears limited in its scope. It only considers groundwater when it is related to surface water, and flowing to a common terminus.

As for its content, the Convention has codified the two main rules of international water law: the equitable and reasonable utilization principle, and the no harm rule, giving supremacy to the first one on the second one (Stephan, 2006). It has also adapted the main principle of international law, the principle of international cooperation, to the topic of international watercourses.

3.2.1.2 The Resolution on "Confined" Transboundary Groundwater

Acknowledging it had left out of its scope other types of groundwater, the ILC adopted in 1994 a "Resolution on confined transboundary groundwater". It is worth mentioning that the ILC does not use here the word "confined" in its hydrogeological meaning. It uses it to designate an aquifer unrelated to surface water while for hydrogeologists, a confined aquifer is an "Aquifer overlain and underlain by an impervious or almost impervious formation" (UNESCO/WMO 1992). This misuse of scientific terms by the legal community has caused a misunderstanding between lawyers and hydrogeologists, or more generally between lawyers and water experts, for years.

In this Resolution, the ILC commends States to be guided by the principles contained in the draft articles (the Convention), "where appropriate", in regulating transboundary groundwater (paragraph 1), and to consider entering into agreements with the other State or States in which the "confined" transboundary groundwater is located (paragraph 2). However, the ILC also recognized in this Resolution that these principles may not be appropriate and it acknowledged "the need for continuing efforts to elaborate rules pertaining to confined transboundary groundwater." In 1998, the Planning Group of the ILC identified shared natural resources as one of the topics for inclusion in the ILC's long-term program of work, (confined groundwater and single geological structures of oil and gas).

3.2.2 State Practice

While State practice is wide over surface waters, it is limited regarding transboundary aquifers.

3.2.2.1 Wide State Practice over Surface Waters

Interstate treaties and agreements concluded on transboundary waters concern in a very large majority international rivers and rarely address transboundary aquifers. Transboundary groundwaters is addressed when it interacts with the surface water body of concern. Table 3.1 below gives few examples of treaties and agreements addressing a surface water body, and giving some consideration to the related groundwaters.

Table 3.1 Selection of treaties on transboundary waters, and the surface and groundwater bodies they cover

Agreement	Surface Water Body	Groundwater Body
Agreement between the Federal Republic of Nigeria and the Republic of Niger concerning the equitable sharing in the Development, Conservation and Use of Their Common Water Resources (Maiduguri, July 18, 1990).	The Maggia/Lamido River basin. The Gada/Goulbi of Maradi River basin. The Tagwai/El Fadama River basin. The lower section of the Komadougou-Yobe River basin (article 1§2).	Groundwater contributing to the flow of surface waters (article 1§3).
Convention on Cooperation for the Protection and Sustainable Use of the River Danube (Sofia, June, 29 1994).	The Danube River.	Groundwater in the catchment area of the river (article 2§1).
Convention on the Protection of the Rhine (Berne, April 12, 1999).	The Rhine.	Groundwater interacting with the Rhine (article 2§a).
Revised Protocol on Shared Watercourses in the Southern African Development Community (2000).	Framework protocol, no Specific surface water body.	No specific groundwater body. Same definition of a watercourse than the UN Watercourse convention.
Convention on the Sustainable Development of Lake Tanganyka (Dar es Salaam, June 12, 2003).	Lake Tanganika (article 3).	"groundwaters that flow into the Lake" (article 1).
The Protocol for Sustainable Development of Lake Victoria basin (Arusha, November 29, 2003).	Lake Victoria.	"underground waters flowing into Lake Victoria" (article 1).

The result of this dominating State practice is the exclusion of a great number of transboundary aquifers from interstate agreements, and the absence of adequate and adapted provisions to cover the specific characteristics of aquifers. Few exceptions exist over transboundary aquifers. One major and remarkable exception is the Arrangement on the protection, utilization and recharge of the Franco-Swiss Genevese aquifer (June 9, 1977) which is the only example of a treaty dealing exclusively with the management of a transboundary aquifer. The

Arrangement creates a Genevese Management Commission with the following wide mandate:

- To propose a yearly aquifer utilization programme;
- To ensure the protection of the resource;
- To remedy possible causes of pollution;
- To inventory all existing water works;
- To manage the artificial recharge of the aquifer.

The Arrangement was negotiated for a period of 30 years. It will end in 2008. Negotiations on the Franco-Swiss borders involving water and other issues will start soon in 2007[1].

The other agreements worth mentioning have a more limited scope and concern two large transboundary aquifer systems with non-renewable groundwaters. They are nevertheless examples of first cooperation over transboundary aquifer systems. The first one is the Nubian Sandstone Aquifer System shared between Chad, Egypt, Libya and Sudan. In 1992, a Joint Authority was established between Egypt and Libya, and the two other States joined later. The Authority is mainly responsible for collecting and updating data and conducting studies. In the frame of the "Programme for the Development of a Regional Strategy for the Utilization of the Nubian Sandstone Aquifer System" conducted by Center for Environment and Development for the Arab Region and Europe (CEDARE), the four States signed two agreements on procedures for data collection, sharing and access to the data system (October 2000).

The second agreement is the North Western Sahara Aquifer System or SASS from its French acronym (Système Aquifer du Sahara Septentrional). The system is shared between Algeria, Libya and Tunisia. In December 2002, the three States agreed on the establishment of a consultation mechanism and approved the maintenance of the database and model of the aquifer, as well as the regular exchange of data and information.

3.2.3 The Recent Developments at the UN International Law Commission

In 2002 the ILC included in its programme of work the topic of "Shared Natural Resources" covering "confined" transboundary groundwaters, oil and natural gas. Ambassador Yamada, the Special Rapporteur appointed on the topic chose a step by step approach, and decided to start by the first item: transboundary groundwaters. He submitted three reports on the topic to the ILC.

---

[1] Oral communication by Sébastien Javogues, from the Communauté des Communes du Genevois, at the meeting on "Gestion partagée des eaux souterraines" organized by the Académie de l'eau, November 24, 2006, Paris, France.

3.2.3.1 From "Confined Transboundary Groundwaters" to Transboundary Aquifers

The first report (2003) (UN A/CN.4/533) is "a very preliminary one, dealing with the outlines of the topic", and its background at the ILC. In this report, the Special Rapporteur defines the scope of the study he has to undertake on groundwater as covering "water bodies that are shared by more than two States but are not covered by article 2 (a) of the UN Watercourse Convention". In the second report (2004) (UN A/CN.4/539) the Special Rapporteur presents seven draft articles intended to provoke substantive discussions at the ILC. However, it is worth noting that in the thinking of the Special Rapporteur an important shift took place. To start with, the Special Rapporteur acknowledges the difference between the meaning of the word "confined" as used by the Commission and its meaning for hydrogeologists (paragraph 13). He thus decides to drop this word. He also decides to "discard the concept of "confined", "unrelated" or "not connected", and to reconsider the assumption announced in his first report to cover "only those groundwaters not covered" by the Watercourse Convention. The Special Rapporteur has therefore widened the scope of his study to include all groundwaters. Moreover, he adopts the concept of aquifer, covering not only the waters but also the geological formation, the container and its content. The Special Rapporteur is meeting here the concerns of the hydrogeological community to which the geological formation is as important as the waters it contains. The seven draft articles include provisions on the scope, definitions, the obligation not to cause harm, the general obligation to cooperate, the regular exchange of data and information and the relationship between different types of uses.

3.2.3.2 A Full Set of Draft Articles on the Law of Transboundary Aquifers

In his third report of 2005 (UN A/CN.4/551), the Special Rapporteur presents a full set of draft articles on the law of transboundary aquifers. The seven previous articles are slightly amended to take into account the comments and suggestions from the debates at the ILC and at the 6th Committee (legal) of the UN General Assembly. The new articles include provisions on the reasonable and equitable utilization with specific factors concerning aquifers and the consideration of non-recharging aquifers, on monitoring, the protection, preservation and management of transboundary aquifers, and on the assistance to developing States. The Commission established then a Working Group to review the draft articles submitted by the Special Rapporteur. In 2005, the Working Group completed the consideration of eight draft articles and was reconvened in 2006 to complete the revision of the draft articles. The revised draft articles were deferred to the Drafting Committee of the ILC and were adopted by the Plenary of the ILC at first reading (Report of the International Law Commission, 2006). The ILC has also decided to request States to provide comments and observations by January 1, 2008.

3.2.4 The Role of UNESCO-IHP

In June 2000, at its 14th Session, the Intergovernmental Council of the International Hydrological Programme (IHP) of UNESCO, recognizing that transboundary aquifer systems are an important source of fresh water (particularly under arid and semi-arid climatic conditions) launched the International Transboundary Aquifer Resources Management project (ISARM) by adopting Resolution No. XIV-12. UNESCO-IHP decided to call other agencies to cooperate in order to establish an inter-agency initiative to promote studies in regard to transboundary aquifers. The objectives of the project are:

- To improve the existing scientific knowledge on transboundary aquifers;
- To compile an international inventory of transboundary aquifers;
- On the long term, to develop a toolkit for a management approach of transboundary aquifers.

Under this project, five focus areas for the proper management of transboundary aquifers were identified: scientific, environmental, socio-economic, legal and institutional (Puri, 2001). Within the framework of its ISARM project, UNESCO-IHP has since the year 2003 provided scientific and technical advices to the Special Rapporteur upon his request, on the issues related to hydrogeology, inviting, co-ordinating and supporting the contributions of international experts, international and national institutions, including centers on groundwater resources such as International Association of Hydrogeologists (IAH), Food and Agriculture Organization (FAO), United Nations Environment Programme/Global Environment Facility (UNEP/GEF), Organization of American States (OAS), International Union for the Conservation of Nature and Natural Resources (IUCN), International Groundwater Resources Assessment Center (IGRAC) and United Nations Economic Commission for Europe (UN ECE). Meetings to prepare the reports were held in Paris, Tokyo and Geneva, and briefings were conducted in Geneva at the ILC and in New York at the 6th Committee of the UN General Assembly to provide guidance on the science of hydrogeology.

3.2.5 Principles and Rules of the Draft Articles

The revised draft articles as adopted by the ILC represent a modified version of the draft articles submitted in the third report. The Working Group had introduced some amendments. For instance, a draft article on sovereignty was added. According to this draft article, an aquifer State has sovereignty over the transboundary aquifer or aquifer system to the extent it is located within its territorial jurisdiction. However, this sovereignty is not absolute. The reference to non-recharging aquifers, as well as their definition had been deleted. Non-recharging aquifers are therefore submitted to the same legal regime as recharging aquifers.

The draft article on the factors relevant to equitable and reasonable utilization, was also amended to give special regard to vital human needs in weighing different utilizations of a transboundary aquifer or aquifer systems. This provision, previously contained in another draft article in the third Report of the Special Rapporteur, was considered more appropriately placed in this draft article, since it sets forth an important principle relevant to determining the appropriate utilization of a transboundary aquifer or aquifer system. Regarding the draft article on the "Prevention, reduction and control of pollution", the Working Group discussions focused on whether more emphasis should be placed on prevention, and in that light, whether the precautionary principle should be placed in an independent article. Given the fragile nature of transboundary aquifers, it was agreed to strengthen the obligation to take a precautionary approach by changing the wording from "are encouraged to take a precautionary approach" to "shall take a precautionary approach". In the draft article on "Management", a last sentence was added mentioning that a joint management mechanism be established wherever appropriate. The Working Group thought that the strengthening of this obligation was particularly important in light of the value placed by groundwater experts on the joint management of transboundary aquifers. However, it also recognized that, in practice, it may not always be possible to establish such a mechanism. The draft articles include finally a provision on the scientific and technical cooperation with developing States which can be provided bilaterally, or through competent international organizations.

3.3 Evolution: Emerging State Practice

While few years ago, State practice over transboundary aquifers was very scarce and almost non-existent, the situation is now showing signs of change. States sharing an aquifer are seeking to establish cooperation mechanisms over their common resource, and are introducing the consideration of a legal component in projects on transboundary aquifers.

3.3.1 Consultation Mechanisms in Projects

3.3.1.1 The Iullemeden Aquifer System (Mali, Niger, Nigeria)

The Iullemeden Aquifer System (IAS) is located South of the Sahara in the arid and semi-arid region of West Africa. It corresponds to a portion of the Niger River basin and integrates Algeria, Benin, Mali, Niger and Nigeria. However, the IAS zone of the project is limited only to Mali, Niger and Nigeria and covers 525,000 km^2 with an exploitable reserve estimated at 2,000 km^3. The system has an important recharge capacity, mainly seasonal, through rivers and ponds. The water is used in majority for irrigation.

A GEF funded project on "Managing Hydrogeological Risk in the Iullemeden Aquifer System" will address the threats and risks on the aquifer system through the establishment of joint mechanism and cooperative frameworks for: (1) identification of transboundary risk and uncertainty issues, (2) formulation of joint risk mitigation and sharing policy; and (3) joint policy implementation through a joint IAS legal and institutional cooperative framework, (http://www.gefonline.org/projectDetails.cfm?projID=2041).

A first temporal consultation mechanism is suggested and is composed of:

- A secretariat;
- A steering committee (political level);
- Technical national committees.

This mechanism would be in charge of improving and maintaining the common database and identifying risks and vulnerable zones. It is expected that this temporal consultation mechanism will develop in a second phase towards a permanent consultation mechanism, which would be in charge of:

- Completing the joint identification of the system;
- Defining and managing the risks;
- Preparing legal and institutional recommendations for the Aquifer States;
- Sustainable development of the IAS.

3.3.1.2 The Coastal Aquifer System of the Gulf of Guinea (Benin, Ivory Coast, Ghana, Nigeria and Togo)

The coastal transboundary aquifer system located along the Gulf of Guinea and shared by five countries (Benin, Ivory Coast, Ghana, Nigeria and Togo) is facing severe deteriorations, such as saline water intrusion, due to poor overall management, and declining groundwater discharge.

A preliminary Transboundary Diagnostic Analysis has identified and prioritized significant threats on the aquifer system, groundwater pollution, wrong land use, saline intrusion, degradation of the coastal ecosystems, recharge reduction, groundwater overexploitation and global climate change impacts. Based on these results, a project on the "Joint Management of Coastal Aquifer in the Gulf of Guinea" has recently been submitted for GEF funding. This project includes a legal component, aiming at the establishment of a cooperative mechanism of technical assistance between the five countries. This mechanism would be conducted by the Project Steering Committee, in charge of supervising the General Coordinator. The General Coordinator would play the role of an implementing organs and would coordinate the activities of the national committee in each State.

3.3.2 Transboundary Aquifers in the EU Water Framework Directive

The topic of transboundary aquifers has also entered the area of specific legal regimes, such as one adopted in the European Union. The EU Water Framework Directive (WFD) (2000/60/EC), which came into force on December 22, 2000 (Official Journal of the European Community 22.12.2000, L 327/1) acts as an umbrella incorporating all water-related elements and topics, based on the concept of integrated river basin management.

The WFD extends to all aquatic systems, including surface waters (rivers and lakes), groundwater and coastal waters. Its objective is to ensure that "good status" is achieved for all surface and groundwater bodies, except for exceptional cases, in 2015. The WFD is based on the concept of integrated river basin management and can thus offer input for best practices in transboundary (surface and ground) water resources management.

Its main provisions regarding transboundary waters include the following requirements:

- Establishment of protected areas in each river basin district (i.e. including international ones), which include groundwater drinking water protected areas;
- Monitoring requirements for transboundary groundwater bodies;
- Requirement for coordination of single international river basin management plan in each international river basin district (hence covering transboundary aquifers);
- And the most important one is the request to delineate and characterize international river basin district and transboundary groundwater bodies (review of the impact of human activity on the groundwater status, and an economic value of water use) (article 5).

Member States of the European Union are requested to comply with the provisions of the directive. So far, eleven Member States with international river basins have assessed transboundary groundwater bodies.

3.3.3 Legal Component in Regional Projects

3.3.3.1 The Regional Development of ISARM in the Americas

The ISARM project was launched in the Americas in 2002, where it is jointly coordinated by UNESCO and the Organization of American States (http://www.oas.org/dsd/isarm/ISARM_index.htm).

In its first phase the programme has focused on the assessment of transboundary aquifers in the continent. One of the most important steps of the programme has been the collection of data on transboundary aquifers. Through preliminary questionnaires sent to the countries, UNESCO and OAS have assessed the prevalence of transboundary aquifers in the American continent in order to identify critical case studies while creating a comprehensive inventory of the transboundary aquifers for the Americas.

At its second workshop held in El Paso, Texas, in 2004, the legal and institutional issue was identified as one of the most important issues to be developed by the project in the future (UNESCO/OAS ISARM Americas Programme, 2005). The third workshop held in Sao Paulo, Brazil, in 2005 discussed the second phase of the ISARM Americas for 2006, and it was decided that the phase will focus on the Diagnostics of Institutional and Legal Framework as well as the socio-economic and environmental aspects. The activities in 2006 were mainly focused on the identification of the legal and institutional issues related to transboundary aquifers in the American countries. A legal questionnaire was thus prepared and circulated to the national coordinators in the participating countries. Its results were discussed and analyzed in a recent workshop held in San Salvador (El Salvador) in November 2006 (on file with author).

3.3.3.2 Capacity Building for Sustainable Utilization, Management and Protection of Internationally Shared Groundwater in the Mediterranean Region (ESCWA, ECE, ECA, in Partnership with UNESCO-IHP [ISARM])

An inter-regional project was formulated on "Capacity building for the sustainable utilization, management and protection of internationally shared groundwater in the Mediterranean region" aiming at strengthening the capacity of water management institutions in the Mediterranean region to implement sustainable forms of utilization, management and protection of internationally shared groundwater resources. Economic and Social Commission for Western Asia (ESCWA), ECE and Economic Commission for Africa (ECA) have forged this inter-regional Mediterranean initiative to address these prevailing issues (project document on file with author). Its main objective is to increase awareness and application by the Mediterranean Development Aid Programme (MEDA) countries of the international norms in the sustainable management of shared aquifers. The shared groundwaters represent a natural resource of critical, strategic social, economic, environmental and political importance to current, medium and long-term development in the water scarce MEDA region. In this sense the shared groundwaters are, at a time, a potential source of conflict as well as a common strategy for cooperation with mutual benefits for sustainable development and social and environmental security in the region. In the MEDA region, the shared aquifer resources dominate and are often the only available water supply to support social and environmental services. The MEDA economies depend largely, directly or indirectly, on the shared coastal aquifers along the Mediterranean to support coastal urban and agricultural development, control land degradation and salinization, and maintain coastal and marine ecosystems in the coastal zone that is of central economical importance to the MEDA.

The Mediterranean partners, comprise originally 10 MEDA countries, Algeria, Egypt, Jordan, Lebanon, Libya, Morocco, Syria, Tunisia and the Palestinian Authority. Since 2006 the MEDA countries are partners of the European Neighborhood Policy (ENP), with Algeria, Armenia, Azerbaijan, Belarus, Egypt, Georgia, Israel, Jordan, Lebanon, Libya, Moldova, Morocco, Palestinian Authority, Syria, Tunisia and Ukraine.

3.4 Conclusion

In the field of transboundary aquifers, State practice was very scarce, even non existent. In his first report, the Special Rapporteur of the ILC expresses his intention to collect State practices with respect to use and management (§25). In this regard, the ILC had prepared a questionnaire sent out to governments and relevant international organizations (UN A/CN.4/555 of April 29, 2005 and Doc. A/CN.4/555/Add.1 of July 15, 2005). Few answers were received. However, during the discussions on the work of the ILC at the 6th Committee of the UN General Assembly, the comments by the official delegates of the member States were globally positive, acknowledging the importance of transboundary aquifers, and the need to adopt rules to govern their management, and favorable to the approach adopted by the Special Rapporteur in dealing with the topic.

Since then, State practice has been slowly emerging, as it appears from the examples presented in the last part of this chapter. The process leading to the establishment of rules of international law is slow. Regarding transboundary aquifers, it seems that this process is now on.

References

Campana, M. E., 2005, Foreword: Transboundary ground water, *Ground Water* **43**(5):646–650.

Foster, S., 1999, Essential concepts for groundwater regulators, in: *Groundwater Legal and Policy Perspectives*, S. M. Salman, ed., Proceedings of a World Bank Seminar, Washington, DC: World Bank.

Puri, S., ed., 2001, *Internationally Shared Aquifer Resources Management, Their Significance and Sustainable Management, A Framework Document*, UNESCO-IHP, Series on Groundwater No.1, (August 4, 2007); http://unesdoc.unesco.org/images/0012/001243/124386e.pdf.

Report of the International Law Commission, 2006, 58th session, A/61/10, p.185.

Stephan, R., 2006, Evolution of international norms and values for transboundary groundwater governance, in: *Governance as a Trialogue: Government-Society-Science in Transition*, A. R., Turton, D. R. M., Claassen, and J. Hattingh, eds., Springer-Verlag, Berlin.

UNESCO/OAS ISARM Americas Programme, 2005, *Third Coordination Workshop*, Sao Paulo, Brazil, Texas, USA, November 30–December 2, 2005, Final report, Washington DC (August 7, 2007); http://www.oas.org/dsd/isarm/Documents/English/ISARMAmericas2005SaoPaulo Reportfeb06final.pdf.

UNESCO/WMO, 1992, *International Glossary of Hydrology*, 2nd revised ed.

World Water Assessment Program, 2003, *Water for People, Water for Life*, The United Nations World Water Development Report.

Yamada, C., 2003a, Special Rapporteur, *Shared Natural Resources: First Report on Outlines*, UN A/CN.4/533.

Yamada, C., 2003b, Special Rapporteur, *Shared Natural Resources: First Report on Outlines*, UN A/CN.4/533/Add.1.

Yamada, C., 2004, Special Rapporteur, *Second Report on Shared Natural Resources: Transboundary Groundwaters*, UN A/CN.4/539.

Yamada, C., 2005, Special Rapporteur, *Third Report on Shared Natural Resources: Transboundary Groundwaters*, UN A/CN.4/551.

Part II
Techniques to Assess the Depletion and Contamination of Shared Groundwater Resources

Chapter 4
Groundwater Geophysics: From Structure and Porosity Towards Permeability?

Klaus Holliger*

Abstract This chapter discusses the potential for obtaining order-of-magnitude-type constraints on the permeability structure of unconsolidated alluvial aquifers through a variety of non-standard geophysical measurements. Knowledge of the permeability distribution within an aquifer is a key prerequisite for reliable predictions of fluid flow and contaminant transport. This information is in turn critical for the effective protection, remediation and sustainable management of the increasingly scarce and fragile surficial groundwater resources in densely populated and/or highly industrialized regions throughout the world. Geophysical constraints with regard to the permeability structure are considered to be particularly valuable because some of them have the potential to bridge the inherent gap in terms of resolution and coverage that exists among traditional hydrological methods, such as core analyses and tracer or pumping tests. Although standard geophysical exploration techniques cannot in general provide direct information on the permeability structure of the probed medium, there are less conventional approaches that are expected to exhibit some more or less direct sensitivity to the permeability structure. Probably the most promising techniques for this purpose are analyses of the attenuation of Stoneley waves and seismic body waves, nuclear magnetic resonance techniques, induced polarization measurements, and gamma logs. Whereas the attenuation of seismic waves in saturated porous media is known to be more or less directly related to the permeability structure of the probed medium, and nuclear magnetic resonance techniques are sensitive to the water content and the pore size, induced polarization measurements and gamma logs exhibit a less direct relation to the permeability structure of unconsolidated sediments via their sensitivity to the grain size and/or the specific surface of the pore spaces.

Keywords Alluvial aquifers, groundwater, geophysics, porosity, permeability, Stoneley waves, seismic attenuation tomography, induced polarization, gamma logs

* Klaus Holliger
Institute of Geophysics, University of Lausanne, CH-1015 Lausanne, Switzerland

* To whom correspondence should be addressed. E-mail: klaus.Holliger@unil.ch

C.J.G. Darnault (ed.), *Overexploitation and Contamination of Shared Groundwater Resources.*

4.1 Introduction

4.1.1 Groundwater and Surficial Alluvial Aquifers

Approximately 70% of the global freshwater reserves is bound in the form of snow and ice and is predominantly concentrated in the Artic and Antarctic regions and hence largely unavailable for human consumption. The overwhelming part of the accessible global freshwater, approximately 97% of the total global reserves, corresponds to groundwater. The amount of freshwater carried by rivers and lakes is largely negligible in comparison. Groundwater is the principal source of drinking water for at least 30% of the world's population. For example, groundwater comprises 50% of the drinking water in North America and up to 80% in many European countries. Even in Switzerland, the country I am living and working in, with its temperate and rather humid climate and its abundance of clean rivers and lakes, more than 80% of the drinking water is drawn from groundwater resources. Here, as in many other parts of the world, unconsolidated surficial alluvial aquifers play a preeminent role. The primary reasons for this are that alluvial aquifers are characterized by very high porosities and permeabilities, are located close the Earth's surface and correspondingly accessible, and have excellent natural cleaning and remediation properties due to the physical filtration of suspended particles and chemical/biological degradation primarily of organic pollutants. Groundwater in general and surficial groundwater stored in alluvial aquifers in particular is thus one of the most widespread and most commonly used natural resources. Most importantly, it is a vital component of the human nutrition chain and in many parts of the world a critical, and often limiting factor for socio-economic development (e.g., Leap, 1999; UNEP, 2000).

In the typically water-rich, developed parts of the world, the unlimited availability of clean water has traditionally been regarded as a common good. For this reason, groundwater is likely to be grossly undervalued in economic terms. This will probably change as the availability of clean groundwater shrinks due to continuously rising demand from households, industry and agriculture, and as supply keeps waning due to the progressive sealing and contamination of aquifers that result from expanding urban settlements and growing industrial and agricultural activity, and due to dramatic trends towards more semi-arid weather regimes in moderate climate zones in the course of rapid global warming (e.g., UNEP, 2000). In response to rapidly rising costs, traditional public water utilities are increasingly susceptible to free-market-type approaches (e.g., The Economist, 2003). A crude, end-member-type estimate of the potential economic value of clean groundwater can be obtained by valuing it at the current market price of bottled table water. This suggests that the value of the yearly global extraction of groundwater is comparable to that of oil and gas at the current high fossil energy prices. Despite its hypothetical nature, this estimate clearly points to the growing need for and the importance of the protection, remediation and sustainable management of groundwater resources in the future. This in turn requires reliable local-scale knowledge of the spatial distribution of the pertinent hydrological properties. In particular, knowledge of the detailed distribution of the permeability within an aquifer is a key prerequisite for reliable predictions of fluid flow and contaminant transport (e.g., Delleur, 1999).

4.1.2 Groundwater Geophysics

Traditionally, the hydrological characterization of aquifers has been based on the analysis of drill cores as well as on the results of tracer and pumping experiments (e.g., Freeze and Cherry, 1979; Rubin, 2003). Core studies can provide detailed local information, but are inherently 1D in nature, whereas tracer and pumping tests tend to capture the gross average properties of a probed region. Due to this inherent gap in terms of resolution and coverage, traditional hydrological techniques may be inadequate for reliably characterizing typical heterogeneous aquifers (e.g., Sudicky, 1986). Modern environmental geophysical techniques have the potential to bridge this gap (e.g., Hubbard and Rubin, 2005). In numerous studies, the corresponding interdisciplinary field of research has provided valuable data at spatial and temporal scales generally not attainable with standard hydrologic measurements (e.g., Hyndman and Gorelick, 1996; Hubbard and Rubin, 2000; Hyndman et al., 2000; Hubbard et al., 2001; Hyndman and Tronicke, 2005; Paasche et al., 2006). In recent years, the terms hydrogeophysics or groundwater geophysics have come to describe these kinds of interdisciplinary research efforts that bridge hydrology and geophysics. The dynamic development of this field finds its evidence, for example, in a rapid growth of papers in the literature as well as publication of several recent review-type monographs on the subject (Rubin and Hubbard, 2005; Kirsch, 2006; Vereecken et al., 2006).

Unfortunately, conventional geophysical approaches do not in general allow for the direct estimation of permeability (e.g., Hyndman and Tronicke, 2005). There are, however, a number of geophysical techniques that are highly sensitive to the water content and thus to the porosity distribution, particularly in alluvial aquifers (e.g., Hubbard and Rubin, 2005). Porosity represents a prerequisite for permeability and hence these two key hydrological properties tend to be systematically and causally linked, albeit on a site- and/or facies-specific basis. One potential approach towards constraining the permeability structure of an aquifer is therefore through the quantitative integration of all available hydrological and geophysical data (e.g., Hyndman et al., 2000; Paasche et al., 2006). This approach is reviewed in Chapter 5. An alternative approach, which we discuss in this chapter, is through less-conventional geophysical approaches that are known or expected to exhibit some sensitivity to permeability or closely related parameters. The methods covered in this chapter are attenuation analyses of Stoneley and seismic body waves, which in porous materials are expected to be more or less directly related to the permeability structure, nuclear magnetic resonance measurements, which are sensitive to the water content and the pore sizes, as well as induced polarization measurements and gamma logs, which are expected to be related to the permeability structure of unconsolidated alluvial materials through their sensitivity to the grain size.

The potential of geophysical approaches for providing constraints on the permeability structure of unconsolidated surficial aquifers is still largely untapped. The corresponding research is novel, inherently trans-disciplinary in nature and hence not without risk. Given the rapidly growing need to manage and protect shallow groundwater resources, we do, however, believe that the potential rewards clearly outweigh the risks and hence also justify the somewhat speculative or forward-looking nature of certain aspects of this chapter. Although the following discussion focuses on saturated alluvial aquifers, it is expected that many of the key arguments can be extended to consolidated sedimentary and fractured "hardrock" aquifers.

4.2 Seismic Attenuation Measurements

The geophysical parameter that probably exhibits the most direct relation to the permeability of fluid-saturated porous media is the inelastic attenuation of seismic waves. Biot's (1962) seminal work clearly demonstrated that pressure re-equilibration between interconnected pore spaces associated with a seismic wave passing through a saturated porous medium may under certain conditions provide a direct link between inelastic seismic attenuation and permeability. Biot's theoretical framework and its various extensions are now generally referred to as poro-elasticity (e.g., Carcione, 2001).

4.2.1 Stonely Wave Logging

Probably the most tangible geophysical application of theory of poro-elasticity is the inversion of Stoneley waves for permeability (e.g., Cheng et al., 1987; Winkler et al., 1989; Tang and Cheng, 1996; Hearst et al., 2000). Stoneley waves are special types of surface or boundary waves traveling along the fluid–solid interface of water or mud-filled boreholes. It can be shown that below a certain threshold frequency, the attenuation of Stoneley waves is strongly dependent on the permeability of the rocks through which they travel (e.g., Hearst et al., 2000). Stoneley wave inversion has reached some degree of maturity in hydrocarbon exploration such that the method is now offered on a commercial basis and considered to be a viable tool for improving permeability models of oil and gas reservoirs, notably also in geologically complex prospects (e.g., Tang and Cheng, 1996; Qobi et al., 2001; Parra et al., 2006a). Unfortunately, the success of Stoneley wave analysis in exploration geophysics has not yet filtered through to hydrogeophysics. In our assessment, the major reasons are the following.

The source frequencies of the standard monopole-type sonic logging tools are too high to effectively excite Stoneley waves in the frequency range in which they are primarily sensitive to permeability (e.g., Taylor et al., 1990; Hearst et al., 2000). Standard sonic tools generally have nominal source frequencies between 10 and 20 kHz, whereas effective Stoneley wave excitation for permeability estimation in seismically slow unconsolidated surficial sediments occurs at frequencies around 1 kHz or less.

Boreholes in hydrocarbon exploration are generally uncased when logged whereas boreholes in surficial unconsolidated sediments tend to be cased with PVC tubes for stabilization. Depending on the final purpose of the borehole, this tubing is either not perforated at all or partially to continuously perforated.

The first issue can be effectively addressed by using sonic logging tools with dipole-type source radiation characteristics and effective source frequency spectra ranging significantly below 1 kHz (e.g., Tang and Cheng, 1996). The anisotropic energy radiation as well as the low source frequencies of such tools should allow for an effective excitation of flexural as well as Stoneley waves in the frequency range that is primarily sensitive to permeability variations even in seismically very slow unconsolidated sediments (Figure 4.1). Such sonic logging tools are indeed becoming increasingly common, as they provide an effective means to determine the shear wave velocity in soft surficial sediments, which is of significant geotechnical importance in its own right.

Addressing the second point is more complex and will probably require additional methodological work to understand the effects of the casing as well as its perforation or lack thereof on Stoneley waves in surficial boreholes and to find means of effectively compensating for such effects. There are indeed reasons to be hopeful as in the oil industry, similar studies have provided effective means for correcting for geometric irregularities of the borehole and the presence of a mudcake (e.g., Tang and Cheng, 1996; Qobi et al., 2001). In this context, it is interesting to note that, in the widest sense, a mudcake along a hydrocarbon well can be regarded as an analogon to a PVC-type casing of a shallow borehole. Moreover, early work by Tubman et al. (1984) indicates that even in the presence of a non-perforated steel casing cemented to the drilled lithological column, Stoneley waves still seem to exhibit some sensitivity with regard to the permeability structure of the undisturbed formation.

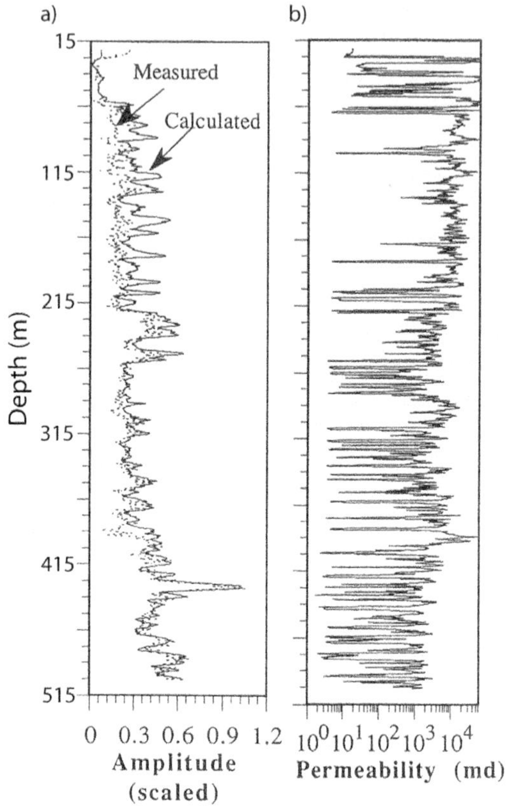

Fig. 4.1 **a** Observed versus modeled Stoneley wave amplitudes, and **b** inferred permeability distribution for a relatively soft and seismically slow sandstone formation (Adapted from Tang et al., 1996 with permission from SEG)

4.2.2 Waveform Inversion of Crosshole Seismic Data

Particularly in recent years, considerable attention has been given to extending the inherent relation between inelastic seismic attenuation and permeability from Stoneley waves to seismic body waves (e.g., Klimentos and McCann, 1990; Yamamoto, 2001; Pride et al., 2003). Although a comprehensive inversion strategy comparable to that existing for Stoneley waves may not yet be within reach, there is general agreement that research in this direction has great potential and that, in concert with suitable complementary information, reliable measurements of the inelastic attenuation of seismic body waves can provide valuable qualitative constraints with regard to the permeability structure and its spatial variability. In particular, recent work has unveiled an unexpectedly strong relation between local variations in permeability and inelastic attenuation of seismic body waves in the frequency range relevant to exploration and environmental geophysics (e.g., Pride et al., 2003). Moreover, Yamamoto (2001) obtained encouraging results estimating the permeability structure of unconsolidated surficial sediments from crosshole seismic attenuation tomograms (Figure 4.2). The corresponding tomographic inversions of the crosshole seismic data were, however, performed using ray-based approaches. Whereas the ray-based tomographic inversion of the first-arrival times is relatively robust and reliable, conventional ray-based attenuation tomograms are based on a number of generally unrealistic assumptions and hence generally prone to errors and artifacts. Moreover, a wave-equation-based inversion approach seems to be indicated because it offers significantly higher resolution and allows for a natural differentiation between scattering attenuation and inelastic attenuation, which represents a major source of uncertainty for essentially all other inversion approaches that are based on some approximation of the governing equations (e.g., Pride et al., 2003). Although a full visco-elastic, or even poro-elastic, waveform inversion approach might be desirable from a point of view of methodological rigor, available evidence suggests that the inversion of the entire seismic wave train recorded in crosshole seismic experiments is not yet practically viable. There are, however, some corresponding examples of visco-acoustic inversions of the first P-wave cycles of the transmitted part of the wavefield (e.g., Watanabe et al., 2004).

Although P-wave attenuation may not be directly translatable into absolute permeability values, it generally is related to changes in permeability. Moreover, the P-wave velocity of saturated clastic sediments is directly related to porosity (e.g., Schoen, 1998). Therefore, systematic deviations of the tomographically

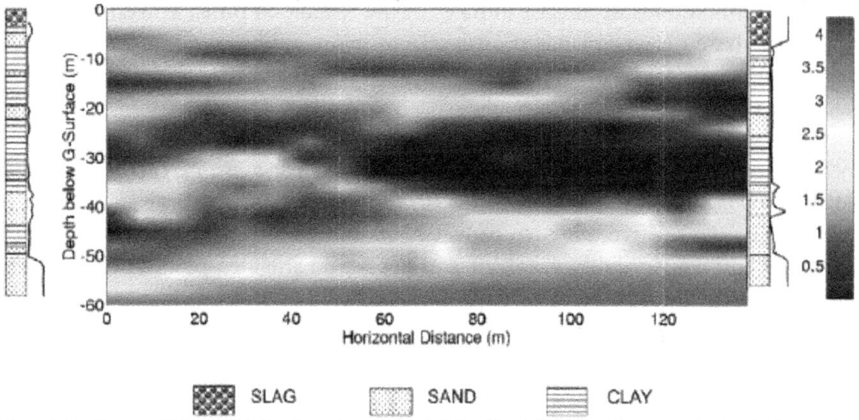

Fig. 4.2 Permeability distribution in a surficial alluvial formation as derived from the tomographic inversion of crosshole seismic traveltime and amplitude measurements (From Yamamoto, 2001 with permission of Elsevier)

inferred velocity and attenuation structures may provide valuable constraints with regard to changes in the relation between porosity and permeability.

4.3 Nuclear Magnetic Resonance Measurements

Nuclear magnetic resonance (NMR), which is based on the observation of secondary magnetic fields generated by precession of the spin magnetic moment of protons in the hydrogen atoms previously excited by external magnetic fields, is a well-established principle in physics (e.g., Levitt, 2002). It is the basis for a wide range of laboratory instruments and analytical techniques throughout the physical, chemical, biological and medical sciences, some of which have truly revolutionized the corresponding fields of research. In the earth and environmental sciences, the corresponding physical concepts have been exploited through the construction of proton procession magnetometers, NMR borehole logging tools and NMR surface sounding equipment (e.g., Hearst et al., 2000; Dunn et al., 2002; Legchenko and Valla, 2002).

To a first approximation, the amplitude of the NMR signal is primarily sensitive to the total amount of hydrogen atoms and thus to the total amounts of water or oil present, whereas the decay of the signal, as quantified by the relaxation time, is sensitive to the pore size, which is defined as the pore volume divided by the pore surface (e.g., Kenyon, 1997; Dunn et al., 2002). Given that, at least in unconsolidated sediments, the latter is expected to be directly related to permeability, this forms a basis for a number of empirical relations relating the relaxation time and/or the NMR-derived porosity to the permeability. These relations are, however, strongly non-linear and site- and/or lithology-dependent, which in turn requires accurate estimates of the relaxation times as well as careful calibration based on independent estimates of permeability of comparable spatial resolution and sensitivity (e.g., Hearst et al., 2000; Lubczynski and Roy, 2003, 2005; Serra and Serra, 2004).

In the hydrocarbon exploration industry, NMR-based logging techniques have indeed proved to be enormously successful (e.g., Hearst et al., 2000; Serra and Serra, 2004). Such measurements now not only represent the basis for accurate estimates of the relative amounts of water, oil and gas filling the pore space, but also for continuous, highly resolved and accurate estimates of the permeability structure along the well bore. A major reason for this success is that the borehole environment allows for an accurate determination of the relaxation times as well as for reliable laboratory-based calibrations based on cores or cuttings. Unfortunately, existing commercial NMR logging tools are in general too bulky for surficial applications, and despite the method's well-proven potential there are at present no commercial slim-hole NMR tools available. It is, however, important to note that the corresponding developments are not prevented by economic considerations but rather by fundamental technical impediments and that the basic potential of NMR logging for groundwater exploration is widely recognized (e.g., Parra et al., 2006b).

Surface-based NMR soundings have proven to be reliable means for determining water content as a function of depth. Difficulties with the accurate determination of the relevant relaxation times as well as the inherent calibration problems faced by all surface-based geophysical techniques have, however, somewhat hampered the method's usefulness as a reliable estimator of the permeability structure (e.g., Legchenko and Valla, 2002; Lubczynski and Roy, 2003, 2005). Recent innovations in hardware, experimental procedures and inversion techniques are, however, expected to overcome many of these problems (Figure 4.3). Moreover, developments are underway to extend the surface NMR method from a purely 1D sounding tool to a spatial imaging technique, which in conjunction with enhancements of the method's sensitivity to permeability has the potential to lead to something like a change in paradigm in certain domains of surface-based geophysical exploration of groundwater resources.

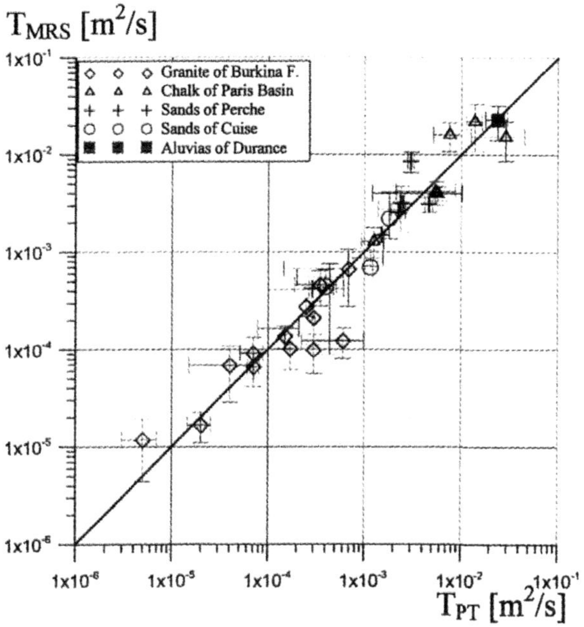

Fig. 4.3 Comparison of hydraulic transmissivities inferred from pump tests TPT and surface NMR measurements TMRS for a variety of lithologies (From Lubczynski and Roy, 2005 with permission of EAGE)

4.4 Grain-Size-Sensitive Geophysical Measurements

A somewhat different and less direct approach towards extracting constraints on the permeability structure of alluvial aquifers is through the sensitivity of certain geophysical techniques to the grain size, which is in turn a key determinant for the permeability of unconsolidated clastic sediments (e.g., Schoen, 1998). Indeed, grain size analysis is one of the oldest, simplest and most robust laboratory techniques used by hydrologists for estimating the permeability of granular materials (e.g., Freeze and Cherry, 1979). This is illustrated by the fact that as early as the late nineteenth century, Hazen (1892) postulated a seemingly universal linear relationship between the logarithm of the permeability and logarithms of the grain size of unconsolidated clastic sediments. The overall validity and universality of Hazen's law and subsequent modifications thereof was rigorously tested and confirmed by Shepherd (1989). He found that although the coefficients of the linear regression varied systematically with the textural maturity of the sediments, the linear correlation coefficients for a large number of widely differing real and synthetic samples were all remarkably high.

In the given context, the grain size is generally defined as a consistently representative measure, such as the mean, mode or median, of the entire grain size histogram of the considered sample (e.g., Freeze and Cherry, 1979; Shepherd, 1989). A widely used measure of grain size is the so-called effective grain size d_{10}, which corresponds to the grain size at which 10% of the sample weight is finer. The primary reason for this is that d_{10} seems to be a particularly versatile and robust measure for a remarkably wide variety of grain size histograms. Moreover, there is both theoretical as well as observational evidence to suggest that the permeability is dominated by the small pore spaces and hence by the small grain sizes present in any given sample, which are likely to be particularly well represented by a measure like d_{10} (Berg, 1970; Schoen, 1998; Slater and Lesmes, 2002). The geophysical techniques that probably exhibit the strongest and most direct sensitivity to the grain size are induced polarization measurements and gamma logs (e.g., Rider, 1991; Schoen, 1998; Hearst et al., 2000; Serra and Serra, 2004), which we shall discuss in the following.

4.4.1 *Induced Polarisation Measurements*

Geoelectric and induced polarization methods are based on the observation of the magnitude and form of the electrical potential resulting from the injection of a quasi-static or low-frequency alternating current into the Earth (e.g., Binley and Kemna, 2005). Of all surface-to-surface geophysical techniques, the electrical resistivity and induced methods are amongst those that can most readily be inverted to provide 2D or 3D tomographic images of the distribution of hydrologically relevant petrophysical parameters. In addition to this, geoelectric and induced polarization measurements can also be acquired in crosshole and borehole logging modes provided that the boreholes are water-filled and either uncased or cased with a continuously perforated, non-conducting PVC tubes. In the past, empirical relations between resistivity and permeability were developed. Although this approach is not entirely without promise in predominantly clay-free environments, it has been found to be extremely site- and facies-specific and hence has largely fallen out of favor. Conversely, laboratory work as well as limited field evidence indicate that order-of-magnitude-type estimates of permeability could be possible from induced polarization measurements (e.g., Boerner and Schoen, 1991; Slater and Lesmes, 2002; Kemna et al., 2004; Binley and Kemna, 2005).

The parameters typically measured in such experiments are the complex electrical conductivity for frequency-domain approaches and the chargeability for time-domain approaches. Induced polarization parameters measured by frequency- and time-domain approaches are known to be closely related. In particular, it can be shown that the complex part of the electrical conductivity observed in frequency-domain experiments is directly proportional to the so-called normalized chargeability derived from corresponding time-domain experiments (e.g., Slater and Lesmes, 2002). The latter simply corresponds to the chargeability divided ("normalized") by the bulk resistivity. Although the mechanisms at play are manifold and complex, surface polarization effects primarily depend on the total grain surface area and thus on the grain size, which can then be related to the permeability using either a simple Hazen-type relation or more complex, but conceptually equivalent measures, such as the specific surface to porosity ratio (Boerner and Schoen, 1991; Slater and Lesmes, 2002; Binley and Kemna, 2005). This work provides strong evidence for a systematic relation between induced polarization effects and grain-size-type measures in unconsolidated sediments ranging from clays to sands (Figure 4.4). The applicability of this concept to coarse-grained and often largely clay-free sand-to-gravel-type sediments that characterize many alluvial aquifers does, however, remain to be tested, both in the laboratory and in the field. Although it is generally assumed that induced polarization effects decrease in magnitude with increasing grain size, the available laboratory and field evidence indicates that the induced polarization effects of sandy to gravely sediments could be strong enough to allow for the detection of any such relationships (e.g., Schoen, 1998).

Finally, it is important to note that in order to convert tomographic induced polarization images into estimates of permeability, the problem of upscaling observations made at the laboratory scale to those made at the field scale must be addressed. In addition to discrepancies related to the differing scales of observations, discrepancies are expected to arise from the fact that resistivity and induced polarization methods are inherently non-unique, that their resolution decreases systematically with distance from the electrodes and that damping and/or smoothing constraints used to stabilize the numerical inversion process artificially limit the bandwidth of the data. One way to effectively bridge the gap in resolution between laboratory measurements and field-scale tomographic measurements could be through induced polarization logging. This approach is relatively well established and commonly used in mineral exploration, where the induced polarization method has its origins (e.g., Hearst et al., 2000). So far, there do, however, seem to be very few applications to hydrological problems (e.g., Ogilvy and Kuzmina, 1972) and hence we believe that the corresponding potential is significant and largely untapped.

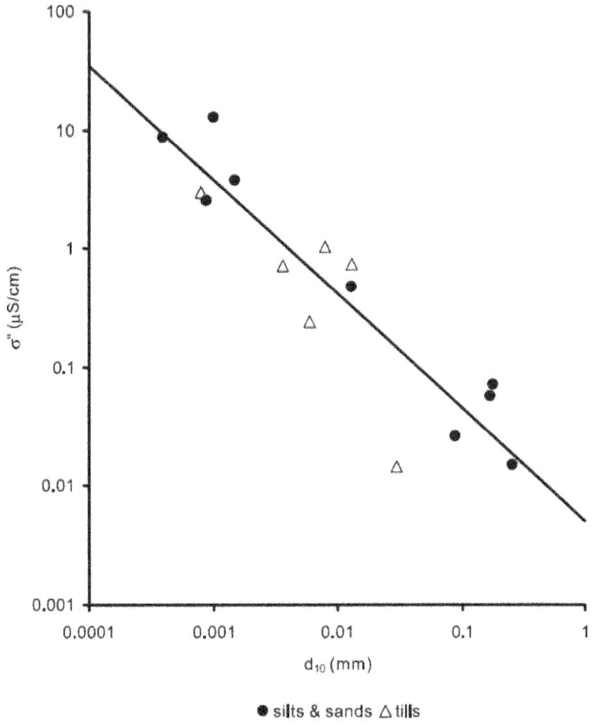

Fig. 4.4 Comparison of the complex part of the electrical conductivity σ" as inferred from laboratory-based induced polarisation measurements with the effective grain size d_{10} of various unconsolidated clastic sediments (From Slater and Lesmes, 2002 with permission of AGU)

4.4.2 *Gamma Logs*

Gamma logs measure the natural radioactivity of the rock column along a borehole, which primarily originates from the decay of the radioactive isotopes of potassium, thorium and uranium. In clastic sediments, the natural gamma radiation is generally dominated by the presence of clay minerals, which contain significant amounts of potassium and tend to selectively enrich mobile thorium. In the hydrocarbon exploration industry, gamma logs are therefore routinely used as quantitative measures of the clay content (e.g., Rider, 1991; Schoen, 1998; Hearst et al., 2000; Serra and Serra, 2004).

In general, there is remarkably good correspondence between the mineralogical and textural definitions of clay in clastic deposits (e.g., Hearst et al., 2000) and hence gamma logs have the potential to provide direct constraints with regard to grain size (e.g., Rider, 1991; Serra and Serra, 2004). The correspondence between the textural and mineralogical definitions of clay does in turn indicates that gamma logs are sensitive to an absolute range of grain sizes rather than to a relative measure of grain size, such as the effective grain size d_{10}, that is indicative of the grain size distribution of the considered sample and is expected to be related to permeability through a Hazen-type relationship (e.g., Freeze and Cherry, 1979; Shepherd, 1989; Klimentos and McCann, 1990; Slater and Lesmes, 2002). Nevertheless, there are good reasons to be hopeful of finding systematic relationships between the gamma log signature and the permeability structure of unconsolidated clastic sediments.

The pronounced sensitivity of gamma logs to grain size is indeed one of the foundations of well-log-based stratigraphy. For example, Rider (1991) and Serra and Serra (2004) report a number of excellent correlations between gamma and grain size logs for a relatively wide range of gamma activities and grain sizes (Figure 4.5). Although the seemingly universal relation between grain size and permeability for unconsolidated sediments is generally invalidated through diagenetic processes and hence not applicable to consolidated sediments (e.g., Schoen, 1998; Helle et al., 2001; Slater and Lesmes, 2002), it is worth noting that, for example, the work of Parra et al. (2006a), which shows a remarkably good qualitative agreement between the gamma log and Stoneley-wave-derived permeability log for a consolidated sandstone-shale sequence.

Both theoretical petrophysical models and laboratory and field evidence indicate that the permeability of unconsolidated sediments is effectively controlled by the small pore spaces and thus by the small grain diameters (e.g., Berg, 1970; Schoen, 1998). Indeed, Slater and Lesmes (2002) argue that it is largely for this reason that the effective grain size d_{10}, which is primarily representative of the smallest grain sizes present in a given distribution, is a particularly reliable and effective measure for relating grain size to permeability in unconsolidated sediments. It is therefore reasonable to assume that there could be systematic relations between the natural gamma activity and the permeability of unconsolidated alluvial deposits, similar to those found for induced polarization effects.

Spectral gamma ray measurements allow for the quantitative determination of the amounts of potassium, thorium and uranium present, which provides direct constraints on the relative amounts and chemical compositions of clay minerals dominating the small grain fractions and/or uranium-rich heavy minerals enriched in coarser grain fractions. Compared to conventional measurements of the bulk gamma ray activity, spectral gamma logs are therefore expected to provide additional compositional and textural constraints, as the dominant radioactive elements tend to be selectively enriched in different parts of the grain size spectrum (e.g., Schoen, 1998).

Finally, it is important to note that of all borehole logging techniques, bulk and spectral gamma logging techniques are probably the ones that are the most robust and the most universally applicable in shallow boreholes. The main reasons for this are that gamma logs are relatively insensitive to borehole filling medium (air, water or mud) or to the presence or absence of a casing as well as its perforation or lack thereof. Provided that the natural gamma radiation is not too weak and adequate corrections are applied, gamma logs can provide reliable results even in the presence of a steel casing (e.g., Hearst et al., 2000).

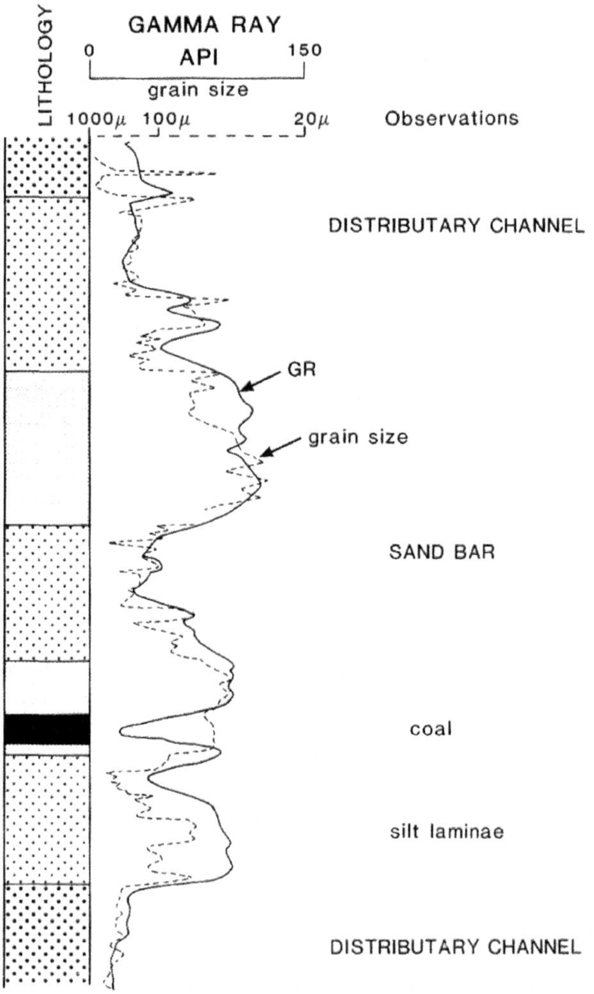

Fig. 4.5 Comparison of the bulk gamma radiation (solid line) and the average grain size (dashed line) for a consolidated clastic sequence (After Rider, 1991)

4.5 Conclusion

We have reviewed and discussed a number of geophysical techniques for groundwater exploration that we believe to have the potential of going beyond the conventional objective of providing constraints on the structure and/or porosity of aquifers. While the importance of these conventional constraints has become quite important in recent times, we believe that the greatest challenges as well as the greatest potential rewards lie in the extraction of constraints on the permeability structure of aquifers from remote-sensing-type geophysical investigations. There are important and far-reaching analogies between hydrocarbon reservoir and groundwater geophysics and it was the advent of geophysical permeability estimates that allowed for bridging the gap between the laboratory and reservoir scales and eventually led to the breakthrough of reservoir geophysics. Indeed, most of the methods discussed in this chapter have already proven their potential for providing permeability constraints in hydrocarbon exploration and are simply waiting to be adapted to surficial environments. One important reason that this has not happened are the widely differing returns on investment in hydrocarbon and groundwater exploration. The increasing tendency towards market-oriented supply and pricing models in the water industry, partially in response to the increasing scarcity and fragility of groundwater resources, could, however, eventually change this situation. While there are good reasons to be confident with regard to the potential of the discussed methods, it is important to realize that most likely none of these techniques will ever prove to be the "silver bullet". In particular, inferred parameter relations will always be site-specific and the need for rigorous calibration with laboratory measurements and/or hydrological data remains paramount. Still, we believe that being able to provide order-of-magnitude type constraints of the permeability structure and/or to changes thereof will greatly enhance the usefulness of groundwater geophysics and thus enhance its acceptance in the hydrological community, which nowadays, and sometimes rightly so, is often still rather skeptical.

References

Berg, R. R., 1970, Method for determining permeability from reservoir rock properties, *Trans. Gulf Coast Assoc. Geol. Soc.* **20:**303–317.

Binley, A., and Kemna, A., 2005, DC resistivity and induced polarization methods, in: *Hydrogeophysics*, Y. Rubin, and S. Hubbard, eds., Springer, Dordrecht, The Netherlands, pp. 129–156.

Biot, M. A., 1962, Generalized theory of acoustic propagation in dissipative media, *J. Acoust. Soc. Am.* **34:**1254–1102.

Boerner, F. D., and Schoen, J. H., 1991, A relation between the quadrature component of electrical conductivity and the specific surface area of sedimentary rocks, *Log Anal.* **32:**612–613.

Carcione, J. M., 2001, *Wave Fields in Real Media: Wave Propagation in Anisotropic, Anelastic and Porous Media*, Pergamon, Amsterdam, The Netherlands.
Cheng, C. H., Jinzhong, Z., and Burns, D. R., 1987, Effects of in-situ permeability on the propagation of Stoneley tube waves in a borehole, *Geophysics* **52:**1279–1289.
Delleur, J. W., 1999, Elementary groundwater flow and transport processes, in: *Handbook of Groundwater Engineering*, J. W. Delleur, ed., CRC Press, Boca Raton, Florida, U.S.A., Chapter 2.
Dunn, K. J., Bergman, D. J., and Latorraca, G. A., 2002, *Nuclear Magnetic Resonance: Petrophysical and Logging Applications*, Elsevier, Kidlington, Oxford, U.K.
Freeze, A. R., and Cherry, J. A., 1979, *Groundwater*, Prentice-Hall, Upper Saddle River, N.J., U.S.A.
Hazen, A., 1892, Some physical properties of sands and gravels, Massachusetts State Board of Health, Annual Report, pp. 539–556.
Hearst, J. R., Nelson, P. H., and Paillett, F. L., 2000, *Well Logging for Physical Properties: A Handbook for Geophysicists, Geologists and Engineers*, Wiley, Chichester, West Sussex, England.
Hearst, J. R., Nelson, P. H., and Paillett, F. L., 2000, *Well Logging for Physical Properties:* 2nd edition, Wiley and Sons, Inc., New York, 492p.
Helle, H. B., Bhatt, A., and Ursin, B., 2001, Porosity and permeability prediction from wireline logs using artificial neural networks: A North Sea case study, *Geophys. Prosp.* **49:**431–444.
Hubbard, S. S., and Rubin, Y., 2000, Hydrogeological parameter estimation using geophysical data: A review of selected techniques, *J. Contam. Hydrol.* **45:**3–34.
Hubbard, S. S., and Rubin, Y., 2005, Introduction to hydrogeophysics, in: *Hydrogeophysics*, Y. Rubin, and S. S. Hubbard, eds., Springer, Dordrecht, The Netherlands, pp. 3–21.
Hubbard, S., Chen, J., Peterson, J. E., Majer, E. L., Williams, K. H., Swift, D. J., Mailloux, B., and Rubin, Y., 2001, Hydrogeological characterization of the South Oyster bacterial transport site using geophysical data, *Water Resour. Res.* **37:**2431–2456.
Hyndman, D. W., and Gorelick, M., 1996, Estimating lithologic and transport properties in three dimensions using seismic and tracer data: The Kesterson aquifer, *Water Resour. Res.* **32:**2659–2670.
Hyndman, D. W., and Tronicke, J., 2005, Hydrogeophysical case studies at the local scale: The saturated zone, in: *Hydrogeophysics*, Y. Rubin, and S. Hubbard, eds., Springer, Dordrecht, The Netherlands, pp. 391–412.
Hyndman, D. W., Harris, J. M., and Gorelick, S. M., 2000, Inferring the relation between seismic slowness and hydraulic conductivity in heterogeneous aquifers, *Water Resour. Res.* **36:**2121–2132.
Kemna, A., Binley, A., and Slater, L., 2004, Cross-borehole IP imaging for engineering and environmental applications, *Geophysics* **69:**97–105.
Kenyon, W. E., 1997, Petrophysical principles of application of NMR logging, *Log Anal.* **38:**21–43.
Kirsch, R., ed., 2006, *Groundwater Geophysics*, Springer, Germany.
Klimentos, T., and McCann, C., 1990, Relationships among compressional wave attenuation, clay content, and permeability in sandstones, *Geophysics* **55:**998–1014.
Leap, D. I., 1999, Geological occurrence of groundwater, in: *Handbook of Groundwater Engineering*, J. W. Delleur, ed., CRC Press, Boca Raton, Florida, U.S.A., Chapter 1.
Legchenko, A., and Valla, P., 2002, A review of the basic principles for proton magnetic resonance sounding measurements, *J. Appl. Geophys.* **50:**3–19.
Levitt, M. H., 2002, *Spin Dynamics—Basics of Nuclear Magnetic Resonance*, Wiley, Chichester, West Sussex, England.
Lubczynski, M., and Roy, J., 2003, Hydrogeological interpretation and potential of the new magnetic resonance sounding (MRS) method, *J. Hydrol.* **28:**319–340.
Lubczynski, M., and Roy J., 2005, MRS contribution to hydrogeological system parameterization, *Near Surf. Geophys.* **131:**131–139.
Ogilvy, A. A., and Kuzmina, E. N., 1972, Hydrogeologic and engineering-geologic possibilities for employing the method of induced potentials, *Geophysics* **37:**839–861.

Paasche, H., Tronicke, J., Holliger, K., Green, A. G., and Maurer, H., 2006, Integration of diverse physical-property models: Subsurface zonation and petrophysical parameter estimation based on fuzzy c-means cluster analyses, *Geophysics* **71**:H33–H44.

Parra, J. O., Hackert, C. L., Xu, P. -C., and Collier, H. A., 2006a, Attenuation analysis of acoustic waveforms in a borehole intercepted by a sand-shale sequence reservoir, *Lead. Edge* **25**:186–193.

Parra, J. O., Hackert, C. L., and Bennett, M. W., 2006b, Permeability and porosity images based on P-wave surface seismic data: Application to a south Florida aquifer, *Water Resour. Res.* **42**:W02415, doi:10.1029/2005WR004114.

Pride, S. R., Harris, J., Johnson, D. L., Mateeva, A., Nihei, K., Nowack, R. L., Rector, J., III., Spetzler, H., Wu, R., Yamomoto, T., Berryman, J., and Fehler, M., 2003, Permeability dependence of seismic amplitudes, *Lead. Edge* **22**:518–525.

Qobi, L., de Kuijper, A., Tang, X. M., and Strauss, J., 2001, Permeability determination from Stoneley waves in the Ara group carbonates, Oman, *GeoArabia* **6**:649–666.

Rider, M. H., 1991, *The Geological Interpretation of Well Logs*, Whittles Publishing, Caithness, Scotland.

Rubin, Y., 2003, *Applied Stochastic Hydrology*, Oxford University Press, New York, N.Y., U.S.A.

Rubin, Y., and Hubbard, S. S., eds., 2005, *Hydrogeophysics*, Springer, Dordrecht, The Netherlands.

Schoen, J. H., 1998, *Physical Properties of Rocks: Fundamentals and Principles of Petrophysics*, Pergamon, New York, U.S.A.

Serra, O., and Serra, L., 2004, *Well Logging: Data Acquisitions and Applications*, Méry Corbon, France.

Shepherd, R. G., 1989, Correlations of permeability and grain size, *Ground Water* **27**:633–638.

Slater, L., and Lesmes, D. P., 2002, Electrical-hydraulic relationships observed for unconsolidated sediments, *Water Resour. Res.* **38**:1213, doi:10.1029/2001WR001075.

Sudicky, E. A., 1986, A natural gradient experiment on solute transport in sand aquifers: Spatial variability of hydraulic conductivity and its role in the dispersion process, *Water Resour. Res.* **22**:2069–2089.

Tang, X. M., and Cheng, C. H., 1996, Fast inversion of formation permeability from Stoneley wave logs using a simplified Biot-Rosenbaum model, *Geophysics* **61**:639–645.

Taylor, K., Wheatcraft, S., Hess, J., Hayworth, J., and Molz, F., 1990, Evaluation of methods for determining the vertical distribution of hydraulic conductivity, *Ground Water* **28**:88–98.

The Economist, 2003, Priceless/Survey: Water, *The Economist* July 17, 2003.

Tubman, K. M., Cheng, C. H., and Toksoez, M. N., 1984, Synthetic full waveform acoustic logs in cased boreholes, *Geophysics* **49**:1051–1059.

UNEP, 2000, *United Nations Environment Programme, Vital Water Graphics*, (August 6, 2007); http://www.unep.org/dewa/assessments/ecosystems/water/vitalwater.

Vereecken, H., Binley, A., Cassiani, G., Revil, A., and Titov, K., eds., 2006, *Applied Hydrogeophysics*, Springer.

Watanabe, T., Nihei, K. T., Nakagawa, S., and Myer, L. R., 2004, Viscoacoustic wave form inversion of transmission data for velocity and attenuation, *J. Acoust. Soc. Am.* **115**:3059–3067.

Winkler, K. W., Liu, H. -L., and Johnson, D. L., 1989, Permeability and borehole Stoneley waves: Comparison between experiment and theory, *Geophysics* **54**:66–75.

Yamamoto, T., 2001, Imaging the permeability structure within the nearsurface sediments by acoustic crosswell tomography, *J. Appl. Geophys.* **47**:1–11.

Chapter 5
Quantitative Integration of Hydrogeophysical and Hydrological Data: Geostatistical Approaches

Klaus Holliger*, Jens Tronicke, Hendrik Paasche, Baptiste Dafflon

Abstract Geophysical techniques can help to bridge the rather broad gap that exists with regard to resolution and coverage for classical hydrological methods. Given the differing sensitivities of various geophysical techniques to hydrologically relevant parameters and their inherent trade-off between resolution and range as well as the inherently site-specific nature of petrophysical parameter relations, the fundamental usefulness of multi-method surveys for reducing uncertainties in data analysis and interpretation is widely accepted. A major challenge arising from such endeavors is the quantitative integration of the resulting generally vast and often diverse databases in order to obtain a unified model of the probed subsurface region that is internally consistent with the entire available database. In this chapter, we review two approaches towards hydrogeophysical data integration that we consider to be particularly suitable and promising as well as largely complementary in their purposes: cluster analysis and Monte-Carlo-type conditional stochastic simulation. Cluster analysis allows for detecting systematic interrelations between various parameters and, based on this information, for producing internally consistent zonations of the target region. Under certain conditions, some of these techniques also allow for a robust and efficient reconstruction of the distribution of the petrophysical target parameters. An entirely different approach to hydrogeophysical data integration is based on Monte-Carlo-type conditional stochastic simulations. These techniques are immensely flexible and versatile, allow for accounting for a wide variety of data and constraints of vastly differing resolution and hardness, and thus have the potential of providing, in a geostatistical sense, highly detailed and realistic models of the pertinent target parameter distributions.

Klaus Holliger, Baptiste Dafflon
Institute of Geophysics, University of Lausanne, CH-1015 Lausanne, Switzerland

Jens Tronicke, Hendrik Paasche
Institute for Geosciences, University of Potsdam, 14415 Potsdam, Germany

* To whom correspondence should be addressed: E-mail: Klaus.Holliger@unil.ch

C.J.G. Darnault (ed.), *Overexploitation and Contamination of Shared Groundwater Resources*.

Keywords Aquifer characterization, cluster analysis, conditional simulation, geostatistics, hydrogeophysics, hydrology, quantitative data integration

5.1 Introduction

Traditionally, aquifer characterization is based on the analysis of drill cores and geophysical borehole logs as well as on the results of tracer and pumping experiments (e.g., Hubbard and Rubin, 2005). Core and logging studies can provide highly detailed local information but are inherently 1D in nature, whereas tracer and pumping tests tend to capture the gross average properties of a probed region. Without complementary information, these techniques may thus be inadequate for reliably characterizing laterally heterogeneous aquifers. The inherent gap in resolution and coverage between core/logging studies and pumping/tracer tests, which generally amounts to several orders of magnitude, can be bridged by high-resolution geophysical techniques (Table 5.1). As a result, the corresponding new field of research, generally referred to as hydrogeophysics or groundwater geophysics, has been evolving strongly, as evidenced by the rapid growth of papers in the literature, as well as the recent publication of several review-type textbooks on the subject (Rubin and Hubbard, 2005a; Kirsch, 2006; Vereecken et al., 2006).

Table 5.1 provides a summary of commonly used geophysical and hydrological techniques for aquifer characterization (e.g., Hubbard and Rubin, 2005). Given the wide gaps in resolution and coverage between traditional hydrological techniques on the one hand and the widely differing sensitivities to hydrologically relevant parameters as well as the notorious trade-off between resolution and range of the various geophysical techniques on the other hand, it is clear that the best results are likely obtained by using multiple methods.

To obtain a detailed and internally consistent aquifer model, the geophysical data thus need to be integrated with all other hydrological data as well any as other kind of relevant *a priori* information (Table 5.1). The quantitative integration of such a diverse database represents a major challenge due to the widely varying physical nature and hardness of the data, their differing scales of resolution and their deterministic or stochastic quantification (e.g., Hyndman et al., 2000; Rubin and Hubbard, 2005b; Paasche et al., 2006).

In recent years, significant progress in the quantitative characterization of hydrocarbon reservoirs has been made by integrating wide ranges of geophysical, petrophysical and production data through geostatistical interpolation, classification and simulation techniques (e.g., Deutsch, 2002; Kelkar and Perez, 2002). Indeed, there are close analogies between hydrocarbon reservoirs and aquifers in general and their geophysical characterization in particular (e.g., Szerbiak et al., 2001). An important difference between the two disciplines is, however, that

Table 5.1 Qualitative classification of hydrogeophysical and hydrogeological exploration techniques with regard to their resolutions and sensitivities to key hydrological parameters

Method	Primary Physical Parameters Inferred (Typical Resolution)	Inferred Hydrological Parameters (Relation to Primary Physical Parameters)
Crosshole georadar tomography	Electromagnetic wave velocity (~1 m)	Water-saturated porosity (strong)
Crosshole seismic tomography	Compressional wave velocity (~1 m)	Water-saturated porosity (relatively strong)
Density logs	Bulk density (~0.1 m)	Bulk porosity (relatively strong)
Sonic logs	Compressional wave velocity (~0.1 m)	Water-saturated porosity (relatively strong)
Neutron porosity logs	Water content (~0.1 m)	Water-saturated porosity (~1:1)
Surface and crosshole electromagnetic sounding	Electrical resistivity (~several meters)	Porosity, permeability, saturation, salinity (weak, ambiguous)
Surface and crosshole geoelectric and induced polarization sounding	Electrical resistivity (~several meters)	Porosity, permeability, saturation, salinity (weak, ambiguous)
Surface georadar profiling	Spatial variations in electromagnetic impedance (~1 m)	Hydrogeological structure (weak, ambiguous)
Surface reflection and refraction seismic profiling	Spatial variations in seismic velocity and impedance (~several meters)	Hydrogeological structure (weak, ambiguous)
Core analyses	Porosity and permeability (for unconsolidated sediments generally >1 m)	Porosity and permeability (~1:1)
Pumping and slug tests	Permeability (~several meters)	Permeability (~1:1)
Tracer tests	Hydraulic velocity and dispersivity (~several meters)	Permeability (strong)
Flowmeter logs	Permeability (<1 m)	Permeability (~1:1)
"Direct-push" measurements	Relative and/or absolute permeability (~0.1 m)	Relative and/or absolute permeability (~1:1)

high-resolution crosshole tomographic methods, which are widely used in hydrogeophysical studies, are usually unavailable for the geophysical characterization of hydrocarbon reservoirs. It is therefore reasonable to assume that the integration of geophysical and hydrological data through suitably adapted geostatistical techniques will eventually prove to be at least as successful as their more established applications in the hydrocarbon industry. Given the rapidly growing need for managing and protecting groundwater resources on a local scale, this clearly is an interesting and exciting prospect, which should be systematically explored and pursued (e.g., Hyndman et al., 2000; Hubbard et al., 2001; Tronicke and Holliger, 2005).

After a brief review of commonly used geostatistical data integration techniques, this chapter discusses two approaches that we consider to be particularly promising and complementary: multivariate statistical data analyses and classifications based on cluster analysis and Monte-Carlo-type conditional stochastic simulations based on simulated annealing. Given the wide variety of techniques available, this choice is inherently subjective and primarily reflects our experience as well as our conviction that these methods are particularly suitable for the reliable structural and/or lithological aquifer zonation on the one hand and the detailed reconstruction of the spatial distribution of hydrological and/or hydrogeophysical target parameters on the other hand.

5.2 Data Integration

The fundamental usefulness of multi-method geophysical surveys for reducing uncertainties in data analysis and interpretation is largely non-debated (e.g., Dannowski and Yaramanci, 1999; Hubbard et al., 2001; Garambois et al., 2002). A common way to integrate multiple and diverse geophysical surveys is to derive independent models, which are then jointly interpreted to obtain a single integrated model of the probed subsurface region. This approach is, however, largely qualitative in nature and hence the outcome depends heavily on the background, experience and preconceptions of the interpreter. Most importantly, this approach also largely excludes a rigorous quantitative assessment of the overall quality and internal consistency of the inferred integrated models.

A more quantitative approach is to link multiple data sets during the inversion and model generation process. As for essentially all quantitative data integration procedures, an important advantage of such joint inversion approaches is that the inherent ambiguity of each data set is reduced due to the additional constraints offered by the other data sets. However, joint inversion techniques usually require far-reaching *a priori* assumptions on the relations between the various parameters (e.g., Gallardo and Meju, 2003; Bosch, 2004; Kowalsky et al., 2005; Linde et al., 2006). The quality and consistency of the resulting multi-parameter models thus depend critically on the reliability of these assumptions and the way they are accommodated in the inversion procedure. Unfortunately, many, if not most, of these parameter relations tend to be non-unique, site-specific and/or spatially variable (e.g., Schoen, 1998). Nevertheless, ongoing algorithmic and computational improvements can be expected to further enhance the attractiveness and applicability of joint inversion approaches.

An entirely different class of approaches to quantitative data integration, which we consider in this chapter, is based on geostatistical principles. In its original definition, geostatistics essentially refers to the interpolation or extrapolation of generally sparsely sampled data based on the observed or inferred covariance structure of these data. The corresponding techniques are referred to as kriging or, if multiple datasets and their interrelations are considered, co-kriging. For convenience and consistence, we shall, however, adopt the considerably broader understanding of this term as practiced in hydrocarbon reservoir characterization, which essentially comprises all multivariate statistical and stochastic techniques that can be used for spatial data analyses (e.g., Deutsch, 2002; Kelkar and Perez, 2002).

Probably the simplest geostatistical approach *sensu stricto* for integrating multiple datasets is indeed through classical co-kriging based on the auto- and crosscovariances of the various parameters and observations (e.g., Cassiani et al., 1998; Gloaguen et al., 2001). This approach can be quite effective for providing smoothed reconstructions of spatial distributions of the target parameters. An inherent shortcoming of all such approaches is, however, that autocovariance functions of the inferred models differ substantially from those of actual data. Moreover, co-kriging assumes the relationships between the various datasets to be linear and unique, which in practice is, however, often not the case.

Bayesian statistics is probably the oldest and most classical technique for the quantitative integration of multiple and/or diverse datasets (e.g., Gelman et al., 2003). This approach requires an initial estimate of the probability distribution of the target parameter, as could be obtained, for example, by kriging or co-kriging of the hydrological database alone. This prior distribution is complemented through estimates of the joint probability distribution of collocated hydrological and geophysical data, which is then used to update the prior distribution in order to obtain the posterior distribution of the hydrological target parameter based on the additional constraints provided by the geophysical data. Bayesian approaches for the integration of hydrogeophysical and/or hydrological data have been successfully used by a number of scientists (e.g., Ezzedine et al., 1999; Chen et al., 2001). A recent comprehensive review of this topic is provided by Rubin and Hubbard (2005b).

In many cases, an important objective of integrated hydrological and hydrogeophysical data investigation is to detect relevant zonations within the probed aquifer as defined, for example, by the interrelations of the pertinent parameters. An effective approach for achieving this is to compare collocated data and use multivariate statistical data classification techniques, such as cluster analysis, support vector engines or neural networks, to divide them into a lithologically and/or hydrologically meaningful number of groups (e.g., Hoeppner et al., 1999; Tronicke et al., 2004; Paasche et al., 2006). Through their mutual relations, these data groups can then be mapped back into the probed aquifer region to define zones characterized by some mutual correlation of the various inferred parameters as well as by corresponding mean values and standard deviations. Cluster-based aquifer zonation and characterization techniques will be discussed in more detail in the following.

Finally, conditional simulations aim at finding models that reproduce all available data, constraints and *a priori* information. Ideally, the data can be characterized by various degrees of resolution and/or hardness and the constraints can be deterministic and/or stochastic in nature. There are basically two avenues in conditional simulation: sequential Gaussian co-simulations and Monte-Carlo-type approaches (e.g., Sen and Stoffa, 1995; Deutsch, 2002; Kelkar and Perez, 2002). Sequential Gaussian co-simulations have been successfully used by Hyndman et al. (2000) to integrate crosshole seismic and hydrological data. An important limitation of Gaussian co-simulations is, however, that, due to their close relation to the classical co-kriging mentioned above, they are based on the assumption of linear relationships between the various parameters. Conversely, Monte-Carlo-type approaches, which will be discussed in more detail in the following, are characterized by a very large degree of flexibility and allow for accounting for non-linear and/or non-unique parameter relations.

5.3 Cluster Analysis

Multivariate statistical methods, including cluster analysis techniques, are powerful tools for exploring and characterizing the relationships between various petrophysical parameters (e.g., Gill et al., 1993; Barrash and Morin, 1997; Bosch et al., 2002). The fundamental principle of cluster analysis is to group data points in a multidimensional space on the basis of their distances from each other. Unlike other multivariate techniques, clustering methods do not assume a specific distribution for the considered variables. It should also be noted that correlated variables do not violate any fundamental assumptions in cluster analysis. Because of their suitability for correlating, integrating and classifying information from a broad range of observations, cluster analysis techniques are widely used throughout the earth sciences (e.g., Dumay and Fournier, 1988; Gill et al., 1993; Hyndman and Harris, 1996; Hammah and Curran, 1998; Bosch et al., 2002; Gueler et al., 2002). A distinction can be made between partitioning and hierarchical clustering approaches. The former partition the observations into a predetermined number of homogeneous groups, whereas the latter produce nested clusters. A further distinction is made between hard or crisp and soft or fuzzy cluster algorithms. The former assign each data point to a distinct cluster, whereas the latter provide information on how strongly a data point is related to all clusters considered. The primary relative advantages of hard and soft clustering techniques are their robustness to outliers and their versatility, respectively. In the following, we briefly review applications of examples of hard and soft clustering to hydrogeophysical data integration.

Because of its conceptual simplicity and algorithmic robustness, the so-called *k*-means approach is one of the most popular and most widely used hard clustering techniques. On the basis of a qualitative assessment of cross plots of the various observed parameters that we wish to integrate and suitably constrained complementary information on the subsurface structures, we first determine the number of clusters. If the number of clusters cannot be specified reliably by *a priori* information or data analysis, statistical criteria can help to constrain the optimal number of clusters (e.g., Kaufman and Rousseeuw, 1990; Everitt, 1993).

Cluster formation then involves iterative regrouping of data points to minimize the variability in each cluster. The thus resulting clusters are characterized by the means and standard deviations of all parameters considered.

We have applied this procedure to a crosshole georadar data set collected at the Boise Hydrogeophysical Research Site (BHRS) near Boise, Idaho (e.g., Tronicke et al., 2004). The corresponding georadar velocity and attenuation tomograms are shown in Figure 5.1 and are expected to contain largely complementary information with regard to hydrological characteristics of the aquifer. Whereas the velocity images are considered to be primarily sensitive to the water content, and thus the water-saturated porosity, the attenuation images are, in the given context, more likely to respond to changes in clay content. While differing significantly in detail, both tomographic images are characterized by predominantly sub-horizontal structures, which is consistent with stratigraphic layering in the gravel and sand deposits at this site (e.g., Barrash and Clemo, 2002). Except for the strong bimodal distribution of values centered about a horizontal line at 11.8 m depth, there does not seem to be a clear and/or systematic relationship between the velocity and attenuation tomograms. Figure 5.2a shows a contoured histographic plot of the estimated velocities and attenuations. Given that this plot is characterized by three distinct and clearly separated maxima, we chose a three-cluster solution for further analysis (Figure 5.2b). Cluster 1 is characterized by higher velocities and lower attenuations than clusters 2 and 3. Clusters 2 and 3 are distinguished from each other only by their velocities. The resulting spatial distribution of clusters in the tomographic plane is shown in Figure 5.2c. Although this procedure has reduced the characterization of the subsurface to only three major petrophysical parameter groupings, the clustered section retains the key structural features of the original tomograms (Figure 5.1).

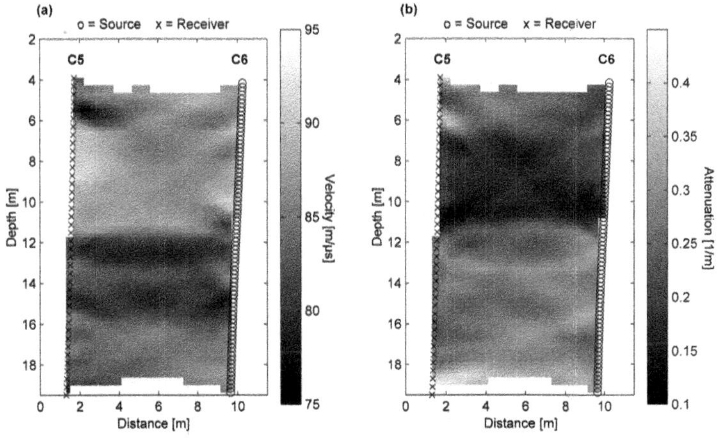

Fig. 5.1 Results of ray-based tomographic inversion of a crosshole georadar data set recorded at the Boise Hydrogeophysical Research Site (BHRS): **a** velocity distribution, and **b** attenuation distribution. Poorly resolved regions are blanked out. The groundwater table is at a depth of 2.96 m (Modified after Tronicke et al., 2004)

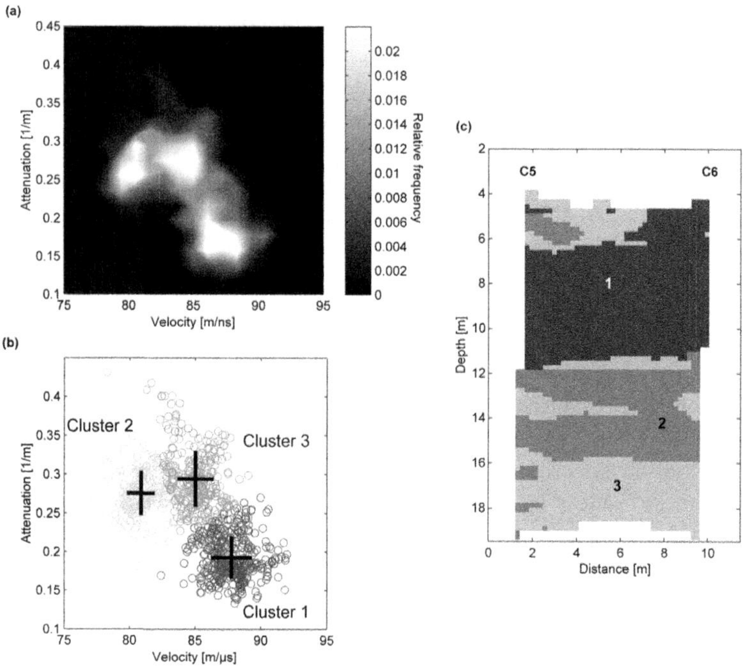

Fig. 5.2 a Contoured histographic plot illustrating trends in the velocity-attenuation relationship for tomograms shown in Figure 5.1. **b** Crossplot of velocity versus attenuation. Crosses delineate cluster centers (mean values), with their dimensions equal to the respective standard deviations. **c** Results of k-means cluster analysis of the BHRS crosshole georadar tomograms visualized as spatial distributions of cluster membership (Modified after Tronicke et al., 2004)

As opposed to hard clustering methods, soft clustering techniques attribute each point partially to each cluster and thus do not only allow for generating an integrated, zoned model, but also for a quantitative assessment of the quality and consistency of this zonation (e.g., de Bruin and Stein, 1998; Knab et al., 2001; Gueler and Thyne, 2004; Bragato, 2004). For the fuzzy *c*-means clustering approach, which we consider in some more detail here, this is achieved by iteratively minimizing the following objective function (e.g., Hoeppner et al., 1999; Gueler and Thyne, 2004; Paasche et al., 2006):

$$J_{FCM}(M,C) = \sum_{i=1}^{c} \sum_{j=1}^{n} m_{ij}^{f} \left\| x_j - v_i \right\|^2 \qquad (1)$$

where M is the membership matrix, C the cluster center matrix, c the number of clusters, n the number of data points, and m_{ij} denotes the degree of membership of data point x_j to cluster i characterized by its cluster center v_i. The "fuzzification" parameter f represents the degree of overlap between the clusters, which is commonly assumed to be equal to 2.

The membership functions obtained through fuzzy *c*-means cluster analysis assign each data point a partial membership to all considered clusters. The corresponding membership values, which quantify the degree of membership of a specific data point to a specific cluster, vary between zero and one: the higher this value, the higher the similarity between a data point and the corresponding cluster center. For any given data point, the sum of all membership values is unity. The results of fuzzy *c*-means clustering can be "defuzzyficated" by assigning each data point to the cluster for which it exhibits the highest membership value. This results in models consisting of homogeneous clusters comparable to those obtained by "hard" clustering algorithms. The resulting spatial distribution of cluster memberships represents a zoned model of the subsurface region probed by the various geophysical techniques. To incorporate information on the consistency of this zonation into the corresponding visualization, one can set the color saturation of each cell proportional to its highest membership value.

An important aspect of fuzzy *c*-means cluster analysis is that this approach not only allows for an internally consistent zonation of the probed subsurface based on all available data, but also for the estimation of the spatial distribution of petrophysical parameters, such as porosity or permeability, which often cannot be directly inferred from the geophysical data. This requires some direct measurements of the petrophysical target parameter in parts of the zoned model as provided, for example, by borehole logs or core samples. Based on these measurements, a corresponding mean value of the petrophysical target parameter is then assigned to each cluster. Using the previously evaluated membership values m_{ij} as weighting factors, we can thus estimate the value of the unknown target parameter at the jth grid cell as a weighted sum over all clusters (e.g., Paasche et al., 2006):

$$p_j = \sum_{i=1}^{c} \overline{p}_i \cdot m_{ij} \tag{2}$$

where $\overline{p}_i$ denotes the mean value of the target parameter p for the ith cluster. The corresponding procedure is conceptually similar to that used for multi-component petrophysical mixing models (e.g., Schoen, 1998). To ensure a meaningful estimation of the spatial distribution of the unknown target parameter, it must exhibit some relation to the geophysical parameters used for the zonation. It is important to note that, in marked contrast to more conventional approaches of petrophysical parameter estimation (e.g., Greaves et al., 1996; Dannowski and Yaramanci, 1999; Garambois et al., 2002), the nature of this interrelation can be rather vague and ill-defined and does not need to be explicitly quantified through theoretical or empirical expressions. It can be shown that the resulting spatial reconstructions are remarkably robust with regard to the number of clusters used in a given analysis. It is also important to note that while the corresponding parameter reconstructions can be expected to faithfully reproduce the large-scale features, they provide an overly smooth picture of the smaller-scale features and inherently underestimate the variance of the parameter distribution (Paasche et al., 2006).

An example of parameter estimation based on fuzzy *c*-means clustering is illustrated in Figure 5.3, which shows reconstructions of the spatial distribution of the natural gamma radiation and the hydraulic conductivity for a shallow alluvial aquifer in northwestern Switzerland (Paasche et al., 2006). Gamma and hydraulic conductivity data were available from borehole logs and direct-push-based slug tests. Parameter estimation was preceded by fuzzy *c*-means cluster analysis of crosshole seismic velocity and crosshole georadar velocity and attenuation tomograms. Despite the limited number of slug tests available, there is a far-reaching similarity between the estimations of the spatial distributions of the two parameters. Given that in alluvial deposits the gamma radiation can be regarded as a proxy for lithology, this result is quite plausible, even expected, which underlines the potential of clustering techniques in general and fuzzy *c*-means cluster analysis in particular for hydrogeophysical and hydrological data integration.

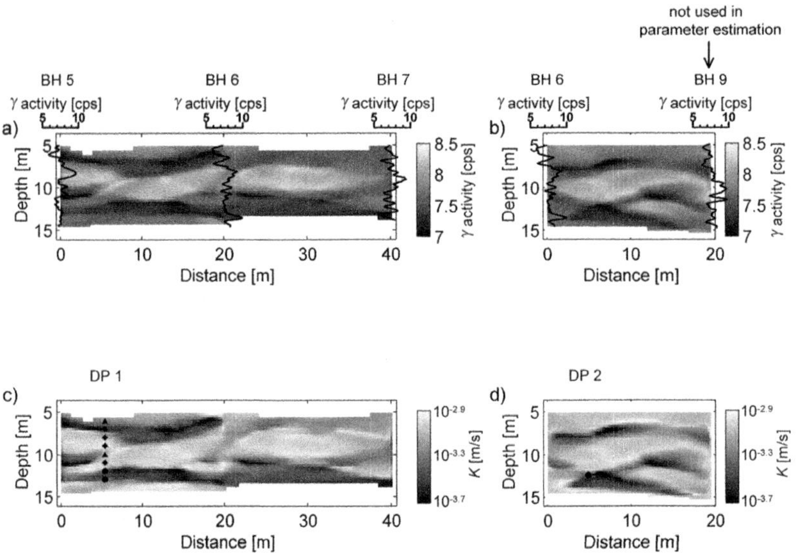

Fig. 5.3 Estimates of the spatial distributions of **a**, **b** the gamma-ray activity and **c**, **d** the hydraulic permeability based on the integration of crosshole seismic and georadar tomograms with natural gamma log and direct-push slug test data using a fuzzy *c*-means clustering approach. In **c** and **d** triangles, diamonds and dots indicate logarithmic hydraulic permeability intervals of –2.8 to –3.0, –3.5 to –3.7, and <-4.0 m/s, respectively (Modified after Paasche et al., 2006)

5.4 Monte-Carlo-Type Conditional Stochastic Simulations

Monte-Carlo-type conditional stochastic simulations aim at producing integrated models that reproduce all available data and constraints, as well as the inferred geostatistical characteristics where no data or deterministic constraints are available (e.g., Sen and Stoffa, 1995; Deutsch, 2002; Kelkar and Perez, 2002). Full-blown Monte Carlo approaches are essentially guaranteed to find the optimal solution but tend to be computationally impractical for multi-method data integration purposes. Instead, directed Monte-Carlo-type approaches, such as genetic algorithms or simulated annealing, which are not only more efficient but also more prone to getting stuck in local minima of the solution space, are being used. While for most intents and purposes genetic algorithms and simulated annealing seem to be largely equivalent, the latter is clearly the preferred approach for conditional stochastic simulations as it is conceptually considerably simpler and easier to parameterize.

The central idea behind simulated annealing is adapted from the thermodynamics of a cooling metallic melt (e.g., Sen and Stoffa, 1995). Atoms can move freely throughout a melt at high temperatures, but as the temperature is lowered, the mobility progressively decreases. Eventually, the system reaches its minimum energy state and the atoms assume fixed positions within a crystal lattice. When applying this idea to conditional stochastic simulations, we require a suitable global objective function that quantifies the overall closeness of the modeled data to the observed data. A common way to achieve this is to formulate the global objective function as a weighted sum of several different components. This objective function is then gradually minimized following the thermodynamic analogy mentioned above. In doing so, the acceptance probability P for a new parameter distribution is calculated as:

$$P = \begin{cases} 1, \text{ if } O_{new} \leq O_{old} \\ \exp\left(\dfrac{O_{old} - O_{new}}{T}\right), \text{ otherwise,} \end{cases} \tag{3}$$

where O_{old} and O_{new} denote the global objective functions before and after the perturbation of the model parameters and T, in reference to the Boltzmann distribution in thermodynamics, is often referred to as the temperature parameter. Thus, all favorable perturbations are accepted, whereas unfavorable ones are accepted according to the exponential probability distribution controlled by T. For deciding whether an unfavorable perturbation is accepted or not, we generate a random number $0 \leq b \leq 1$ and compare it with P. The perturbation is accepted if $b < P$, otherwise it is rejected. Lowering T results in lower probabilities for the acceptance of unfavorable perturbations. How and when T is lowered is controlled by the so-called annealing scheme or "cooling path", the choice of which may have a significant impact on the computational efficiency and/or the convergence characteristics of the simulation process (e.g., Sen and Stoffa, 1995; Deutsch, 2002).

Traditionally, simulated-annealing-based conditional simulations start off with a random field of values based on the inferred/assumed probability density function of the target parameter. This initially uncorrelated structure is then gradually "organized" by repeated random swapping of values to meet the structural and/or petrophysical constraints imposed by the available geophysical data. Although this conventional approach is quite robust and flexible, it also suffers from a number of shortcomings. In particular, deterministic information with regard to the larger-scale structure, as provided by geophysical data, is difficult to incorporate into the constraints imposed on the stochastic simulation process and hence the lateral continuity tends to be systematically underestimated by the resulting models (e.g., Tronicke and Holliger, 2005). This problem can be alleviated (i) by linking each value of the larger-scale geophysical models, as provided, for example, by georadar and/or seismic crosshole tomography, to a conditional distribution of the target parameter through a Bayesian-type approach and (ii) by using a variogram based on an inferred parametric model at shorter lags while retaining the structural information contained in the geophysical data at larger lags.

The potential of this approach is illustrated on a realistic porosity model for a heterogeneous alluvial aquifer, which is characterized by a scale-invariant layered structure with a horizontal-to-vertical aspect ratio of ~10 and a power spectrum decaying as ~$1/f$ with f denoting the spatial frequency (Figure 5.4a). This aquifer model represents a particularly challenging test case, as it is statistically highly non-stationary, exhibiting pronounced structural complexity both at the small and large scales (Figure 5.4a). It is probed by crosshole georadar data and porosity logs with boreholes located at 0, 10, 20, and 30 m lateral distance. Synthetic cross-hole georadar data are generated using a finite-difference solution of Maxwell's equations. After picking the traveltimes of the first arrivals, these data are then tomographically inverted for the spatial distribution of the electromagnetic velocity (Figure 5.4b). Together with the synthetic porosity logs, this represents the basis for the reconstruction of the detailed porosity structure using our new conditional simulation approach outlined above. To this end, we first use a Bayesian-type approach to infer the relation between the porosity logs and the georadar velocity along the boreholes to establish a probability distribution of possible porosity values for each georadar velocity present in the tomogram. In a next step, we need to determine the horizontal and vertical target variograms. We use the porosity logs to infer a parametric variogram model at shorter lags and combine it with the larger-scale structural information contained in the tomograms. The resulting conditional simulation (Figure 5.4c) does indeed compare quite favorably with the original porosity model in terms of its small- and large-scale structures as well as their lateral continuity with regard to the overall distribution of the porosity values. Finally, we have repeated the same stochatistic simulation using a more conventional simulated annealing approach (e.g., Tronicke and Holliger, 2005). A comparison of these results (Figure 5.4c, d) clearly illustrates the strength and potential of our new approach and thus the importance that must be attributed to the incorporation of the larger-scale scale deterministic information into simulation process as well as the estimation of the underlying variogram model.

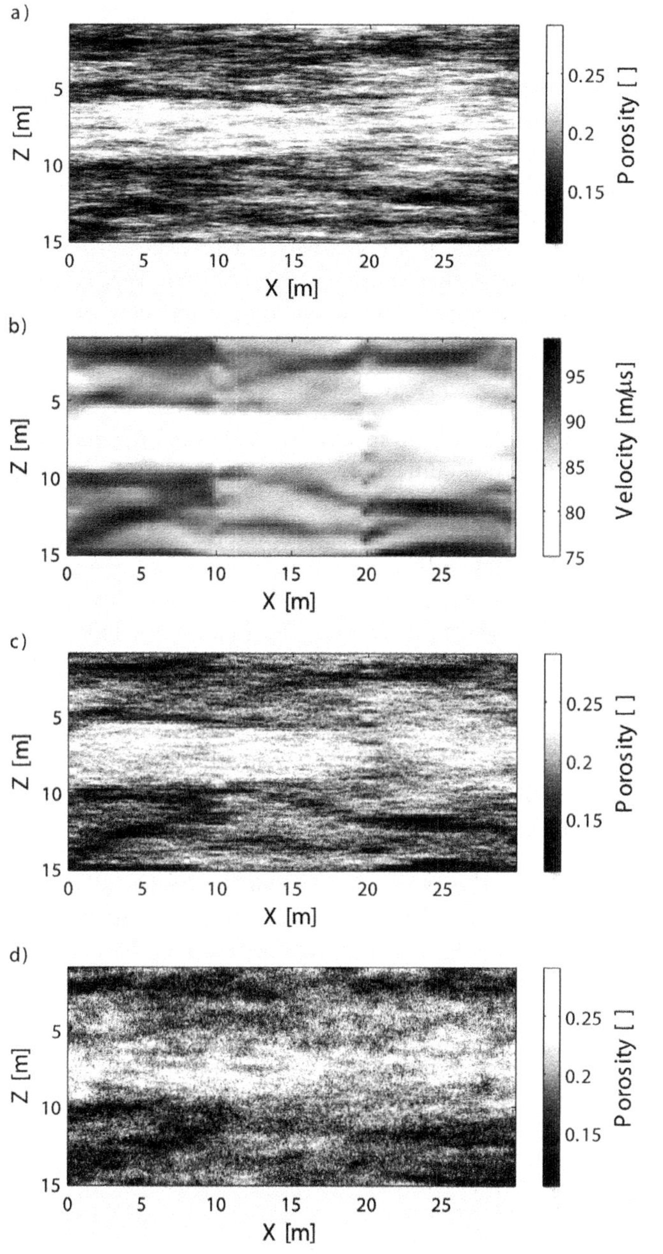

Fig. 5.4 a Model of porosity distribution within a heterogeneous aquifer. **b** Result of tomographic inversion of synthetic crosshole georadar data acquired between boreholes located at 0, 10, 20 and 30 m. Stochastic realizations of porosity distribution constrained by georadar tomogram, porosity logs and inferred parametric variogram model using **c** a novel Bayesian-type and **d** a conventional simulated annealing approach (e.g., Tronicke and Holliger, 2005)

5.5 Conclusion

We have reviewed two largely complementary techniques that we consider to be useful and effective means for the quantitative geostatistical integration of typical hydrogeophysical and/or hydrological databases: cluster analysis and Monte-Carlo-type conditional stochastic simulations. These methods are largely complementary and highly effective for their respective purposes, but their choice is necessarily subjective and there are alternative approaches, which under certain conditions could prove to be more suitable and/or effective. That said, it is important to realize that all quantitative data integration techniques are fundamentally limited by the quality, resolution and hydrological sensitivity of the hydrogeophysical input data. Improvements in geophysical data acquisition and interpretation provide more realistic images of the subsurface, which in turn allow for the development of more reliable site-specific petrophysical relations with available hydrological data and thus ultimately for more improved integrated aquifer models. Whereas, for example, the inversion of potential or diffusive field data have already been based on the governing partial differential equations for quite a while, wavefield-type data, such as crosshole seismic or georadar data, are still routinely inverted based on asymptotic approximations. Given that crosshole georadar and seismic data arguably have the highest potential resolutions and hydrological sensitivities of common hydrogeophysical techniques, the emerging trend towards wave-equation-based inversion approaches could thus result in something like a change of paradigm in local-scale aquifer characterization. Similar improvements can be expected from novel uses of existing geophysical techniques. For example, surface-based georadar and seismic techniques have so far almost exclusively been used for the structural imaging of aquifers. While this information is undoubtedly useful, equivalent studies in exploration geophysics indicate that surface seismic data have the potential to provide reliable and relatively highly resolved constraints on the velocity and impedance distributions within the probed subsurface regions. In summary, we believe that there are reasons to be hopeful that innovations in data acquisition and interpretation in concert with ongoing improvements in quantitative integration techniques will continue to enhance the effectiveness and broaden the use and acceptance of hydrogeophysical aquifer characterization techniques.

References

Barrash, W., and Morin, R. H., 1997, Recognition of units in coarse, unconsolidated braided-stream deposits from geophysical log data with principal components analysis, *Geology* **25:**687–690.

Barrash, W., and Clemo, T., 2002, Hierarchical geostatistics and multifacies systems: Boise Hydrogeophysical Research Site, Boise, Idaho, *Water Resour. Res.* **38:**1196, doi:10.1029/2002WR001436.

Bosch, M., 2004, The optimization approach to lithological tomography: Combining seismic data and petrophysics for porosity prediction, *Geophysics* **69:**1272–1282.

Bosch, M., Zamora, M., and Utama, W., 2002, Lithology discrimination from physical rock properties, *Geophysics* **67:**573–581.

Bragato, G., 2004, Fuzzy continuous classification and spatial interpolation in conventional soil survey for soil mapping of the lower Piave plain, *Geoderma* **118:**1–16.

Cassiani, G., Boehm, G., Vesnaver, A., and Nicolich, R., 1998, A geostatistical framework for incorporating seismic tomography auxiliary data into hydraulic conductivity, *J. Hydrol.* **206:**58–74.

Chen, J., Hubbard, S., and Rubin, Y., 2001, Estimating hydraulic conductivity at the South Oyster site from geophysical tomographic data using Bayesian techniques based on the normal regression model, *Water Resour. Res.* **37:**1603–1613.

Dannowski, G., and Yaramanci, U., 1999, Estimation of water content and porosity using combined radar and geoelectrical measurements, *Eur. J. Environ. Eng. Geophys.* **4:**71–85.

De Bruin, S., and Stein, A., 1998, Soil-landscape modelling using fuzzy c-means clustering of attribute data derived from a digital elevation model (DEM), *Geoderma* **83:**17–33.

Deutsch, C. V., 2002, *Geostatistical Reservoir Modeling*, Oxford University Press, New York, N.Y., U.S.A.

Dumay, J., and Fournier, F., 1988, Multivariate statistical analyses applied to seismic facies recognition, *Geophysics* **53:**1151–1159.

Everitt, B. S., 1993, *Cluster Analysis*, Edward Arnold, London.

Ezzedine, S., Rubin, Y., and Chen, J., 1999, Bayesian method for hydrogeological site characterization using borehole and geophysical survey data: Theory and application to the Lawrence Livermore National Laboratory Superfund site, *Water Resour. Res.* **35:**2671–2684.

Gallardo, L. A., and Meju, M. A., 2003, Characterization of heterogeneous near-surface materials by joint 2D inversion of dc resistivity and seismic data, *Geophys. Res. Lett.* **30:**doi: 10.1029/2003GL017370.

Garambois, S., Sénéchal, P., and Perroud, H., 2002, On the use of combined geophysical methods to assess water content and water conductivity of near-surface formations, *J. Hydrol.* **259:**32–48.

Gelman, A., Carlin, J. B., Stern, H. S., and Rubin, D. B., 2003, *Bayesian Data Analysis*, 2nd ed., Chapman & Hall/CRC, Boca Raton, Florida, U.S.A.

Gill, D., Shomrony, A., and Fligelman, H., 1993, Numerical zonation of log suites and logfacies recognition by multivariate clustering, *AAPG Bull.* **77:**1781–1791.

Gloaguen, E., Chouteau, M., Marcotte, D., and Chapuis, R., 2001, Estimation of hydraulic conductivity of an unconfined aquifer using co-kriging of GPR and hydrostratigraphic data, *J. Appl. Geophys.* **47:**135–152.

Greaves, R. J., Lesmes, D. P., Lee, J. M., and Toksoez, M. N., 1996, Velocity variations and water content estimated from multi-offset, ground-penetrating radar, *Geophysics* **61:** 683–695.

Gueler, C., and Thyne, G. D., 2004, Delineation of hydrochemical facies distribution in a regional groundwater system by means of fuzzy c-means clustering, *Water Resour. Res.* **40:**doi: 10.1029/2004WR003299.

Gueler, C., Thyne, G. D., McCray, J. E., and Turner, A. K., 2002, Evaluation of graphical and multivariate statistical methods for classification of water chemistry data, *Hydrogeol. J.* **10:**455–474.

Hammah, R. E., and Curran, J. H., 1998, Fuzzy cluster algorithm for the automatic identification of joint sets, *Int. J. Rock Mech. Min. Sci.* **35:**889–905.

Hoeppner, F., Klawonn, F., Kruse, R., and Runkler, T., 1999, *Fuzzy Cluster Analysis: Methods for Classification, Data analysis and Image Recognition*, Wiley, Chichester, West Sussex, England.

Hubbard, S. S., and Rubin, Y., 2005, Introduction to hydrogeophysics, in: *Hydrogeophysics*, Y. Rubin, and S. S. Hubbard, eds., Springer, Dordrecht, The Netherlands, pp. 3–21.

Hubbard, S. S., Chen, J., Peterson, J. E., Majer, E. L., Williams, K. H., Swift, D. J., Mailloux, B., and Rubin, Y., 2001, Hydrogeological characterization of the South Oyster bacterial transport site using geophysical data, *Water Resour. Res.* **37:**2431–2456.

Hyndman, D. W., and Harris, J. M., 1996, Traveltime inversion for the geometry of aquifer lithologies, *Geophysics* **61:**1728–1737.

Hyndman, D. W., Harris, J. M., and Gorelick, S. M., 2000, Inferring the relation between seismic slowness and hydraulic conductivity in heterogeneous aquifers, *Water Resour. Res.* **36:**2121–2132.

Kaufman, L., and Rousseeuw, P. J., 1990, *Finding Groups in Data: An Introduction to Cluster Analysis*, Wiley, Hoboken, New Jersey, U.S.A.

Kelkar, M., and Perez, G., 2002, *Applied Geostatistics for Reservoir Characterization*, Society of Petroleum Engineers, Richardson, Texas, U.S.A.

Kirsch, R., ed., 2006, *Groundwater Geophysics*, Springer, Germany.

Knab, M., Appel, E., and Hoffmann, V., 2001, Separation of the anthropogenic portion of heavy metal contents along a highway by means of magnetic susceptibility and fuzzy c-means cluster analysis, *Eur. J. Environ. Eng. Geophys.* **6:**125–140.

Kowalsky, M., Finsterle, S., Peterson, J., Hubbard, S., Rubin, Y., Majer, E., Ward, A., and Gee, G., 2005, Estimation of field-scale soil hydraulic and dielectric parameters through joint inversion of GPR and hydrological data, *Water Resour. Res.* **41:**W11425, doi:10.1029/2005WR004237.

Linde, N., Finsterle, S., and Hubbard, S., 2006, Inversion of tracer test data using tomographic constraints, *Water Resour. Res.* **42:**W04410, doi:10.1029/2004WR003806.

Paasche, H., Tronicke, J., Holliger, K., Green, A. G., and Maurer, H., 2006, Integration of diverse physical-property models: Subsurface zonation and petrophysical parameter estimation based on fuzzy c-means cluster analyses, *Geophysics* **71:**H33–H44.

Rubin, Y., and Hubbard, S. S., eds., 2005a, *Hydrogeophysics*, Springer, Dordrecht, The Netherlands.

Rubin, Y., and Hubbard, S. S., 2005b, Stochastic forward and inverse modelling: The "hydrogeophysical" challenge, in: *Hydrogeophysics*, Y. Rubin, and S. S. Hubbard, eds., Springer, Dordrecht, The Netherlands, pp. 3–21.

Schoen, J. H., 1998, *Physical Properties of Rocks: Fundamentals and Principles of Petrophysics*, Pergamon, New York, U.S.A.

Sen, M., and Stoffa, P. L., 1995, *Global Optimization Methods in Geophysical Inversion*, Elsevier, Amsterdam, The Netherlands.

Szerbiak, R. B., McMechan, G. A., Corbeanu, R. M., Forster, C., and Snelgrove, S. H., 2001, 3-D characterization of a clastic reservoir analog: From 3-D GPR data to a 3-D fluid permeability model, *Geophysics* **66:**1026–1037.

Tronicke, J., and Holliger, K., 2005, Quantitative integration of hydrogeophysical data: Conditional geostatistical simulation for characterizing heterogeneous alluvial aquifers, *Geophysics* **70:**H1–H10.

Tronicke, J., Holliger, K., Barrash, W., and Knoll, M. D., 2004, Multivariate analysis of cross-hole georadar velocity and attenuation tomograms for aquifer zonation, *Water Resour. Res.* **40:**doi:10.1029/2003WR002031

Vereecken, H., Binley, A., Cassiani, G., Revil, A., and Titov, K., eds., 2006, *Applied Hydrogeophysics*, Springer, Dordrecht, The Netherlands.

Chapter 6
Hydrogeological Settings in Dobrudja Area and Groundwater Monitoring Networks in Transboundary Aquifers

Marta Machkova

Abstract This chapter presents the geology and hydrogeology of transboundary aquifers in Dobrudja area as well as the corresponding monitoring network as part of the integrated management of transboundary groundwater between Bulgaria and Romania. The physicochemical characteristics of the Upper Jurassic—Valanginian aquifer (Deep Aquifer) and the Sarmatian aquifer (Upper Aquifer)—are reviewed in details. The groundwater monitoring network developed in the transboundary aquifers in Dobrudja area as part of the requirements of Article 8 of the EC Water Framework Directive 2000/60/EC (WFD) is assessed and its scheme selection approach is explained.

Keywords Transboundary groundwater, carbonate aquifers, pollution, over-exploitation, monitoring network

6.1 Introduction

Within the framework of the project "Integrated Management of Transboundary Groundwater between Bulgaria and Romania in Dobrudja/Dobrogea Area" (proposed for financing by the Phare Bg-Ro CBC Program), an actualization and proposal for enlargement of groundwater monitoring network in Dobrudja area were implemented. The area is located in Northeast Bulgaria and has a surface of about 5,500 km^2. The western boundary coincides with the watershed of the Suha

Marta Machkova
National Institute of Meteorology and Hydrology - Bulgarian Academy of Sciences, 66 Tsarigradsko Shosse Blvd., 1784 Sofia, Bulgaria

C.J.G. Darnault (ed.), *Overexploitation and Contamination of Shared Groundwater Resources.*

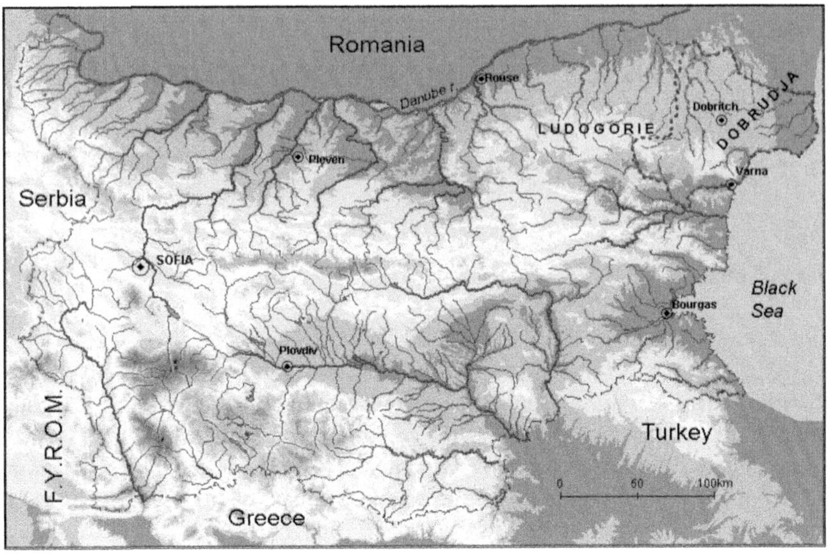

Fig. 6.1 Location of the Dobrudja area

River valley. The southwestern border is the watershed of the Karamandere River, watershed of the Provadiiska River. The eastern border is the Black Sea and the northern one is the State border with Romania. The two basic aquifers are spread to the north, crossing the State border with Romania (Figure 6.1).

Considering the country's administrative division, the region belongs to the Dobrich district and partially to the Varna district. The biggest settlements are the Dobrich town and the third biggest town in Bulgaria—Varna. The continuous growing industrial activities are concentrated in both towns. From an economic point of view, the area is predominantly agricultural but the irrigated arable land occupies a relatively small part of the territory. A significant number of large stock-breeding farms (mainly pork and cattle farms) have been established there. At present, most of them are closed. The tourist activities are well developed in the coastal area where there are a number of modern resorts.

The northwestern half of the area corresponds approximately to the hydrographic basin of the Danube River. The rest of the area is drained by various small rivers flowing into the Black Sea. The area is quite flat with altitudes from 0 m a.s.l. along the Black Sea to 380 m in the southwest part of the region. This monotonous flatness is only interrupted by narrow valleys with depths of up to 200 m from the altitude of the plateau, especially in the southern sector (Batova River) and in the west (Suha River).

6.1.1 Geology

The geology is relatively well known due to the boreholes drilled for different purposes. The boreholes generally reach significant depths (more than 400 m) but can penetrate into much greater depths (exceeding 4000 m).

The area is part of the Moesian Platform which occupies the territory to the north of the domain of the Balkan Mountain Ridge and to the south of the domain of the South Carpathians. This platform has two main complex structures, one superimposed over the other: a folded and fractured Paleozoic basement and a non-folded cover which can nevertheless be somewhat fractured, with sedimentary rocks dating from the Permian-Triassic to the Quaternary (Iovtchev, 1976).

The Quaternary materials (Figure 6.2) are made up of Aeolian Pleistocene sediments (loess), clayey towards the top and the southern sector, and sandy or mixed with alluvial sediments towards the bottom and the northern outcrops. These materials cover most of the study area except for the canyon floors which are the high-quality cultivated areas.

The Miocene is built up of various groups of formations. The upper group is basically built up of two carbonate formations of Sarmatian age, between which lies a formation of banded marls and clays having a thickness of 45–80 m. The most recent carbonate formation is Karvuna (10–50 m), which is built up of organogenic and conchiferous limestone, sometimes compact and sometimes highly

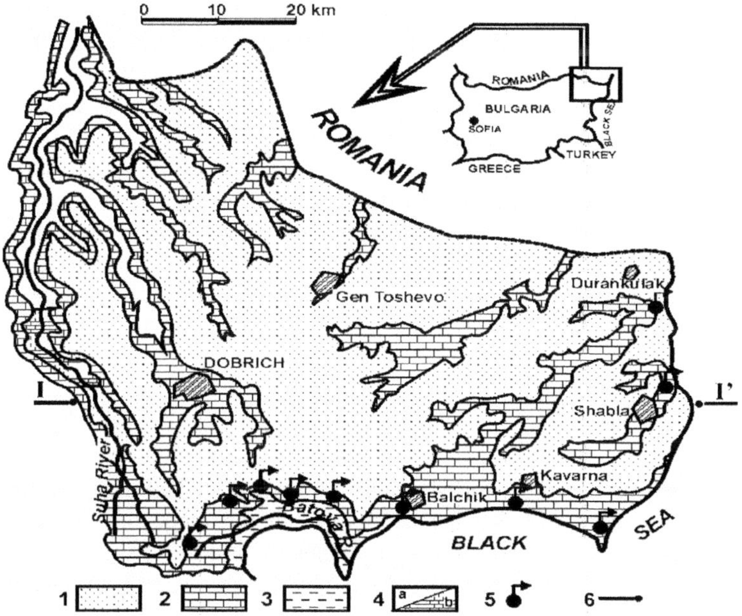

Fig. 6.2 Hydrogeological scheme of the Dobrudja area

porous due to the dissolution of fossil shells. The lower carbonate formation is presented by organogenic, conchiferous, oolithic and detrital limestones, and more rarely, clays or marls having a thickness of 30–120 m.

The Lower Miocene group comprises four formations with complex facies. It is built up of sands and clays or marls with a few intercalations of organogenic limestones. In the southern sector the thickness is greater, ranging from 150 to 200 m, while towards the north it decreases being less than 10 m. Miocene materials are discordantly superimposed over a Paleogenic group which practically does not outcrop in the area. The drilling data indicate that the group is present only in the southeastern half of the study area, ranging from 40 to 740 m in thickness.

The Paleogene is built up of clay and marl, organogenic and marly limestone, sand and sandstone. Paleogene materials lie discordantly over a group from the Hauterivian to the Upper Cretaceous. This group outcrops along the Suha River valley, varying in thickness from 30 to 400 m (increasing towards the West and Northwest). This is a series of alternating marly limestones and limestone marls, more terrigenous or with limestone containing chert and conglomerates. Concordantly with the lower part of the previous series are materials from the Jurassic-Valanginian age forming a thick package of limestones, dolomite limestone and dolomites. These materials are present throughout the region, but outcrop only in the western sector, in the Suha River valley and some of the most western tributaries. The outstanding formation in these materials is Kaspichan (Cheshitev and Kânchev, 1989), slightly folded (dipping by less than 10°) by adaptation to the play of fault systems. The thickness of these materials ranges from 400 to 900 m, increasing generally towards the Southeast. The Middle Jurassic is built up of clayey formations, sandstones, and/or limestone marls, with thickness ranging from 5 to 150 m, sometimes absent in the northwestern sector. The thickness increases generally to the Southeast.

The Permian-Triassic materials fill mainly the Hercinian paleorelief. The thickness is extremely variable (50–4000 m) and can even be absent in some sectors. The Triassic is built up of alternating carbonate and detrital materials or marly materials. The Permian materials consist of a volcano-sedimentary complex, sometimes with intercalations of carbonate and anhydrite materials.

6.1.2 Hydrogeology

The whole North Bulgaria, from the Balkan Mountain to the Danube River and the terrestrial boundary with Romania, falls within the range of a large platform structure continuing into the territory of Romania. From a hydrogeological point of view, it is a large artesian reservoir named "Lower Danube artesian basin" or "Moesian artesian basin" in Bulgaria. This is the largest hydrogeological structure in the country of particular hydrogeological and economic significance (Figure 6.3).

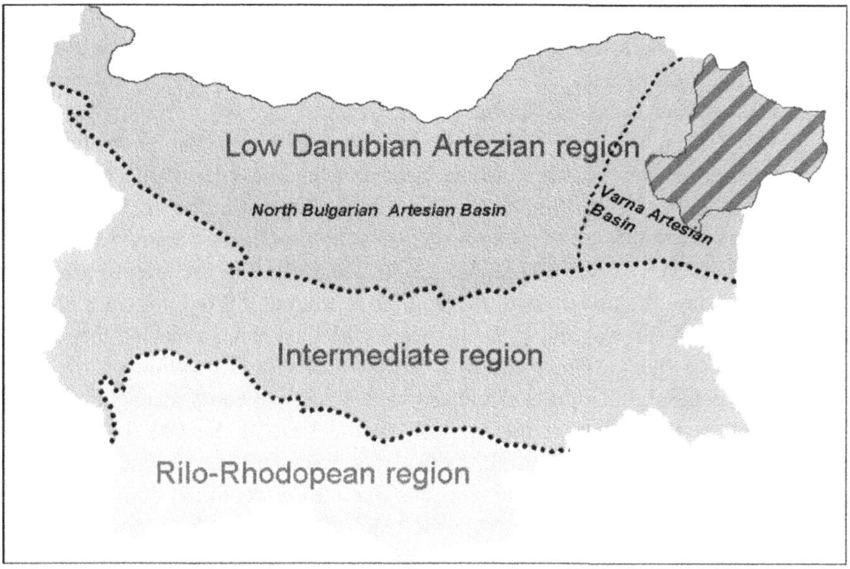

Fig. 6.3 Varna artesian basin

This structure is divided into two parts: the West part is called the North Bulgarian artesian basin and the East part is the Varna artesian basin resulting from a tectonic arched uplift in the region West of the Dobrich town. The water-bearing rock complex extends widely in a horizontal direction and has a varying lithological composition and stratigraphic vertical range. The groundwater is formed in overlying rocks with a heterogeneous lithological composition. A hydraulic contact exists due to tectonic disturbances in some isolated areas. The aquifers are characterized by hydrodynamic, hydrochemical and hydrogeothermal zoning in a horizontal and vertical direction. The groundwater in the recharge zone is fresh with low temperature and total dissolved solids (TDS) values. Dipping groundwater increases the temperature and changes the chemical composition.

6.1.2.1 Upper Jurassic–Valanginian (Deep Aquifer)

The Deep Aquifer is widespread and is the largest and most important aquifer in Bulgaria. It is built up of a thick carbonate complex. The carbonate layers have radial dipping patterns. The complex comprises several formations of different ages from Callovian to Valanginian. It is divided into three zones: dolomite zone (at the base of the complex) built up of various dolomites; intermediate zone, presented by alternating limestone and dolomites; and limestone zone at the top.

The Deep aquifer is confined under younger materials. The aquifer scarcely outcrops with the exception of some local areas along the river valleys (Suha

River valley and more along its tributary Karamandere River). This is the sector where the aquifer is closest to the ground surface; the highest parts lie at altitudes between 0 and 100 m a.s.l. The slope of the aquifer declines gently towards the North, while towards North-Northeast it appears to be sub-horizontal (Pulido-Bosch et al., 1996a, b). To the Northwest of the area, within the Romanian territory, the aquifer declines towards the North and reaches a thickness of up to 1500 m. There are localized outcrops near the Danube River in the Romanian Plain. To the North of Bucharest, it is confined at a depth of over 2000 m (Iurkiewicz and Oraseanu, 1997). However in Southeast Romania (South Dobrudja), in the area to the north of Mangalia Lake, the thickness of the aquifer is around 350 m and it lies at 150–200 m b.s.l. (Feru and Capota, 1991). Towards the East and Southeast, the aquifer is separated into blocks which are thrown down or rise depending on the deep fractures. In general, the aquifer declines progressively towards the Southeast and extends, still confined, under the seabed. To the East of the coastline, seismic studies of the seabed (Dachev et al., 1988) have revealed that the aquifer remains confined beneath more recent sediments and declines to depths of 2000 m b.s.l. In the continental area, the greatest depths are reached in two troughs: in the sector of the Batova River and Balchik town at nearly 1000 m b.s.l.; and in the Shabla-Bulgarevo trough, at more than 700 m b.s.l. The greatest change in the level of the aquifer occurs in the Dobrich fracture zone. Some 15 km to the South of Cape Kaliakra, offshore oil wells penetrate the carbonate aquifer at a depth of around 1000 m, confined under Cretaceous and Paleogene sediments. This sector coincides with another horst that is evident in the seismic studies (Robinson et al., 1996).

Figure 6.4 shows a schematic map of the piezometric surface obtained from the general mean data; despite the imprecision, some inferences can be made on a regional scale. The first noteworthy feature is a clear differentiation of the aquifer in two sectors: in the more western part, there is a high piezometric level reaching 100 m a.s.l.; in the more eastern part, the levels vary from 15 to 20 m a.s.l. The zone of highest levels coincides with a recharge sector of the aquifer since it outcrops along the Suha River and its tributaries; in addition, this sector shows intense fracturing which appears to aid the surface recharge of the aquifer (Dobrich fracture zone). Such a clear difference between the two sectors indicates that these are hydraulically somewhat disconnected; this would not be a complete disconnection since practically the only source of inflow possible in the eastern sector is precisely from the surface recharge in the Suha River sector. Groundwater flow would occur in a form that is virtually radial and divergent from such a piezometric dome, indicating an influx of groundwater towards Southeast Romania (which agrees with the observations of Davidescu et al., 1991), towards the East of the study area and in the direction of the Devtnia springs to the South. In the Northeast sector of the area, the piezometric lines indicate the existence of a flow towards the North-Northeast, approximately in the direction of Mangalia Lake (Romania), where there is known to be an important discharge of thermal waters from the carbonate Deep Aquifer (Davidescu et al., 1991). The hydraulic head here is known to be similar to that in the coastal area of the Bulgarian sector.

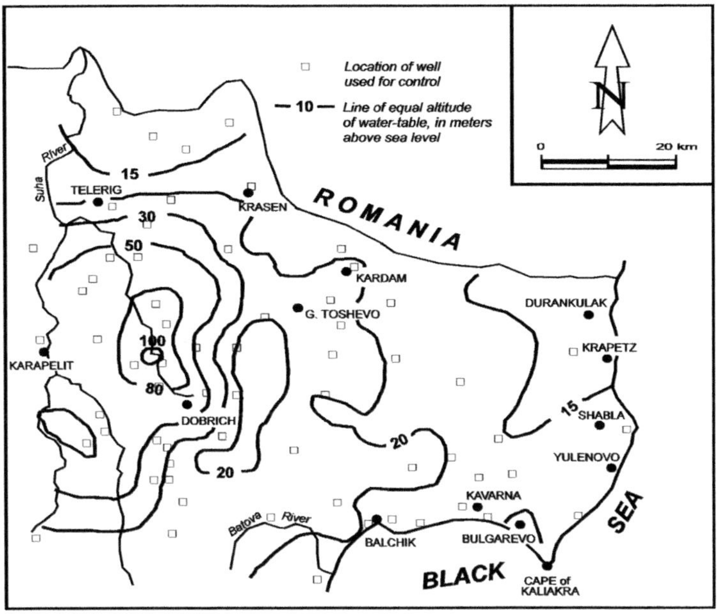

Fig. 6.4 Schematic map of the piezometric surface of the Deep Aquifer

The hydraulic gradient toward East, North and West ranges from 0.0075 to 0.002. The filtration characteristics are rather varied due to alterations in the karsting and cracking of sediments. Local permeability values vary from 8–10 m^2/d to more than 2000 – 3000 m^2/d, with an average rate of 200–600 m^2/d (Danchev et al., 1981; Velikov et al., 1989). The permeability coefficient is predominantly between 2 m/d and 5 m/d. The values of specific yield are between 0.01 and 0.10.

The aquifer is characterized by a vast groundwater resource (for the Varna artesian basin, the natural resources are about 13–14 m^3/s) and it is the main source used for different purposes including water supply in the region.

Physicochemical Characteristics

The water temperatures show a spatial distribution in accordance with the geological structure of the area. In the western and central parts temperatures are 14–17°C—quite similar to those of the Upper aquifer. The temperatures increase towards the East due to the effect of the geothermal gradient reaching 32°C near Balchik town, 38°C near Cape Kaliakra and 41°C in Shabla. The highest temperatures ranging from 45°C to 52°C have been measured in Varna and neighboring resorts. Changes in the general tendency are definitely related to disturbances induced by the rises and falls of blocks limited by faults. The pH values (7.1–7.4) do not significantly vary within the area, although pH values have a slight tendency to increase towards the coastal zone. The groundwater is fresh (TDS 0.4–0.5 g/l) around the recharge zone as well as in the intermediate zone. The groundwater hydrochemical type, being bicarbonate-calcium (magnesium) and chloride-sodium, changes from West to East. The conductivity of the water in the western half of the aquifer is less than 650–700 μS/cm. There is a steady rise in this parameter towards the East exceeding 3000 μS/cm at certain points of the coastal border (Machkova et al., 1997). These changes are also related to a certain increase in the total mineralization (mainly for Na^+, Cl^-, but also for K^+ and H_4SiO_4) due to the longer water-rock contact and the corresponding dissolution and leaching processes taking place. HCO3 concentrations vary from some 500 mg/l on the western border to 300 mg/l on the eastern border with scattered values of over 600 mg/l. This steady decrease in the direction of the groundwater flow could be related to the distribution of the CO_2 itself, combined with the calcite precipitation processes. In this sense, the concentration of SO_4 in the water, ranging between 20 and 45 mg/l in most of the aquifer, decreases steadily towards the eastern border to values of less than 10 mg/l. The odor of H_2S is evident near the overflowing coastal wells. In Southeast Romania, controversy exists over the origin of this H_2S (concentrations of 2–30 mg/l, according to Feru and Capota, 1991), which is present in all the formations found between the Paleozoic and the Miocene (Marin and Nicolescu, 1993). According to Feru (1993), the origin of the H_2S lies in the reduction of sulfates by anaerobic microorganisms in the presence of organic matter rather than the oxidization of sulfur. Chloride contents are less than 50 mg/l in the western third of the aquifer, but rise notably near the Black Sea to over 1500 mg/l. Mixing processes of freshwater and saline water (mainly fossil or connate) are apparently the origin of this spatial distribution of the concentrations. Since the aquifer is confined over practically the entire area, the NO_3^- contents are low, between 4 and 12 mg/l. Some local values greater than "the contamination threshold" i.e. 30 mg/l would be due to the mixing of water from the Upper Aquifer and/or possible pollution through the annular space of the wells (Pulido-Bosch et al., 1996a, b, 1997). The anoxic environment that characterizes this aquifer also facilitates the reduction of nitrogenous species. The presence of ammonium ions can not be an indication of recent organic pollution as is the case with the Sarmatian. The genesis of the ammonium ions in the Deep groundwater is closely related to the presence of a very large amount of natural organic substances in the water-bearing rocks (Monahova and Kostova, 1989) and the existence of coal and petroleum deposits in the South-Southeast part of the region (Monahova and Vakarelska, 1982). Due mainly to a vertical migration, the mixing of allochtone hydrocarbons with the Deep groundwater causes a change in the oxidation–reduction properties of the aquifer especially in its deeper confined parts.

The Deep aquifer is recharged directly by the infiltrated precipitated water in the outcropping areas of the Kaspichan formation but is considerably fed more by rivers than by overlying aquifers. The aquifer is discharged naturally by Zlatina and Devnya springs (Figure 6.5), the main discharge points.

Fig. 6.5 One of the Devnya springs "Valsheben"

The Deep Aquifer is subject to intense exploitation as it is a primary source of fresh and drinking water in Northeast Bulgaria. As a result, a number of water-supply wells have been established which are distributed unequally over the territory. The water consumption is more considerable in the regions around the larger settlements and the resorts. Along the Black Sea coastal border from Krapetz to Balchik and Varna, there is a number of boreholes for self-discharge thermal water which is used for sport and recreation activities as well as a source for thermal energy for the needs of hotels. Some of them have no closure systems, so the water flows continuously without regulation. Although this latter effect has not been firmly established, it is clear that this situation represents a steady bleeding of the aquifer as well as a waste of great quantities of resources. Thus, in some wells drilled in the Krapetz (the eastern zone), which were originally artesian, the piezometric level has fallen by over 4 m for the last 20 years. The long-term observation on the groundwater levels in the western part of the area also shows steady downward trends (Figure 6.6).

6.1.2.2 Sarmatian Aquifer (Upper Aquifer)

The Upper Aquifer represents the upper part of a common heterogeneous water-bearing complex with different water-transmitting and water storage properties. The bottom part of the aquifer is built up of unconsolidated sands. Above them lie the detritus and shelly limestone separated locally by carbonate clays. The total

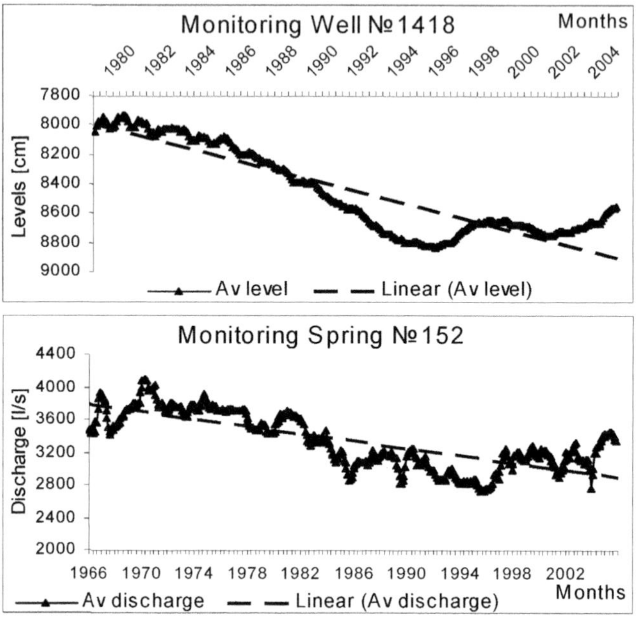

Fig. 6.6 Downward trends (Deep aquifer)

thickness of the complex is more than 240 m. The layers are almost horizontal with a slight dipping of 1–5° from East toward Southeast and Northeast. The rocks have a high porosity resulting from the spaces between sand grains and organogenic remains. The limestones are significantly karstified. Depending on the density of the limestone and their clay content, the volume of the caverns is 10–30% of the total volume. The Evksinovgrad formation clay is the bottom aquiclude for a large part of the area. The water-bearing rocks are almost 80% covered by loess, diluvia and diluvia-alluvial deposits of varying thickness. The aquifer is mainly unconfined.

In horizontal plan, the aquifer can be divided into two parts. In the west part - up to the Suha River valley, it is cut by the river network, while in the eastern part the aquifer is uninterrupted. The depth to groundwater levels depends on the relief, varying from 4–10 m up to 90 m from the surface. In the eastern part, the direction of groundwater flow almost coincides with the relief and the water flows toward East and Northeast. In the west part, the groundwater flows toward the river network.

The filtration properties of the aquifer are rather varied. The permeability coefficients range from 1 to 15 m/d (in the western and central parts) to 20–160 m/d in the East. The transitivity values vary from 5–90 m^2/d up to 200–600 m^2/d in the central part reaching 1500–2000 m^2/d in the eastern part. (Danchev, et al., 1981; Velikov, et al., 1989).

Rainfall infiltration and surface water recharge the aquifer. In outcropped areas, the infiltration coefficient varies from 8–10% up to 22–23%.

The main natural discharge of the aquifer occurs on the eastern and southeastern borders, both along the Black Sea coast and the Batova River, where the largest springs are located. Seasonal fluctuations of the spring discharge are very low. A significant part of groundwater is drained by the lakes Durankulak, Shabla - Ezeretz, Shablenska Tuzla, near the village of Vaklino. Part of the groundwater flow continues toward the territory of Romania. Along the western border, there are several smaller springs whose discharge ranges from 1–2 to 20 l/s.

Physicochemical Characteristics

The predominant type of water is magnesium bicarbonate, and to a smaller extent, calcium bicarbonate; this is determined by the lithology of the aquifer and by a Quaternary detritus cover being rich in magnesium. Only on the eastern coast, where there is widespread contact with the Sea, the water type changes to sodium chloride. The spatial evolution of the groundwater temperature shows the same distribution pattern as the mean air temperature ranges between 12°C in the recharge area (Dobrich sector) and 15 to 17°C in the discharge areas. This temperature increase might be due to the existence of vertical flows from the Deep aquifer with a greater hydraulic head. In general, in the central sector of the region (the main recharge area), higher values have been registered for the concentrations of HCO_3^- (500 mg/l), Ca^{2+} (100 mg/l) and Mg^{2+} (100 mg/l).

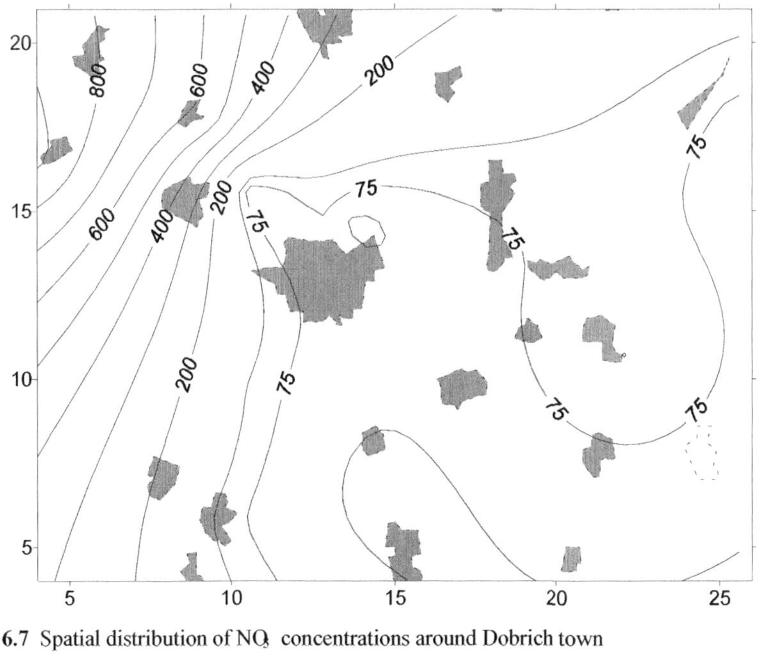

Fig. 6.7 Spatial distribution of NO_3 concentrations around Dobrich town

The values of these ions decrease towards the discharge zones, suggesting subsaturation in carbonate minerals in these directions. The groundwater sulfate content is generally very low in the entire region (less than 80 mg/l). Only near Cape Kaliakra, higher values of approximately 112 mg/l have been recorded. The distribution of nitrates and chlorides is related to three main processes of pollution: scattered discharge of urban liquid waste and unpurified animal waste; excess fertilizer applied in cultivated areas; and incipient sea-water intrusion on the eastern coastal strip (Machkova, et al., 1997; Velikov, et.al., 1996). The nitrate content exceeds 200 mg/l in extensive areas of the aquifer, with maxima at certain points such as the central sector (Figure 6.7), where values reach over 600 mg/l. In general, the maximum values are registered in the principal infiltration area of the entire aquifer, coinciding with developed agricultural sectors and some of the major urban agglomerations. In the southern and eastern coastal sectors, except near the discharge area of the Lake Durankulak, the nitrate values are generally below 30 mg/l, thus indicating a major dilution of the polluting discharge along the flow. The highest chloride contents are closely related to local processes of sea-water intrusion near the eastern coast. Mean values for most of the aquifer are around 50 mg/l, which are considered normal for this region. Along a 5 km strip, parallel to the eastern coastline, values exceed 500 mg/l and reach almost 1000 mg/l. The spatial variation of water conductivity shows that this parameter represents properly the sectors most affected by one or more of the pollution processes described above. Conductivity is about 600 to 700 µS/cm in the discharge areas of the Suha and Batova Rivers, increasing in the contact zone between the aquifer and the Black Sea to values of more than 1500–2000 µS/cm (to the North of Cape Kaliakra). All this can be interpreted as a consequence of the incipient process of sea-water intrusion (Chapter 9). In the central sector the conductivity reaches values of 1300 µS/cm, probably due to the agricultural contamination of the groundwater.

The natural flow of the whole aquifer is about 7.5 to 9 m^3/s. A considerable part of them are exploited.

The intensive use of groundwater has caused a decrease in the water level by 30–50 m and higher around the town of Dobrich and northeast of the village of General Toshevo, as well as the abovementioned local processes of sea-water intrusion near the eastern coast.

Within the area of the Sarmatian aquifer are located several wet zones of international significance, protected by Ram Sara Convention (1971). These are as follows: the Lake complex Durankulak, Shabla - Ezeretz and Shablenska Tuzla. The existence and maintenance of this complex is in close relationship with the Sarmatian groundwater.

6.2 Monitoring Scheme

The main goal of the proposed groundwater monitoring scheme in the transboundary aquifers in Dobrudja area is to apply the requirements of Article 8 of the EC Water Framework Directive 2000/60/EC (WFD). Taking into account the significance of both aquifers for the sustainable development of the settlements in the area, it is essential to preserve the groundwater resources and use them in a sustainable manner. The groundwater monitoring should facilitate the process of their management and contribute further to determining and reversing the negative qualitative and quantitative tendencies. This is related to the calculation of water balance, establishment of databases and geographic information systems (GIS) including groundwater modeling, all of which would not be possible without reliable specific monitoring data.

6.2.1 Groundwater Monitoring Network Assessment

The analysis of the present monitoring network shows that it is not sufficient to provide reliable information on natural or affected groundwater abstraction, irrigation and land use conditions in both aquifers, as well as on the problems related to groundwater pollution. The present scheme can not be used for water management in the transboundary aquifers between Bulgaria and Romania.

This leads to the need for increasing the density of the observation points and improving their spatial distribution, especially in the areas characterized by transboundary groundwater transfer, as well as in the areas facing a high risk of pollution and overexploitation.

The territory belongs entirely to the Black Sea River Basin Directorate (BSRBD), occupying its northern part to the East of the underground watershed between the Varna and North Bulgarian artesian basins. The National Institute of Meteorology and Hydrology (NIMH) deals with the groundwater-surface water quantitative monitoring whereas the qualitative one is done by the Executive Environmental Agency (EEA).

The quantitative monitoring network comprises boreholes, shaft wells and springs. In most cases, the groundwater levels are observed using suitable wells constructed for some other purposes, but not used at present (water supply, exploration, etc.). In a few cases, shallow water is observed using wells dug in villages. The NIMH permanent technical staff measures once or twice per month the water levels by using portable instruments (contact level gauge). Discharges are measured 12 times per year (rarely more frequently) by current meters (rod measurement methods). Small springs and artesian wells are measured by the volume method. Water temperature is measured in all observation points.

The chemical species are measured with different frequencies (1–4 times per year). Information about major components exists whereas data for trace components are much more limited.

The present quantity monitoring network related to the Deep aquifer consists of:

- 19 observation points—boreholes located in the south-southwestern part of the recharge zone (Figure 6.8). Their depth ranges from 325 to 700 m while water levels vary from 25 to 150 m below the surface. Six of these are located in the study region. The groundwater level measurements were started between 1980 and 1983;

- 9 observation points—artesian wells are located on the Black Sea coast (Figure 6.8). Their depth ranges from 1798 to 2123 m, uncovering and discharging thermal water. The measurements of the artesian wells were started in 1986, but were canceled in 2002. The reason for that is the continuously increasing demand for thermal water during the last several years, which makes the access to some of them difficult and even restricted. At present the observations have been restored to three of the boreholes.

Two monitoring stations were established at the main group of springs discharging Deep aquifer in 1966. They are the Devnia group consisting of 15 springs and the Zlatina group consisting of 3 springs which are not being measured at present.

The present quantity monitoring network related to the Upper Aquifer consists of:

- 15 observation points (11 shaft wells and 4 boreholes) are monitored at present. Their depth ranges from 10 to 77 m, while the water levels vary from 1.7 to 65 m below the surface. All monitoring wells are located along the main groundwater flow directions to the East and to the South- Southeast. Two of these (wells in Kochmar and Dobrich) are located in the western part where the aquifer is discharged by the Suha River and its tributaries. The measurements

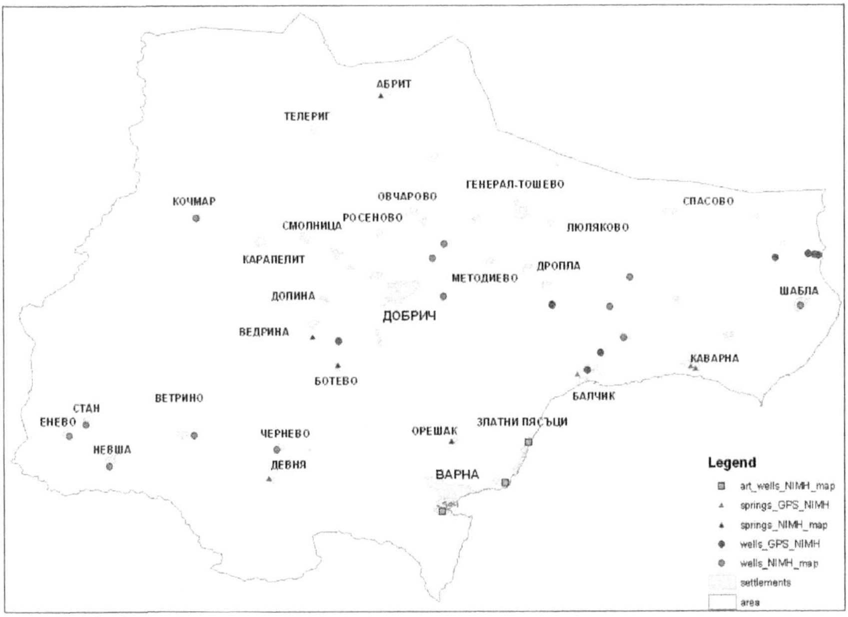

Fig. 6.8 Existing monitoring points

of groundwater levels and temperatures were started in the 1959–1962 period and in the 1980s in some wells;

- 7 springs discharging the Upper Aquifer have been monitored since the 1960–1967 period. One of them, spring No. 148a (Abrit), is located along the groundwater flow line to Romania. Springs No. 149 (Novo Botevo) and No. 150 (Botevo) discharge into the Suha River, while all the rest discharge into the Black Sea. The measurements of springs No. 803 (Kavarna) and No. 806 (Balchik) were canceled for some years but were restored in 2004.

A qualitative monitoring network, maintained by the EEA, consists of 24 observation points. Most monitoring points (boreholes, shaft wells and drainages) are located in pumping stations for drinking water supply. Raw water taken as water samples is analyzed by the local laboratories at the EEA. The water samples of 14 observation points are analyzed according to a base program (macro-components) including some heavy metals. Pesticides are analyzed in the remaining ten points.

It is suggested that the current network may be used as a basis for the draft monitoring scheme which is to be expanded with new observation points. Table 6.1 shows data on the number of the points in which monitoring activities are being carried out.

Based on the analyses of the available information, it could be concluded that the existing monitoring networks for both quantitative and qualitative monitoring are insufficient. The observation points are unequally distributed as the density is far below the required basic one: 22 wells for the Upper and Deep Aquifers, instead of at least 80–90 according to the population density, groundwater vulnerability and exploitation. The available monitoring points are insufficient for drawing hydrodynamic maps, for model validation, etc. In most cases, their technical status is unsatisfactory for a lack of shielding equipment. In addition, the equipment for manual measurements does not guarantee sufficient accuracy of the primary data whereas there is no equipment for automatic measurements. The groundwater tables and spring discharges are measured with outdated instruments produced in the former USSR. Moreover, the access to most monitoring points is difficult due to their locations, frequently far from villages and roads. The roads are often in a bad condition, making their crossing difficult particularly during rainy or winter periods. The monitoring of water balance components (recharge parameters) like precipitation and discharge of the few rivers is in a similar condition.

The main task at this stage is to identify the appropriate points for the future transboundary monitoring scheme. The proposed draft scheme will be expanded,

Table 6.1 Type and number of the exiting monitoring points

Aquifers	Quantitative Monitoring			Qualitative Monitoring
Deep Aquifer	Springs	Boreholes	Artesian wells	
	15	6	3	8
Upper Aquifer	Springs	Boreholes	Shaft wells	
	7	4	11	18

elaborated into details and shaped into its final version during the next stage of the project. All activities envisaged by the WFD, related to the revision and detailed outlining of the water bodies, the completion of their description, the anthropogenic impact assessment and risk assessment will be conducted during the next stage as well. On this basis and further on the proposed expanded draft monitoring scheme, a final transboundary monitoring scheme will be established. The purpose of the current monitoring scheme is to facilitate this process, to provide the equipment needed for manual and automatic measurements as well as to avoid the construction of new and expensive monitoring wells.

The proposed scheme covers the territory which overlaps with the geographic concept of the Dobrudja area, delineated by the surface watershed of the Suha River (including its tributary Karamandere River) and its continuation to the State border between Bulgaria and Romania to the West, the Black Sea to the East, and the Provadiiska River to the South.

At this stage, no extension of the surface water (precipitation and rivers) is proposed, although it is of particular importance for the quality and quantity of the groundwater. An additional supply of equipments for the river monitoring points and meteorological stations is proposed. Additional laboratory equipments for the Regional Laboratories in Varna and Shumen are required for the analytical methods in compliance with ISO standards.

6.2.2 *Scheme Selection Approach*

In accordance with the requirements of the WFD, the monitoring system should permit the assessment of the water balance components, the groundwater resource and the establishment of decision support through flow modeling activities which do not exist at present, but are obviously required to avoid water shortages across the border and to allow effective and sustainable joint monitoring and management of the transboundary waters. The qualitative monitoring does not meet the requirements needed for the basic evaluation of the groundwater body status. The number of sampling points in the Upper and Deep Aquifers are not well distributed while at least 50 points should be sampled at both aquifers. The hydrochemical evaluations types are not sufficient as well. Important "on-site" evaluations should be introduced for gases like oxygen, hydrogen sulfide, carbon dioxide, and trace elements as well.

6.2.2.1 Characteristics of the Scheme Proposed

The main considerations that should be taken into account when selecting the monitoring scheme are the intense exploitation and the water quality.

Groundwater is used through boreholes and shaft wells as well as through springs along the rivers and the Black Sea coast.

The planned monitoring scheme should have a complex character and permit the solution of the problems related to groundwater use, protection and management. It should include the following dimensions:

- Monitoring groundwater levels/discharge, temperatures including water quality;
- Monitoring and collecting meteorological and hydrological data that are relevant for water balance and quantification of groundwater resources. In this case, the monitoring points are the already existing meteorological stations including hydrometric stations established on the main rivers and tributaries;
- Collecting water use data such as pumping rates and volumes of pumped water implemented by the local companies responsible for water supply and irrigation;
- Taking into account the location of point and diffuse sources of pollution including qualitative and quantitative data.

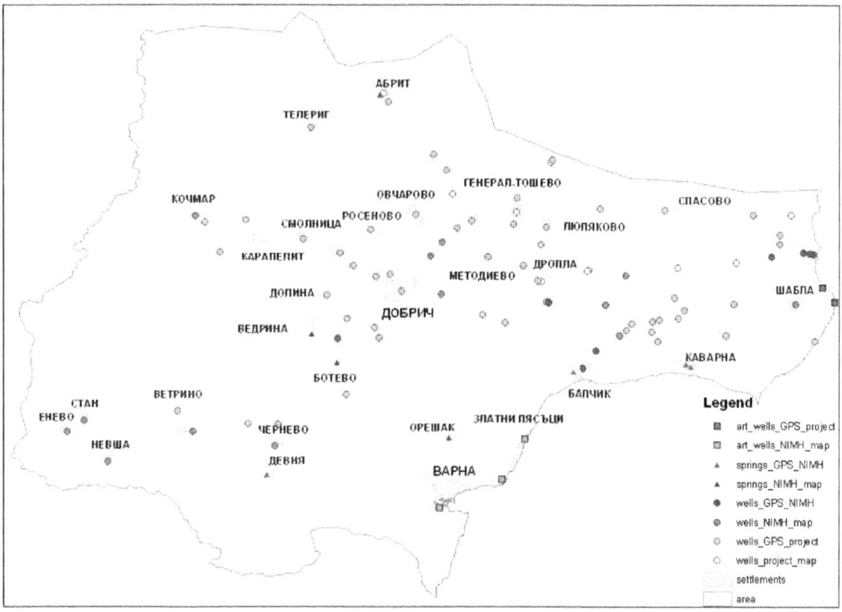

Fig. 6.9 All observation points

In order to implement the future requirements, the draft monitoring scheme should include representative groundwater points where groundwater quality and quantity will be monitored. For these purposes, shaft wells and boreholes may not be used for other purposes. As far as the monitoring is concerned, no drilling of new boreholes is foreseen. Based on archival materials, potential points, predominantly boreholes have been identified that are appropriate for monitoring. Through several field campaigns 45 observation points—boreholes (with a depth ranging from 164 to 3496 m)—in the Deep Aquifer, and 20 points—boreholes and shaft wells (with a depth from 6.10 to 199 m)—in the Upper Aquifer were chosen. Their accessibility and actual technical status were also assessed. The potential monitoring points are presented in Figure 6.9 together with the existing monitoring points. Those for which a possibility exists for electric supply will be equipped with automatic equipment for measurement, data storage and data transmission. All borehole outsets should be repaired and adapted for these purposes in order to protect the equipment which will be installed as well as the groundwater against barbarism etc. The preparatory works including a ferroconcrete cover of the well head and a lockable iron lid providing access to the well are obligatory before the equipment installation.

6.2.2.2 Frequency of Measurement of Parameters and Indicators

The frequency of measurements of the quantity and quality parameters is among the most important components of the monitoring programs. It should be defined with regard to the aquifer types and groundwater fluctuation, variability of the parameters observed, and the effects of various stresses on the aquifers including the WFD requirements.

In the launch of the subsequent project, it is recommended to perform a sampling campaign over the whole territory analyzing a range of chemical parameters which will clarify the actual ecological status of the groundwater. This, as well as the analyses of historical data, will be helpful to identify human effects and risk assessment. After the anthropogenic impact assessment is performed, the indicators for qualitative monitoring will be identified. The monitoring frequency will be defined according to the results of risk assessment in pursuit of the requirements of the WFD.

The initial base for the future transboundary groundwater monitoring scheme will be the current monitoring points of EEA and NIMH plus the observation points proposed in the draft scheme. Elements of the groundwater monitoring scheme are boreholes, self-discharging boreholes, shaft wells, springs and drainages. The elements related to the groundwater quality and quantity are precipitations measured by meteorological stations, hydrometric stations at the rivers, pollution sources including locations and quantity, and pumping rates and volume of pumped groundwater.

6.2.2.3 Indicators and Parameters for Quantitative and Qualitative Groundwater Monitoring

- Water levels in boreholes and shaft wells, water temperature, discharge of springs and artesian boreholes (or pressure in some artesian boreholes), major chemical components (base program), minor and some trace components including pesticides, mandatory components according to the WFD;
- The qualitative monitoring will be implemented by "on-site" (portable) and standard laboratory equipment for evaluation of chemical components characterizing the groundwater status. The measurement and analysis of water samples including data processing will be performed according to the requirements of the WMO and ISO standards.

Acknowledgement This report has been elaborated based on the results obtained in the framework of the Copernicus project ERB-CIPA-CT93-0139 financed by the EC as well the results obtained within TA of PPF Phare CBC Programme financed by the EC.

References

Cheshitev, G., and Kânchev, I., eds., 1989, *Geological Map of P.R. Bulgaria scale 1:500,000*, Supreme Technical Council, Sofia.

Feru, M. U., 1993, Considérations sur la genèse et la circulation des eaux thermominérales karstiques de la zone de Mangalia (Roumanie), *Theoretical and Applied Karstology* **6**:189–197.

Feru, M. U., and Capota, A., 1991, Les eaux thermominérales karstiques de la zone de Mangalia (Roumanie), *Theoretical and Applied Karstology* **4**:143–157.

Iurkiewicz, A., and Oraseanu, I., 1997, Karst terrains and major karst systems in Romania, in: *Karst Waters and Environmental Impacts*, G. Günay, and A. I Johnson, eds., Balkema, Rotterdam, pp. 471–478.

Dachev, C., Stanev, V., and Bokov, P., 1988, Structure of the Bulgarian Black Sea area, *Bollettino di Geofisica Teorica ed Applicata* **30**:79–107.

Danchev, D., Damyanov, A., Nikolov, E., Benderev, A., and Gushkov, I., 1981, Regularities in the distribution of groundwater in the southeast of Dobrudja and Ludogorieto region, *Annals of the High School of Mining and Geology* **21**:140–153 (in Bulgarian).

Davidescu, F. D., Tenu, A., and Slavescu, A., 1991, Environmental isotopes in Karst Hydrology. A lay-out of problems with exemplifications in Romania, *Theoretical and Applied Karstology* **4**:77–86.

Iovtchev, I., ed., 1976, *Tectonic Map of P.R. Bulgaria scale 1:500,000*, Main Authority of Geodesy and Cartography, Sofia.

Machkova, M., Dimitrov, D., Pulido-Bosch, A., Calaforra, J. M., Calvache, M. L, Lopez-Chicano, M., Pentchev, P., and Velikov, B., 1997, Main hydrogeological problems in the karstic aquifers of NE Bulgaria, in: *Karst Waters and Environmental Impacts*, G. Günay, and A. I., Johnson, eds., Balkema, Rotterdam, pp. 53–58.

Matchkova, M., Pulido Bosch, A., Dimitrov, D., Velikov, B., and Lopez-Chikano, M., 1997, Evaluation of the hydrochemical characteristics of the karst waters in the Bulgarian—Romanian Dobrudja region, *BJMH* **8**(3–4).

Marin, C., and Nicolescu, T., 1993, The geochemistry of groundwater from southeastern Dobrogea, Romania, *Travaux de l'Institut de Spéologie "E. Racovitza"* **32**:229–247.

Monahova, L., and Kostova, M., 1989, Main results of the study of water-soluble organic substances in the sediment of North Bulgaria, *Petroleum and Coal Geology* **26**:32–44.

Monahova, L., and Vakarelska, M., 1982, New data on the hydrogeochemistry of Dobrudzha coal deposit, *Petroleum and Coal Geology* **17**:31–48.

Tenu, A., and Davidescu, F., 1993, Relations entre le Lac Techirghiol (Dobroudja du sud, Roumanie) et les aquifères adjacents, *Theoretical and Applied Karstology* **6**:163–174.

Pulido-Bosch, A., Lopez-Chicano, M., Calvache, M. L., Calaforra, J. M., Machkova, M., Dimitrov, D., Velikov, B., and Pentchev, P., 1996a, Principales características de los acuíferos carbonatados de la región de Dobrich (Bulgaria), *Revista Sociedad Geológica España* **9**(3–4):143–156.

Pulido-Bosch, A., Lopez-Chicano, M., Calvache, M. L., Calaforra, J. M., Machkova, M., Dimitrov, D., Velikov, B., and Pentchev, P., 1996b, Consideraciones sobre el flujo en el acuífero kárstico profundo de la región de Dobrich (NE de Bulgaria), in: *Contribuciones a la Investigación y Gestión del Agua Subterránea*, I. Morell, and J. R. Fagundo, eds., Colecció Summa, Ciències experimentals, Ed. Universitat Jaume I, **4**:67–85.

Pulido-Bosch, A., Machkova, M., Lopez-Chicano, M., Calvache, M. L., Dimitrov, D., Calaforra, J. M., Velikov, B., and Pentchev, P., 1997, Hydrogeology of the Upper Aquifer, Dobrich Region (Northeastern Bulgaria), *Hydrogeology Journal* **5**(2):75–85.

Robinson, A. G., Rudat, J. H., Banks, C. J., and Wiles, R. L. F., 1996, Petroleum geology of the Black Sea, *Marine Petrol. Geol.,* **13**:195–223.

Velikov, B., Damyanov, A., and Pentchev, P., 1989, Hydrogeological conditions and hydrochemical characteristics of sea water intrusion in aquifers near the Black Sea coast of Bulgaria, *Proceedings 40 years Department of Hydrogeology and Engineering Geology*, UMG, Sofia, pp. 102–112 (in Bulgarian).

Velikov, B., Dimitrov, D., Machkova, M., Pulido-Bosch, A., Lopez-Chicano, M., and Calvache, M. L., 1996, Hydrogeochemical regime observations on the groundwater in the area of Dobrich - Northeast Bulgaria, *Geology and Mineral Resources* **8**:3–7.

Part III
Tools to Predict/Forecast the Future Depletion and Contamination of Shared Groundwater Resources

Chapter 7
Coupled Surface/Subsurface Flow Systems: Numerical Modeling

Rachid Ababou*, **Ahmad Al-Bitar**

Abstract This chapter contains a summarized presentation of novel modeling approaches for coupling hydrological flows, where the "coupling" involves porous aquifers as well as open water bodies. First, basic concepts on partially saturated flow systems are presented (mostly without equations). A brief review of a fully 3D modeling approach, involving the concept of a variably saturated macroporous medium, is given. The rest of the chapter focuses on a 2D approach, based on vertically hydrostatic plane flow assumptions, for dealing with coupled aquifer-stream flow in an alluvial river valley and floodplain. This approach assumes that the water tables of stream and aquifer are connected. Preliminary simulation results are shown for an actual test site. Overall, this chapter is meant to provide a quick overview on some environmental flow modeling problems requiring retroactive surface/subsurface coupling, based on some of the methods presented in the corresponding NATO-ASI lecture.

Keywords Porous media, groundwater, water table, hydrogeology, permeability, Darcy-Forchheimer, transmissivity, Boussinesq-Dupuit, hydraulics, kinematic wave, diffusive wave, Saint-Venant, macro-porous, finite volume 2D–3D, numerical simulation, variably saturated, partially saturated, streamflow; flooding

7.1 Introduction, Summary and Objectives

This chapter builds upon a NATO-ASI lecture given by the first author in 2006 in Varna (Bulgaria) on the modeling of coupled surface/subsurface flow and transport phenomena in hydrological systems. Here, we provide a very brief

Rachid Ababou, Ahmad Al-Bitar
Institut de Mécanique des Fluides de Toulouse, Allée du Professeur Camille Soula, 31400 Toulouse, France

*To whom correspondence should be addressed. E-mail: ababou@imft.fr

C.J.G. Darnault (ed.), *Overexploitation and Contamination of Shared Groundwater Resources*.

overview of some of the topics, summarizing the lecture, and focusing mostly on hydrological flow problems involving retroactive interactions between stream flow and aquifer flow in an alluvial floodplain.

For more details on the flow modeling approaches, the reader is invited to peruse the cited references at the end of this chapter. For instance, the numerical modeling approaches implemented in the 3D BIGFLOW code are described in detail in Ababou et al. (1988), Ababou et al. (1992), Ababou and Bagtzoglou (1993) and Trégarot (2000). The latter also describes the various BIGFLOW 2D/3D approaches to partially saturated and coupled flow problems, including the new 2D BIGFLOW module for plane flow modeling. Finally, the reader can consult Al-Bitar et al. (2005) concerning an extension of the 2D BIGFLOW module to deal with seawater intrusion in a freshwater aquifer via the sharp interface approach (with or without coupling to surface flow).

7.2 Basic Concepts

In this chapter's summary presentation, we choose to skip the equational formulation of the models, in order to focus on example simulations. Let us however recall here a few basic concepts—most of which have been defined, possibly differently, in other reference text books such as Freeze and Cherry (1979), Marsily (1986), Bear (1972, 1988), among others.

- *Saturated flow.* Water flow in saturated porous media, with the porosity being entirely filled with water;
- *Unsaturated flow.* Water flow in unsaturated porous media, the porosity being only partially filled with water, and the rest with air; resistance to air flow is neglected; air pressure equilibrates rapidly to atmospheric pressure. The classical equation governing unsaturated flow (without partial saturation) is the Richards equation (Richards, 1931);
- *Variably saturated and partially saturated flow.* Water flow can occur either in fully saturated mode, or unsaturated, or both. Several saturated/unsaturated zones may coexist, and their spatial pattern can evolve with time, e.g.: partially saturated involving unsaturated infiltration towards groundwater's free surface. For early studies on the latter topic, see Freeze (1971), and Vauclin et al. (1979). To model partially saturated flow with a single equation and single domain, the Richards equation must be cast in a conservative mixed form, in terms of two state variables, namely, water content (storage term) and pressure head (flux divergence term): see Ababou et al. (1988, 1992) and Celia et al. (1990) among others;
- *Darcy's law, permeability, and hydraulic conductivity.* Darcy's law expresses proportionality between flux density q (m/s), and hydraulic gradient. Note that water velocity is $V = q/\theta$, where θ is porosity (or water content). Permeability

(m^2) expresses the inverse viscous resistance to flow, intrinsic to the porous medium. Hydraulic conductivity (m/s) expresses the same but depends on the fluid, i.e. water under standard conditions. In unsaturated media, Darcy's law retains the same form: it is quasilinear, with hydraulic conductivity a function of pressure head or water content. Historical references: Darcy (1856, Appendix D); Buckingham (1907); Richards (1931);

- *Macroporous media.* A macroporous medium can be an open flow domain, a surficial layer of thickly vegetated soil, a pebble bank, etc. It is viewed here, for modeling purposes, as a medium with coarse pores and/or large permeability (as a limit case, infinite permeability and 100% porosity can be used in subdomains). This leads to two different effects to be accounted for in a model (for details, see Ababou et al., 1996, 1998, 2002, 2006):

  1. *Dynamic effects*: due to larger Reynolds numbers, the head loss law is no longer linear, inertial effects become important, and Darcy's law must be replaced with Ward-Forchheimer's quadratic velocity law (Ward, 1964).
  2. *Kinetic effects*: the water retention curve is a quasi-step function, leading to instantaneous wetting or drainage of the porosity at any point (x,y,z) where a mobile free surface exists.

- *Dupuit-Boussinesq plane flow.* Assuming vertically hydrostatic/quasi-plane flow, the Darcy-based equations can be vertically averaged to obtain the Dupuit-Boussinesq flow equations in the (x,y) plane. The variables are the vertically averaged hydraulic head H (m), and the 2D specific discharge rate vector Q (m^2/s). The latter is obtained by vertical integration of Darcy's velocity (or flux density vector) q (m/s). It should be noted that the unsaturated zone flux is neglected (assuming instantaneous response of water table to wetting/drying processes). The storage coefficient is the aquifer's "effective porosity". Historical references: Dupuit (1863); Boussinesq (1904).
- *Saint-Venant equations.* The 2D Saint-Venant equations are depth-averaged approximations of the Navier-Stokes equations (conservation of mass and momentum) for the case of free surface hydraulics over a canal, river bed, flood plain, or coastal area with known bathymetry. The variables are the 2D water velocity vector V (m/s), and the water elevation "H" or the water depth "h". There exists also a 1D system of Saint-Venant equations for describing section-averaged flow along a river or canal. Historical reference: Saint Venant (1871).
- *Kinematic-diffusive wave equation.* The 2D kinematic-diffusive wave equation is a further simplification of the Saint-Venant (and Navier-Stokes) equations, valid only for slowly varying velocity and negligible inertial terms. There exists a well known 1D version of this equation for canals (x), but it is the 2D version that is of interest here (x,y). The idea in this work is to couple a surface "layer" and a subsurface "layer" based on groundwater hydraulics (Boussinesq) and surface hydraulics (kinematic-diffusive wave). For classical presentations of the kinematic-diffusive wave, and the required "roughness" parameters, see Chow et al. (1988) or Bedient et al. (2002).

- *Anisotropy*. Porous and macroporous media can be strongly anisotropic, although the correct amount of anisotropy should depend on model resolution (anisotropy may not be needed if fine heterogeneities are resolved explicitly). Darcy's law may be formulated with anisotropic permeability as 2nd rank symmetric tensor. In our numerical model, the tensor is diagonal but the reference system may be rotated with respect to gravity. More generally, the model accommodates anisotropy in various generalized forms of Darcy's law. Thus, for plane flow, both aquifer transmissivity and roughness coefficients can be anisotropic. For 3D macroporous media, with large Reynolds number, Ward-Forchheimer's quadratic law can be made anisotropic as described in Trégarot (2000), after Knupp and Lage (1995).

7.3 The General 2D/3D Flow Model Implemented in the BIGFLOW Code

A general mathematical model that incorporates the above-mentioned features and properties, is implemented numerically in the finite volume 2D/3D BIGFLOW code, aimed at simulating variably saturated and coupled hydrological flows in either 2D (x,y) or 3D (x,y,z).

The 3D module includes a generalized Darcy law for variably saturated or partially saturated, porous or macroporous media, including Forchheimer's inertial term. Its capabilities were studied and demonstrated, for instance, in Ababou et al. (2002).

The 2D module implements the coupled plane flow models of interest here and in the remainder of this chapter. It is based on vertically averaged flow equations, assuming vertically hydrostatic plane flow. When surface water is present, the Boussinesq equation of groundwater flow is coupled with a mass conservative form of the kinematic-diffusive wave equation. The coupling is performed implicitly, without recourse to decoupling/recoupling strategies, within a single generic equational model which has spatially distributed parameters in the (x,y) plane.

7.4 The Coupled Stream/Aquifer Plane Flow Model (2D)

This section gives a summary presentation and illustration of the 2D plane flow approach for modeling coupled aquifer-stream flow. Note that an application to the Garonne river valley will also be presented further below, based on the actual geometry and topography of the study site. The objective is to test the coupled model, and *in fine*, to contribute to the development of a model for simulating flow and solute transport in wet hydrosystems (e.g. floodplains) with surface/ subsurface interactions.

As mentioned earlier, the numerical model used to achieve these objectives is the finite volume 2D/3D BIGFLOW code for variably saturated and coupled hydrological flows in 2D (x,y) or 3D (x,y,z). The full capabilities of the BIGFLOW code are described in several references given at the end of this chapter. For instance, Ababou et al. (2002) presents macroporous flow tests with the 3D module, while Al-Bitar et al. (2005) presents the 2D module coupling Boussinesq freshwater flow with seawater intrusion via a sharp interface approach. However, in the remainder of this Chapter, we focus on freshwater surface/subsurface flow problems without seawater intrusion.

Let us now focus on the equational model used for dealing with coupled surface/subsurface plane flow in an alluvial floodplain (without seawater intrusion). The corresponding "2D module" of BIGFLOW consists of:

- The transient Boussinesq flow equation in (x,y,t);
- The transient kinematic–diffusive wave equation for surface flow in (x,y,t);
- The implicit coupling relation between the two flow equations;
- The spatially distributed parameters of the previous equations.

The flow equations, and the implicit coupling involved in the 2D module, are briefly presented in Figures 7.1 and 7.2. The coupling is based on the assumption of perfectly connected water table between surface water and aquifer: thus, H(x,y,t) is continuous in the (x,y) plane. The parameters include permeability, porosity, bed roughness, and topography/bathymetry, i.e.:

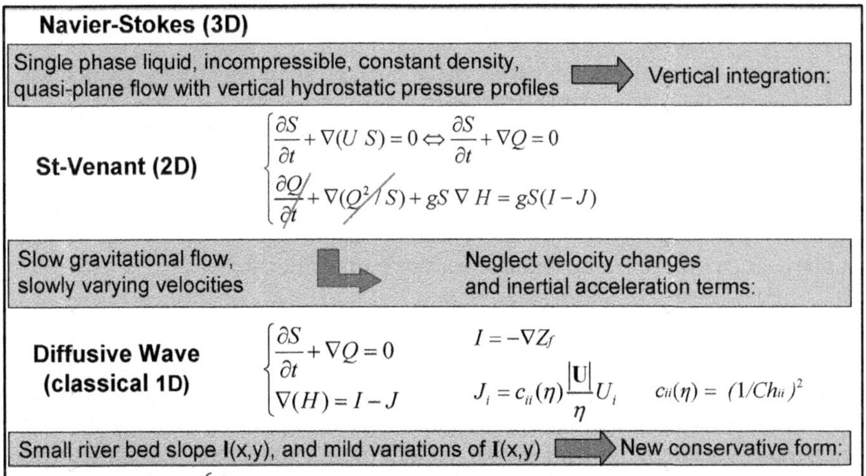

Fig. 7.1 Equations of the surface flow model. The 2D diffusive wave equation is implemented in BIGFLOW's 2D plane flow module, with or without coupling to groundwater flow (see below)

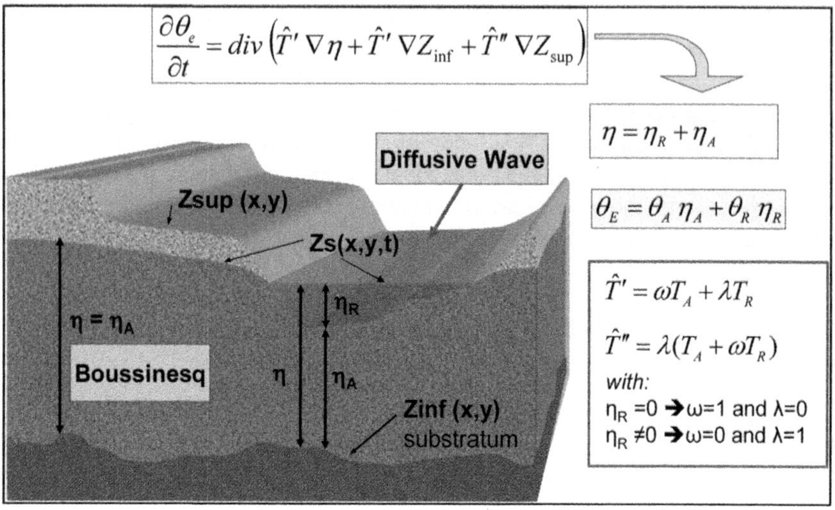

Fig. 7.2 Coupled aquifer/stream flow equations in BIGFLOW's current version of the 2D coupled flow module (2007). The Dupuit-Boussinesq equation governs groundwater flow, and the diffusive wave model is used for stream or floodplain hydraulics. The two are strongly coupled in a single equation governing water table elevation H(x,y,t) or water depth η(x,y,t). Note that water storage θ(x,y,t) [m^3/m^2] is another variable, related to water depths

- Elevation of ground surface and river bed, $Z_{SUP}(x,y)$;
- Elevation of aquifer substratum, $Z_{INF}(x,y)$.

Note. It should be noted that the coupled flow simulations shown in this *Chapter* were obtained with an earlier version of the coupling scheme (which was developed in Trégarot, 2000). The earlier version of the coupling scheme is equivalent to our current scheme (described by the equations shown in Figure 7.2) in some particular cases involving mild slopes, weak coupling, *etc.*

Test simulations. Several preliminary tests (*not shown here*) were successfully performed on simple geometries, with either aquifer or streamflow, without coupling. The purpose was to validate various parts of the model, e.g.:

- 1D aquifer flow on a sloping substratum, and
- 1D surface flow on a sloping impervious bed,

for which analytical solutions are available.

Another series of tests, like the one in Figure 7.3, focused on simplified 2D geometries, but this time, with full stream-aquifer coupling. The example in this figure concerns the passage of a "flood" in a 2D coupled stream-aquifer system with a rectangular meander. Given the lack of analytical solutions, the results were assessed by inspecting mass balance and by observing a few qualitative features of the simulated water table, like the delayed response of the aquifer, the bypassing of the meander by groundwater flow, *etc.*

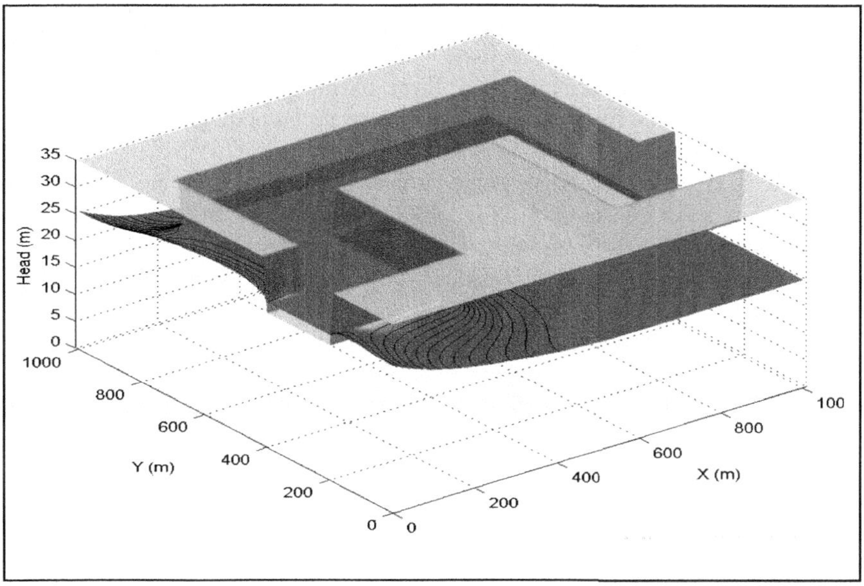

Fig. 7.3 Simulation of a simplified stream/aquifer flow system with the 2D coupled plane flow module of BIGFLOW: qualitative test with simplified geometry (a single rectangular meander)

7.5 Application: Coupled Aquifer/Stream Flow in the Garonne Valley

In contrast with the previous cases, the simulations presented in this section concern a coupled stream-aquifer system with realistic (x,y) geometry and topography, including a curved river meander and a pebble bank islet. However, it should be cautioned that the fully coupled model is still being tested at this stage. Therefore, the results presented below should only be seen as a preliminary assessment of the coupled model's equations and parameters, for a realistic, geometrically complex stream-aquifer flow system.

7.5.1 Stream-Aquifer Exchanges in the Garonne River

The context of this modeling is an eco-hydrological study of the Garonne river valley, aimed at understanding hydro-bio-chemical exchanges in the wet zone of the Garonne valley and other similar alluvial systems. In particular, the main objective is to quantify by in situ measurements and numerical modeling, the hydraulic and mass transport interactions in the stream-soil-aquifer system, and in particular the roles of the river's hyporheic zone, pebble banks and islets. In this work, we focus on hydraulic interactions.

Figure 7.4 illustrates schematically these exchanges in a short meandering stretch of the Garonne river. The meander shown in the photograph, located near the town of Monbequi, is easily recognizable with its pebble bank islet (small island). It is a part of the larger computational domain used in the simulations shown further below. The arrows in Figure 7.2 highlight water and solute exchanges through the hyporheic zone and the pebble banks, and also, stream-aquifer interactions through river bed and river banks. The schematic is adapted from Ababou et al. (2006).

On a larger scale, this study zone is a subdomain of a longer 75 km stretch of the alluvial plain of the *Garonne* river, between the cities of Toulouse and Moissac (France). The physical parameters of the stream-soil-aquifer system ("wet zone") were studied in order to analyze hyporheic water exchanges, and their effects on biochemical nutrient uptake. In this context, there arose a need for a modeling approach to analyze retroactive (two-way) interactions between surface water and groundwater via the stream's hyporheic zone. This should help quantify exchanges between the various compartments of the hydrosystem.

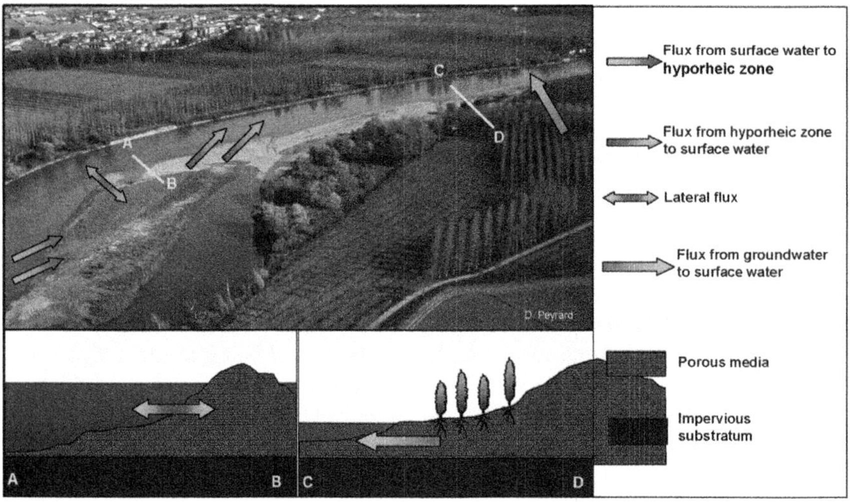

Fig. 7.4 Schematic representation of flow-transport interactions in a meander of the Garonne river, involving the hyporheic zone, the pebble banks, and the soil-aquifer system (Photo credit: D. Peyrard (ECOLAB))

Modeling was conducted at mainly two different scales: "meso-scale" (short river reach ~ 4–8 km), and "macro-scale" (longer stretch of Garonne river floodplain, ~75 km long and a few kilometers across). The meso-scale site is located on the inner side of a river mander near the town of Monbequi. It is instrumented as follows: about 20 piezometers were installed, water levels were recorded continuously during at least 2 years, and concentrations in conservative elements like chloride were measured in each piezometer under various hydrologic conditions. See Weng et al. (2003) concerning *in situ* data, as well as (uncoupled) groundwater flow modeling and calibration. Their calibrated hydraulic conductivities were used in the simulations presented in the next section.

7.5.2 Simulations of Coupled Stream-Aquifer Flow in 2D

The remaining figures in this Chapter show preliminary simulations of coupled aquifer/stream flow conducted on the actual topography of the Garonne river bed and its alluvial floodplain, at the scale of a meander.

The topographies of ground surface and river bed, shown in these figures, were obtained from several data sources with extremely different spatial resolutions, and were then interpolated and merged on a single grid using a GIS in the process. We propose to call the numerical map resulting from *data fusion* calculations an *Integrated Digital Elevation Model (IDEM)* (Figure 7.5).

The flow simulations are displayed in plane view at three times labeled t(1), t(2), t(3) in Figures 7.6–7.8. A perspective view is also shown, for intermediate time t(2), in Figure 7.9.

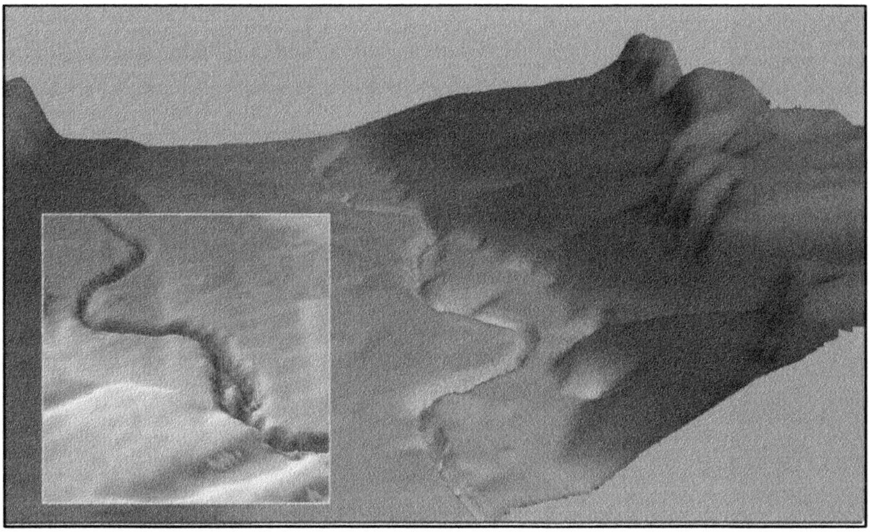

Fig. 7.5 Perspective view of the alluvial Garonne river valley, displayed with a GIS. This topography/bathymetry is an Integrated Digital Elevation Model (IDEM) obtained by data interpolation and data fusion. Two scales are show here. The zoomed area (inset) displays a sub-domain comprising the central meander and its island. The coupled aquifer-stream flow simulations shown in the accompanying figures were conducted for this sub-domain

It should be noted that the flow in the river (and in the companion aquifer) was obtained by "forcing" and *ad hoc* condition at the *upstream* boundary, in order to mimick qualitatively the passage of a "synthetic" flood in the river. At a later phase of this work, it is planned that actual time series of river stages and discharge rates (limnigraphs and hydrographs) will be used to enforce observed flood events coming from upstream. Transient boundary conditions can be accommodated by our flow code in the form of time series inputs, with constant as well as variable time step.

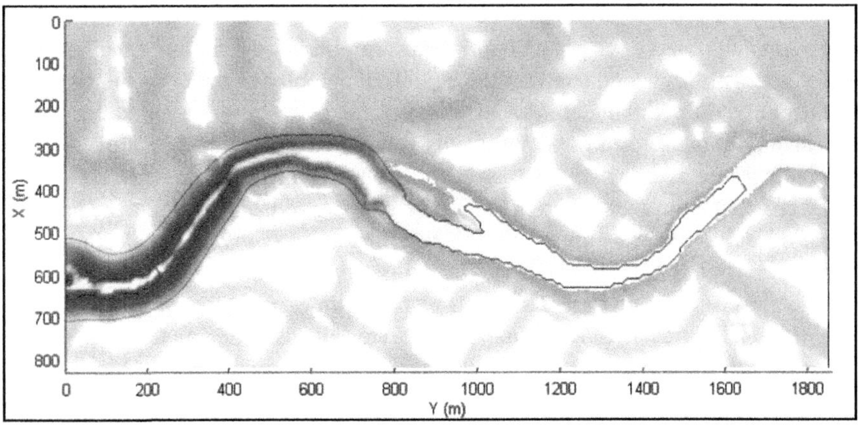

Fig. 7.6 Simulation of coupled stream/aquifer plane flow with the BIGFLOW code, during a synthetic flood forced at the left boundary, in a meandering stretch of the Garonne river: early time t(1)

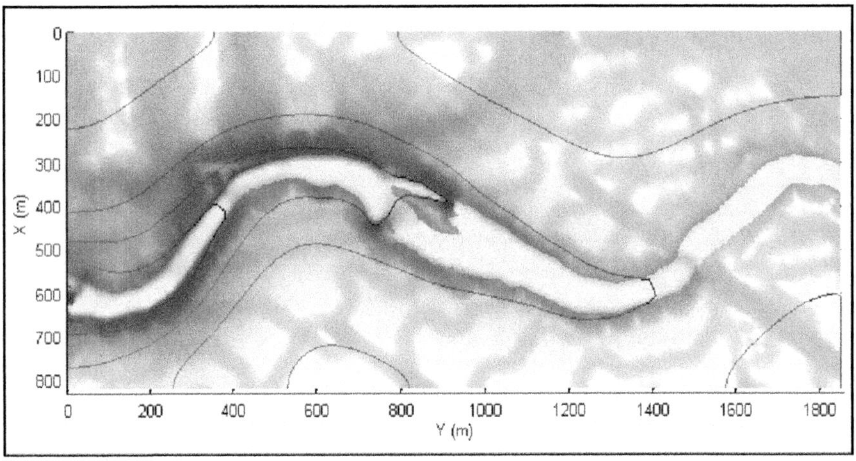

Fig. 7.7 Simulation of coupled stream/aquifer plane flow with the BIGFLOW code, during a synthetic flood forced at the left boundary, in a meandering stretch of the Garonne river: intermediate time t(2)

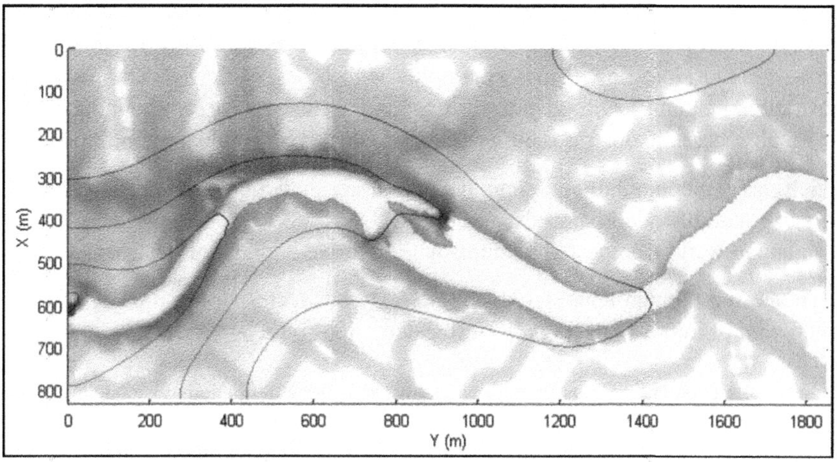

Fig. 7.8 Simulation of coupled stream/aquifer plane flow with the BIGFLOW code, during a synthetic flood forced at the left boundary, in a meandering stretch of the Garonne river: later time t (3)

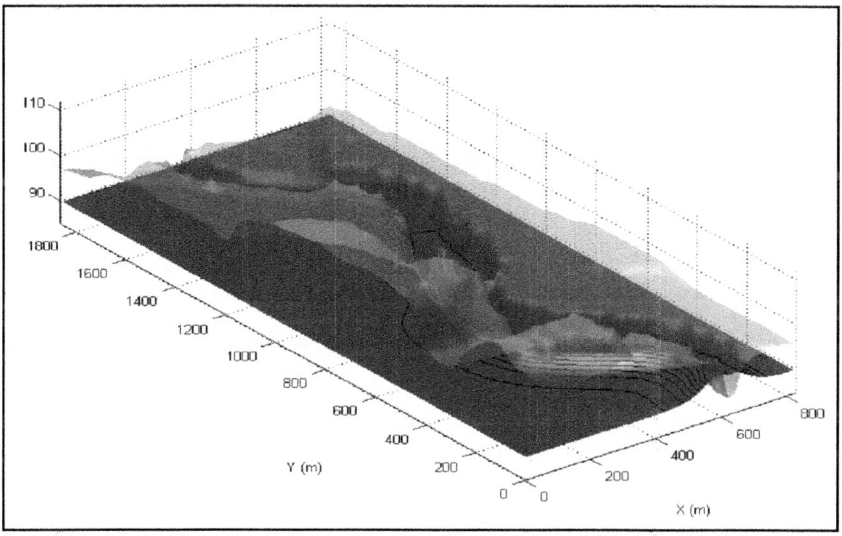

Fig. 7.9 Perspective view of the coupled stream/aquifer plane flow simulation performed with the BIGFLOW code, for a synthetic flood "event" forced at the left boundary and propagating in the meandering Garonne river and its companion aquifer (intermediate time t(2))

On the other hand, there may be a difficulty with *downstream* conditions, depending on available data and scale. In the absence of downstream data, there may be a need to design "transparent" boundary conditions downstream.

For instance, a "reasonable" downstream condition may be to set the gradient of water depth (η) to zero in both the stream and aquifer. This is equivalent to setting a null "diffusive" flux, in terms of hydraulic diffusion. However, this condition makes sense only if there exists a significant "advective" flow component, that is, gravitational flow. This occurs only if the slopes of river bed and aquifer substratum are significant, i.e. non zero, at least at the downstream boundary.

Finally, here are other possible methods to deal with boundary conditions:

- Setting up a computational domain much larger than the domain of interest;
- Using spatially adaptive grids; and/or,
- Calculating flow on nested domains with iterative boundary updates.

The nested domain approach could alternate between fine mesh calculations in the domain of interest, and coarse mesh calculations on a larger domain.

Acknowledgments The authors gratefully acknowledge the contributions to modeling results described in this work, under various forms and at different times, by several collaborators or coworkers, in particular, doctoral students G. Trégarot (2000) and D. Peyrard (currently), and researchers at the Ecolab laboratory in Toulouse (J. M. Sanchez-Perez, S. Sauvage, P. Vervier). Some of the modeling results were obtained under funding by the following French programs and institutions: GIS ECOBAG; Programme National Zones Humides; and Programme National de Recherche en Hydrologie (GDR 2290, CNRS).

References

Ababou, R., and Bagtzoglou, A. C., 1993, *BIGFLOW: A Numerical Code for Simulating Flow in Variably Saturated, Heterogeneous Geologic Media (Theory & User's Manual, Version 1.1)*, Report NUREG/CR-6028, U.S. Nuclear Regulatory Commission, Government Printing Office, Washington D.C., U.S.A., 139 pp.

Ababou, R., and Trégarot, G., 2002, Coupled modeling of partially saturated flows : Macro-porous media, interfaces, and variability, in: *Computational Methods in Water Resources,* Proceedings CMWR, 23–28 June 2002, Delft, The Netherlands, 8 pp.

Ababou, R., Gelhar, L. W., and McLaughlin, D., 1988, *Three-Dimensional Flow in Random Porous Media*, Report No. 318, Ralph Parsons Laboratory for Water Resources & Hydrodynamics, Massachusetts Institute of Tech., Cambridge, MA 02139, March 1988 (2 vols., 833 pp.).

Ababou, R., Sagar, B., and Wittmeyer, G., 1992, Testing procedures for spatially distributed flow models, *Advances in Water Resources* **15**:181–198.

Ababou, R., Trégarot, G., and Bouzelboudjen, M., 1996, Variably saturated subsurface flow with layers and interfaces: Perched water tables and stream-aquifer connection, in: *ModelCare'96,* Proc., International Groundwater Modeling Center (IGWC), GWMI Series No.96-OX, Colorado School of Mines, 10 pp., 25–27 Sept. 1996.

Ababou, R., Trégarot, G., and Larabi, A., 1998, Partially saturated hydrological flows: Numerical experiments and analyses, in: *Computational Methods in Water Resources,* XII International Conference on (CMWR) Proceedings, 15–19 June 1998, The Chersonese, Crete, Greece, 8 pp.

Ababou, R., Al-Bitar, A., Peyrard, D., Quintard, M., Sanchez-Perez, J. M., Sauvage, S., Vervier, P., and Weng, P., 2006, Modeling coupled surface/subsurface flow interactions: Implementation and comparison of three models based on Darcy, Boussinesq/Saint Venant, and Boussinesq/diffusive wave, with application to the Garonne floodplain, Midi-Pyrénées, France, in: *Groundwater Hydraulics in Complex Environments,* Proceedings IAHR-GW 2006, Toulouse, France, 12–14 June 2006 (to be published).

Al-Bitar, A., and Ababou, R., 2005, Random field approach to seawater intrusion in heterogeneous coastal aquifers: Unconditional simulations and statistical analysis, in: *GeoENV: Geostatistics for Environmental Applications*, P. Renard, H. Demougeot-Renard, and R. Froidevaux, eds., Springer.

Bear, J., 1972, *Dynamics of Fluids in Porous Media*, Elsevier, New York, 764 p.

Bear, J., 1988, *Dynamics of Fluids in Porous Media*, American Elsevier, N.Y. (1972), republished in 1988 by Dover Publications Inc., N.Y., 764 pp.

Bedient, P. B., and Huber, W. C., 2002, *Hydrology and Floodplain Analysis*, 3rd Ed., Prentice-Hall Publishing Co., Upper Saddle River, NJ, 763 p.

Boussinesq, J., 1904, Recherches théoriques sur l'écoulement des nappes d'eau infiltrées dans le sol et sur le débit des sources, *Journal de Mathématiques Pures et Appliquées*, **10**:5–78 and 363–394.

Buckingham, E., 1907, *Studies on the Movement of Soil Moisture*, U.S. Dept. Agric., Bureau of Soils, Bulletin No. 8.

Celia, M. A., Bouloutas, E. T., and Zarba, R. L., 1990, A general mass conservative numerical solution for the unsaturated flow equation, *Water Resources Research* **26**(7):1483–1496.

Chow, V. T., Maidment, D. R., and Mays, L. W., 1988, *Applied Hydrology*, Mc Graw-Hill International Editions, Civil Engineering Series, 572 pp.

Darcy, H. P. G., 1856, *Les Fontaines Publiques de la Ville de Dijon, Exposition et Application des Principes à Suivre et des Formules a Employer dans les Questions de Distribution d'Eau*, Victor Dalmont (publisher), Paris.

Dupuit, J., 1863, *Etudes Théoriques et Pratiques sur le Mouvement des Eaux dans les Canaux Découverts et à Travers les Terrains Perméables*, 2nd edition, Dunod, Paris.

Freeze, A., 1971, Three dimensional, transient, saturated-unsaturated flow in a groundwater basin, *Water Resources Research* **7**(2):347–366.

Freeze, A., and Cherry, J. A., 1979, *Groundwater*, Prentice-Hall, Englewood Cliffs, NJ.

Knupp, P. M., and Lage, J. L., 1995, Generalization of the Forchheimer-extended Darcy flow model to the tensor permeability case via a variational principle, *Journal of Fluid Mechanics* **299**:97–104.

Marsily (de), G., 1986, *Quantitative Hydrogeology (Groundwater Hydrology for Engineers)*, Academic Press, New York, 440 pp.

Richards, L. A., 1931, Capillary conduction of liquids through porous medium, *Physics* **1**:318–333.

St Venant, Barré de., 1871, Theory of unsteady water flow, with application to river floods and to propagation of tides in river channels, *French Academy of Science* **73**:148–154, 237–240.

Trégarot, G., 2000 (in french), *Modélisation Couplée des Ecoulements à Saturation Variable avec Hétérogénéités, Forçages, et Interfaces Hydrologiques*. Doctoral thesis, Institut National Polytechnique de Toulouse: No. 1663 (dir. R. Ababou). Institut de Mécanique des Fluides de Toulouse, France, May 2000, 496 pp.

Vauclin, M., Khanji, D., and Vachaud, G., 1979, Experimental and numerical study of a transient two-dimensional unsaturated-saturated water table recharge problem, *Water Resources Research* **15**(5):1089–1101.

Ward, J. C., 1964, Turbulent flow in porous media, *Journal of the Hydraulics Division*, Proceedings of the ASCE. Vol. HY 5, pp. 1–12.

Weng, P., Sánchez-Pérez, J. M., Sauvage, S., Vervier, P., and Giraud, F., 2003, Hydrological modelling to characterize the riparian wetland of a large alluvial river (Garonne river, France), *Hydrological Processes* **17**:2375–2392.

Chapter 8
Quantitative Stochastic Hydrogeology: The Heterogeneous Environment

Rachid Ababou*

Abstract This chapter is devoted to quantitative hydrogeology, and particularly, to flow in heterogeneous geologic formations, with particular focus on two main points: the quantitative representation and generation of heterogeneous geologic media (continuous or discontinuous) in a statistical framework, and the appropriate governing equations to model groundwater flow in such media. The growing interest in "*stochastic hydrogeology*" is due to the fact that complex spatial patterns can be efficiently represented by statistical methods, whereby the heterogeneous environment is represented via random variables and random fields. Along these lines, various topics will be discussed: the generation of purely "synthetic" random media; bayesian estimation and geostatistical approaches to incorporate data; continuous *versus* discrete geologic media; and related flow modeling and upscaling issues in 2D and 3D.

Keywords Porous media, groundwater, hydrogeology, stochastic methods, Darcy-Forchheimer, permeability, random fields, random media, boolean media, fractured media and networks, Bayesian estimation, conditioning, implicit finite volumes (3D), conjugate gradients, numerical simulations, unsaturated flow, variably saturated flow

8.1 Introduction, Summary and Objectives

This chapter builds upon a NATO-ASI lecture given by the author in 2006 in Varna (Bulgaria) on "quantitative stochastic hydrogeology", and particularly, on numerical and stochastic approaches to heterogeneous geologic media.

Rachid Ababou
Institut de Mécanique des Fluides de Toulouse, Allée du Professeur Camille Soula, 31400 Toulouse, France
* To whom correspondence should be addressed. E-mail: ababou@imft.fr

C.J.G. Darnault (ed.), *Overexploitation and Contamination of Shared Groundwater Resources*.

Hydrogeology is the study of groundwater flow in geologic formations. The title and subtitle of this chapter stem from the fact that the *environment* of hydrogeology is made up of *heterogeneous* geologic media (soils, sedimentary aquifers, topographic surfaces, strata, beds and layers, fractures, macropores, etc). The growing interest in *stochastic* hydrogeology and groundwater (GroundWater, denoted GW for short) flow modeling, can be explained by the fact that very complex spatial patterns can be efficiently represented based on a *stochastic* approach. The heterogeneous environment is represented via random variables, or random fields, based on classical probability concepts (random variables) and on more specialized concepts like random functions of space (random fields), bayesian estimation and/or geostatistics.

However, before exploring the specific topic of stochastic hydrogeology, we describe in Sections 8.2–8.4 various "qualitative" aspects of groundwater hydraulics and hydrogeology. Section 8.5 is more "quantitative": it develops the concepts needed for porous media flow modeling, including a discussion of Darcy's law and related equations for different types of hydrogeologic flows (see also, along the same lines: Bear, 1972; Freeze and Cherry, 1979; Marsily, 1986). Here are some related questions and issues:

- What are the different types of geologic media and aquifer systems?
- What are the relevant types and configurations of groundwater flow systems (vertically, horizontally, in three dimensions)?
- What is a geologic porous medium (for all practical purposes)?
- What is Darcy's law? For isotropic media? Anisotropic media? Unsaturated media? Fast flows? And how is Darcy's law related to other fluid dynamics equations (Stokes, Navier-Stokes)?
- What is the typical permeability of a coarse sand and gravel aquifer? Of a multilayered medium? Of a fractured rock? And (how) does it depend on scale?

Finally, Section 8.6 develops in more detail the central topic of "*stochastic hydrogeology*", including the required statistical concepts and tools needed to generate and to characterize a random geologic medium, continuous or discrete, in 2D/3D. It is pointed out that the geologic medium needs to be adequately parameterized, e.g. in terms of granulometry, porosity, permeability. The remaining "task" is then to represent the spatial pattern in a statistical framework. Here are a few questions and issues related to this topic:

- What is a "continuous" random medium, or random field F(x,y,z)?
- What is a discrete or "boolean" model, or boolean field B(x,y,z)?
- Why is the PDF of permeability often assumed log-normal? What is a log-normal PDF? (PDF: Probability Density Function).
- What are the relevant statistical characteristics of a spatially distributed function, or random field F(x,y,z)? What does the PDF represent for a random field, and what does it fail to capture?

- What is a spatial covariance function? A2-point structure function?
- How do the properties of a random medium vary with scale?

The last part of Section 8.6 is devoted to flow modeling in 2D/3D randomly heterogeneous geologic media. A specific numerical tool (the BIGFLOW 2D/3D code based on implicit finite volumes) is used in some cases, to illustrate high resolution simulation of groundwater flow systems in randomly heterogeneous media. Overall, the examples discussed cover a number of cases: unsaturated soils, sedimentary aquifers, and fractured or fissured rocks.

A natural extension of this chapter would concern the fate of tracers and contaminants, migrating in statistically heterogeneous flow systems. The relevant transport mechanisms are (i) pure advection and (ii) subgrid diffusion or hydrodynamic dispersion. However, due to lack of space, this aspect will not be treated here, except for a few references to particle tracking models.

8.2 Overview on Global Water and Groundwater Cycles

The "global" water cycle is an average over global space-time scales, i.e.:

- At the spatial scale of the entire globe (the planet), or only 30% of its surface if the focus is on continental surfaces (mountains, plateaus, plains, valleys, lakes);
- At the time scale of the solar year (yearly fluxes and water budgets).

The global water cycle is only briefly examined here, for the sake of comparison between subsurface and surface water.

Table 8.1 displays the total volume and also the mean travel time (or residence time) in oceans, fresh surface waters, and subsurface waters (Groundwater). The latter constitutes only 0.27% of the total water. The mean GW travel time for active (exploitable) aquifers is about 330 years, to be compared to 0.033 years ≈ 12 days for streams.

Table 8.1 Global water volumes and residence times on earth

	Volume (*thousands of km*^{3})	Fraction of Total Volume (*dimensionless*)	Mean Travel Time (*years*)
Oceans	1,370,323	0.9393	3,000
Groundwater (all kinds)	60,000	0.0412	5,000
Groundwater (active)	4,000	0.0027	330
Lakes	230	0.00016	10
Streams	1.2	0.000001	0.033

Table 8.2 displays the global water budget on continents, on a yearly basis, according to the following budget equation:

$$P = \text{Qsurf} + \text{Qsubsurf} + \text{Evap} \quad [\text{mm/year}] \tag{1}$$

where P is the yearly rainfall rate, *Qsurf* is the specific discharge rate of surface water, *Qsubsurf* is same for subsurface waters (GroundWater), and *Evap* is evaporation flux density.

Table 8.2 Global water budget (continents)

P	=	Qsurf	+	Qsubsurf	+	Evap
730 mm	=	171 mm	+	81 mm	+	478 mm

It can be seen that the total flow of GW on continents is far from negligible: it is about half the total flow of the continents' surface waters. A final note on physical units may be useful here. Specific discharge rate «q» is defined as:

$$q\ [(liter/s)/km^2] = 1000 \times Q\ [m^3/s]/A\ [km^2]$$

and we have the equivalence:

$$1\ (liter/s)/km^2 = 31.536\ mm/year$$

For example, on a basin of area $A = 100\ km^2$, with a discharge rate $Q = 1\ m^3/s$, we obtain $q = 10\ (l/s)/km^2 \approx 315\ mm/year$. Now, according to the above table, the specific discharge rate of surface waters is *171 mm/y* or equivalently $5.4\ (l/s)/km^2$, and that of groundwater is *81 mm/y*, or equivalently $2.6\ (l/s)/km^2$, this being an average over the entire continental surface of the earth.

8.3 Groundwater Flow Systems Configurations, Layouts and Scales

This section offers a brief qualitative description, and site-specific examples, of typical groundwater flow systems with various spatial configurations (layouts), at different scales of analysis relevant for hydrogeologic applications (e.g. local *vs.* regional, vertical cross section *vs.* plane view, *etc.*).

8.3.1 Basic Concepts and Definitions (Porous Media Flow)

Before starting, it may be necessary to pre-define a few essential concepts (to be developed in more detail in subsequent sections).

- *Saturated flow.* In saturated porous media, the porosity is entirely filled with water.
- *Unsaturated flow.* In unsaturated media, porosity is only partially filled with water (and the rest with air). Resistance to air flow is neglected. Air pressure equilibrates rapidly to atmospheric pressure.

- *Partially saturated, or variably saturated flow.* Water flow occurs either as fully saturated, or unsaturated, or both. There may be several saturated/ unsaturated zones, evolving with time.
- *Permeability, or hydraulic conductivity.* This parameter is the inverse of viscous resistance to flow. It is the term of proportionality between flux and hydraulic gradient in Darcy's law.

8.3.2 Flow Models, Space-Time Scales and Coupling

The main point of this section is that, depending on the domain scale of interest, the relevant fluid dynamics and the corresponding equational model may be drastically different.

For example, in a karstic system, one may have to study the detailed Navier-Stokes flow of air/water in a single large conduit, or else, to resort to 0-dimensional input/output reservoir modeling for the entire karstic basin. In the case of a sedimentary aquifer, one may use Darcy's law with vertical averaging assuming quasi-plane flow (Dupuit-Boussinesq), or else, require a more complete model coupling 3D aquifer flow with 1D stream flow and with 2D watershed runoff at the scale of a large hydrologic basin.

Accordingly, hydrological models may be ranked as follows, in order of increasing complexity:

- Lumped input/output models (time-dependent reservoirs or compartments);
- Spatially distributed models based on Darcy's law (2D plane or 3D flow);
- Spatially distributed models with more general governing equations, and/or with coupling between Darcy's law and other mechanisms such as surface flow, vapor flow (evaporation), heat flow (thermo-convection) and density-dependent flow (saltwater in arid zones and in coastal aquifers).

Figure 8.1 illustrates these categories with specific models used in the literature (*among many others*). An example of a fully lumped model for surface hydraulics is the *Muskingum* model for flow in a canal stretch (*not shown here*). Some hybrid models (like the "TOP Model") are based on an astute combination of spatially distributed and partially lumped "reservoir" models. Other models like SHE, MOD-HMS, or BIGFLOW include a strong degree of coupling between different parts of the flow system: saturated aquifer, unsaturated soil, stream flow. In the current version of BIGFLOW-Py (2006), the three sub-systems can be fully integrated within a single numerical module, based on a generic "single equation" model with strongly implicit coupling.

Various numerical simulations of subsurface flow in heterogeneous environments will be shown later in this Chapter (Section 8.6.3, Section 8.6.4, etc.). See also, in this book, Chapter 7 (*"Coupled surface/subsurface flow systems: Numerical modeling"*).

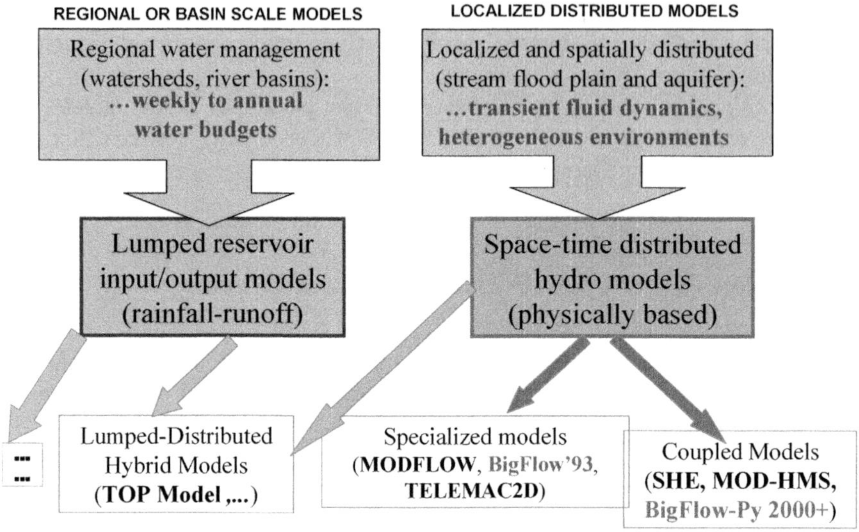

Fig. 8.1 Types of hydrologic models: space-time scales, lumping, distributing and coupling

8.3.3 Aquifers, Flow Systems, Piezometry (3D or Cross-section)

In this section, we inspect some typical groundwater flow configurations in (x,y,z) or in vertical cross-section (x,z), with emphasis on flow geometry in local or regional systems. We will also introduce, by the same token, the notion of piezometric head, hydraulic head and hydraulic gradient.

8.3.3.1 Stream-aquifer Flow Systems

To begin with, we examine the situation in a stream-aquifer system, which gives just a glimpse at the complexity of a real hydrologic flow system.

Figure 8.2 illustrates schematically a 3D flow system at the intermediate scale of a river stretch including a "slice" of watershed drained by the river stretch (only the left bank is shown). Three main flow paths can be distinguished:

a) Overland flow or direct runoff (which may be modeled as sheet flow).
b) Subsurface storm flow (or "hypodermic" flow).
c) Infiltration into the saturated zone *via* the unsaturated zone.

All three flow paths eventually reach the river, but at different time scales; note that the flow paths (a,b,c) are ranked here from "fast flow" to "slow flow".

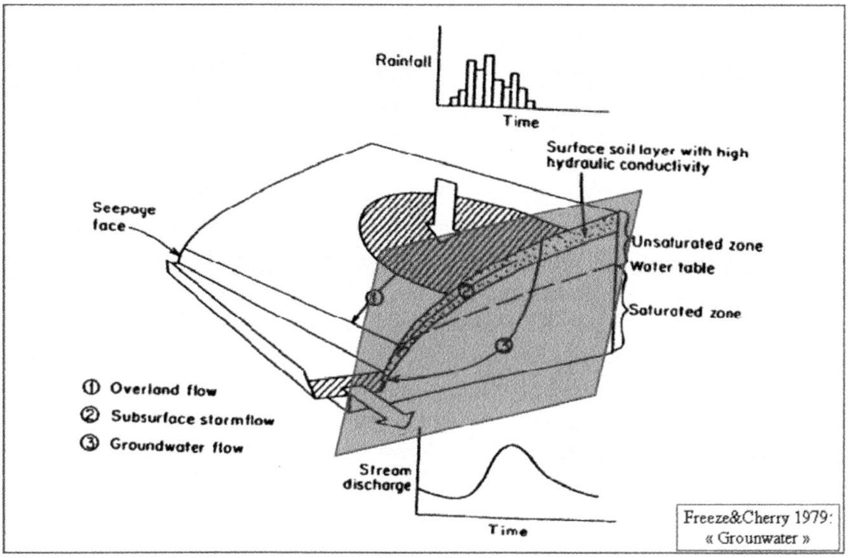

Fig. 8.2 Schematic illustrating hydrologic flow at the scale of a watershed connected to a river: the "slice" of watershed is drained by the river stretch (only the left bank is shown here). The three flow paths are discussed in the text (Modified from Freeze and Cherry, 1979)

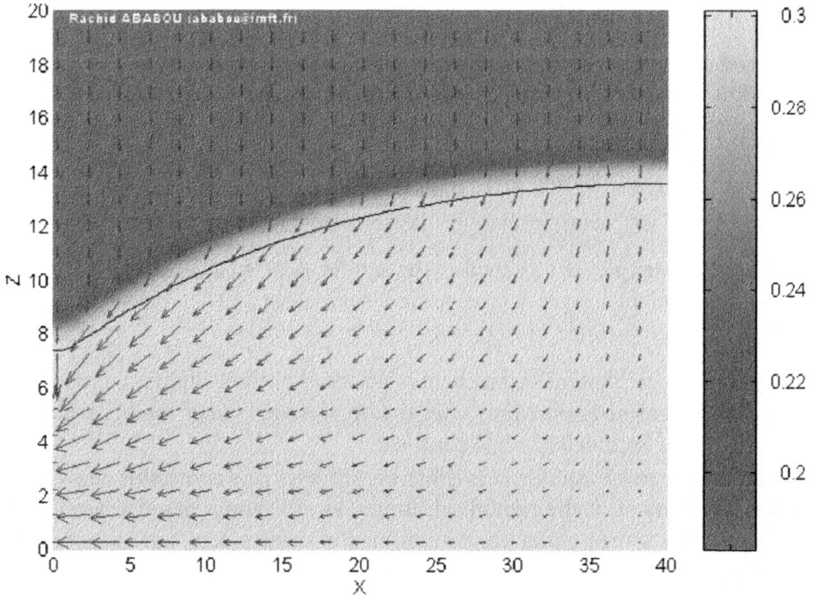

Fig. 8.3 Simulated discharge of a half-aquifer in a stream (located at left) during rainfall infiltration, in vertical cross-section. The thick continuous line is the water table (atmospheric pressure isosurface). Arrows represent Darcy velocity, or flux density q [m/s]. Colors (or gray scales) represent volumetric water content θ [m^3/m^3]. Porosity is 0.30 m^3/m^3

Figure 8.3 shows the detailed flow field in a vertical cross section (such as the cross-sectional plane displayed in the 3D stream-aquifer system of Figure 8.2). The simulation was carried out using the BIGFLOW code with the option of partially saturated flow in a 3D domain (reduced to a thin vertical slice). This simulation accounts for infiltration through the unsaturated zone, as well as saturated groundwater flow; the free surface is calculated as the atmospheric pressure isosurface. Stream-aquifer exchange flux is calculated without accounting for feed-back coupling effects, such as stream level changes. Indeed, the stream level is artificially fixed in this example (see Chapter 7 in this book for a more elaborate model of *fully coupled* stream-aquifer flow).

It can be seen in Figure 8.3 that saturated flow is quasi-horizontal, at least far enough from the stream-aquifer boundary. On the other hand, unsaturated flow is nearly vertical. This interesting effect is due in part to boundary conditions, but also to the fact that unsaturated permeability, being moisture-dependent, is horizontally stratified in this case (the porous medium itself is homogeneous and isotropic). The resulting effect is that flow is orthogonal to the unsaturated "permeability strata", a situation analogous to the refraction law in "ray optics".

8.3.3.2 Vertical and Horizontal Circulations in Multiple Aquifer Systems

In order to understand groundwater circulations, it is essential to account for vertical as well as horizontal circulations, for two reasons:

- Local and shallow aquifers are usually embedded in larger and deeper groundwater flow systems, delimited by various recharge and discharge zones at the topographic surface;
- Aquifer systems often have a "sandwich" structure, with permeable layers separated by less permeable ones, with possible communications between them (case of "leaky" or "semi-permeable" layers, also named "aquitards").

A schematic example of a typical multi-aquifer flow system in vertical cross-section is provided in Figure 8.4, modified from Bear (1972, 1988). The most important features illustrated in Figure 8.4 are:

- The hydrogeologic system shown here is made up of three aquifers, the top one (A) being unconfined (phreatic) with a free surface "water table", while the others (B,C) are confined or semi-confined;
- The recharge area of aquifer B (visible on the left) is a permeable impluvium surface where most of the rainfall infiltrates locally into a phreatic aquifer; as can be seen, the same aquifer can be confined in some places and unconfined in other places;
- There is a local groundwater body "perched" on a clay lens (to the right); this perched groundwater may be ephemeral (e.g. seasonal);

- Layers A/B and B/C are leaky layers ("aquitards"); leakage flux is respectively upwards at A/B and downwards at B/C, as indicated by local piezometric observations;
- The piezometer near center taps into aquifer B and flows naturally because the piezometric surface of aquifer B is above ground surface (case of artesian wells that extract water without pumping);
- The "true" piezometers in the figure are hydraulic observation tubes (e.g. aluminum or PVC) *punctured at one end* in order to measure *locally* the total hydraulic head, or piezometric head.

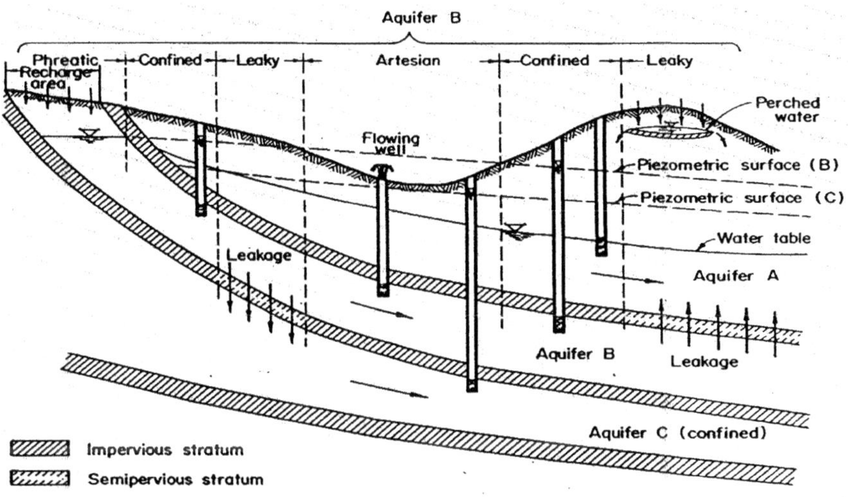

Fig. 8.4 Schematic vertical cross-section of a multi-aquifer system (Modified from Bear, 1972)

8.3.3.3 Piezometric or Hydraulic Head in Confined and Unconfined Aquifers

The piezometric head concept used just above can be made more precise as follows. Piezometric head, or hydraulic head H [*meters*], can be defined as:

$$H = \frac{p - p_{ATM}}{\rho g} + z \tag{2}$$

where axis "z" is vertical pointing upwards. For unconfined aquifers, H can be interpreted as the elevation of water table surface $z = Z_{WT}(x,y,t)$, defined as the surface where $p(x,y,z,t) = p_{ATM}$. Thus, applying the above equation yields:

$$Z_{WT} = H\left(x, y, Z_{WT}, t\right) \tag{3}$$

and equating the result with the vertically averaged head (as measured in fully perforated piezometers), we have the interpretation:

$$\bar{H}(x,y,t) \approx Z_{WT}(x,y,t) \tag{4}$$

For confined aquifers, in the absence of a free surface, hydraulic head H can be interpreted as the total energy of pore water per unit weight; similarly, the total pressure p + ρgz, where ρgz is gravitational head, can be interpreted as the total energy of pore water per unit volume:

$$p+\rho gz \ \text{Joules/m}^3 \quad \text{(Water Energy/m}^3\text{)} \tag{5}$$

Note that a pair of "true" piezometers aligned vertically can indicate the direction of vertical flow (from high to low piezometric head). In contrast, the usual "integrating" piezometer is punctured from end to end. It can only measure a vertically averaged piezometric head. A group of three "integrating" piezometers can be used to detect flow direction in the (x,y) plane of a sub-horizontal aquifer layer, but this device cannot be used to detect vertical fluxes.

8.3.3.4 A Laboratory Scale Groundwater Flow System in Vertical Cross-Section

Figure 8.5 shows a photograph of a physical model of groundwater flow in a heterogeneous geologic porous medium.

This reduced scale model is a thin vertical slice, filled with several porous materials, about 60 cm long and 40 cm high. The true scales that the model attempts to reproduce (albeit qualitatively) should be, typically, a few kilometers horizontally and ten to a hundred meters vertically. Aquifer heterogeneity is relatively simple, and "deterministic". The thick top layer is coarse sand, but it also includes a thin, localized clay lens. The thick bottom layer is made of fine sand, and it contains a long clay layer which almost (but not completely) separates the system into two distinct aquifers.

The general flow is from top right downwards (recharge zone) to top left upwards (discharge zone). This "groundwater system" may be viewed as a shallow unconfined aquifer overlaying a deeper confined aquifer, with some communications between the two. Perched water can occur above the clay lens, while the flow is partially confined beneath the clay lens. Also, depending on recharge conditions, the piezometric head in the bottom confined aquifer may be higher than soil surface, thus producing artesian flow in one of the piezometric tubes at left.

As it can be seen in Figure 8.5, a yellow fluorescent dye tracer is injected in the physical model, at a chosen depth, in one of the tubes located at right. Injection is fast and localized, mimicking instantaneous point source pollution. The migration of the tracer plume is influenced by the imperfectly layered structure of the aquifer system. It is observed that the plume first moves from left to right in the fine sand,

Fig. 8.5 Photograph of a vertical slab physical model of groundwater flow in a heterogeneous aquifer system. This scale model was made for qualitative demonstration of flow paths and contaminant dispersion. It was designed in 2006 at the Institut de Mécanique des Fluides de Toulouse (R. Ababou, D. Bailly, E. Milnes, M. Marcoux, et al., Porous Media Group, IMFT, Toulouse, France) in collaboration with J. Langot at Science Animation (Centre de Culture Scientifique, Technique et Industrielle de Midi-Pyrénées, Toulouse, France)

while dispersing isotropically (vertically and horizontally). However, when the plume touches the coarse/fine sand interface, it separates in two parts, an upper plume and a bottom plume:

- The upper plume continues to disperse isotropically (diffusively) as it moves slowly, horizontally, towards the surface discharge zone;
- The bottom plume moves faster in the coarse sand, and disperses longitudinally in stream tube fashion, before emerging upwards into the fine sand, where it becomes again more diffusive (isotropic). Eventually, it exits the system at the surface discharge zone, and this, a long time before the arrival of the slower upper plume.

8.3.4 Regional Circulations and Piezometric Maps (Plane View)

While most groundwater flow systems are in fact 3D, they can be studied in plane view based on vertically averaged approximations of flux density, velocity and hydraulic head (H).

For instance, in the Paris basin, the vertical thickness of a single confined aquifer (*Sables Albiens*) is just a few meters, while the horizontal scale of the flow system is several hundreds of kilometers, corresponding to the diameter of the subsurface basin. Actually, this layered system of confined aquifers is slightly curved, bowl-shaped, and it outcrops at ground level around the perimeter of the basin. But in spite of the 3D geometry of the aquifer, regional flow can be represented as a plane flow system, vertically averaged, whereby vertical circulations are neglected. This seems justified in the case at hand, given the high contrast of horizontal/vertical scales (five orders of magnitude).

The hydraulic head distribution can then be represented as a piezometric map in the (x,y) plane, as shown in Figure 8.6. The piezometric contours shown in this figure were obtained by plane flow simulations, with parameter fitting procedures based on field data (Grenet et al., 1996). It is clear that groundwater flow is converging towards Paris (deep pumping wells) and away from surface recharge areas located around the basin (aquifer outcrops are particularly visible on the south-east border). In such "plane flow" models, groundwater extraction by pumping wells and recharge by rainfall infiltration are all taken into account as 2D "sink/source" terms in m/s (m^3/s per m^2 in the plane).

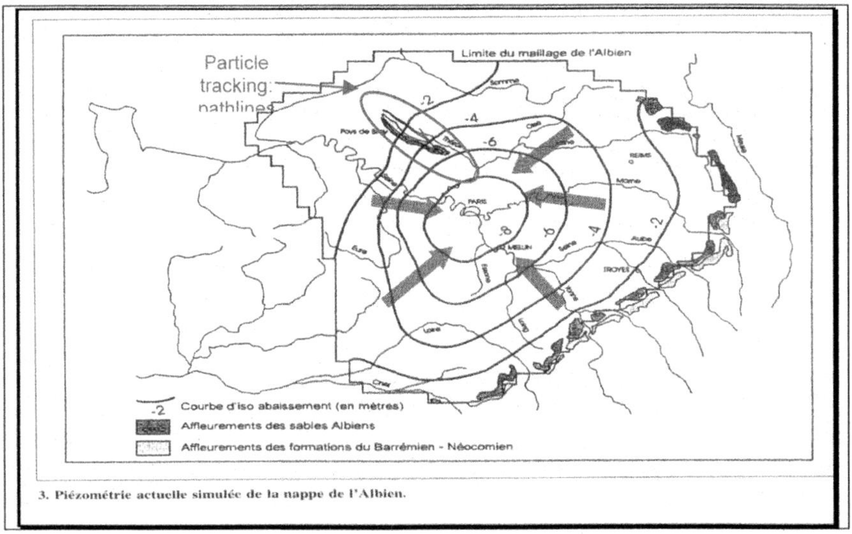

Fig. 8.6 Confined aquifer of the *Sables Albiens* (Albian Sands) in the Paris Basin: simulated piezometric map (Modified after Grenet et al., 1996)

8.3.5 Inlets and Outlets (Recharge Areas, Springs)

Without going into details, here is a short list of different types of inlets and/or outlets (sometimes both). What follows is only a qualitative description of how water flows in and out of subsurface flow systems.

- A natural well in a karstic formation can function as a possible natural inlet for surface runoff; it can also function as an outlet (Fontaine du Vaucluse in France); the hydraulics of natural wells are quite complex, especially in highly dynamic regimes, as the geologic medium surrounding the natural well may be highly fissured and/or connected with other wells and galleries.
- Outcrops of permeable aquifer layers can form a distribution of more or less localized recharge areas at ground surface (e.g. South-East border of Paris Basin); in other cases they can function as outlets (causing surface flooding by groundwater exfiltration);
- Large parts of watersheds (basins) can function directly as recharge areas (e.g. plateau impluvium with high infiltration capacity);
- Isolated springs constitute localized outlets for groundwater (in the form of "emergence springs", "artesian springs", "overflow springs", etc.); springs can also function as "lines of springs" running along a hillside parallel to the main river valley; and finally, spring outflow can be highly variable, possibly intermittent;
- Stream/aquifer interactions cause water to be exchanged to or from shallow groundwater systems; for example, phreatic aquifers help sustain surface stream flow during low flow periods (in this case the stream is an "outlet" for groundwater);
- Groundwater can outflow to the sea (shoreline, seafloor);
- Water is extracted through the soil-plant system by evapo-transpiration (direct evaporation and plant transpiration).

Some of the above phenomena will be discussed later in more quantitative fashion, e.g. infiltration in permeable soils, stream/aquifer interactions, or flow in fissured media. Next, we examine another important "outflow" mechanism that is man-made rather than natural: groundwater extraction by pumping wells.

8.3.6 Groundwater Extraction by Pumping Wells

Let us single out one of the most important anthropogenic (man-made) causes of groundwater "outflow", that is: groundwater extraction by "pumping wells" or "groups of wells".

The hydraulic functioning of pumping wells can be studied by considering either a single isolated well, a group interacting wells, or even, a continuous distribution of wells. Indeed, the spatial distribution of pumping wells can be so dense that they are sometimes represented as a continuous "well field", and modeled like a distributed "sink term".

Our goal in this section is to discuss practical examples, rather than develop the equational models of pumping wells. We discuss below two real-world examples of fairly dense "well fields" located in some of the most intensively exploited areas of the world. The two examples are extremely different in terms of agricultural, industrial and urban development.

a) **First example: the Gaza Coastal Aquifer.** The Gaza Coastal Aquifer (365 km^2) supplies freshwater for 1.3 million inhabitants (population density: 3,560 persons/km^2) via at least 4,000 pumping wells. In 1998, the annual water production rate of the wells was about *155* Mm^3 (155 millions of cubic meters). Groundwater quantity and quality are linked, and they are both affected negatively by overexploitation. Head losses due to overexploitation enhance seawater intrusion, among other sources of pollution. For details on the Gaza Coastal aquifer in relation to pumping and seawater intrusion, see Qahman (2004) and Qahman et al. (2005).

b) **Second example: California's Central Valley aquifers.** The San Joaquin and Sacramento rivers form the Central Valley system, located in a large structural trough in Central California. It extends about *400 miles* long and *20–70 miles* wide (over *20,000 sq miles*). It is one of the most important agricultural regions of the world, at least for comparable area. According to USGS reports, by the end of the 1960s, water-levels in the confined aquifer system of the San Joaquin Valley had declined by more than *400 ft* due to decades of intensive groundwater extraction. However pumping decreased later on, in the 1970s, owing to long-distance imports of surface water. However the system remains fragile. For instance, according to a survey by California's Department of Water Resources, the number of water wells drilled each year (as reported by drillers) was about 15,000 drills/year on average, but increased to as much as 25,000 drills/year during the 1987–1992 "drought" period. The Central Valley groundwater system is studied in the USGS report by Bertoldi et al. (1991).

8.4 Heterogeneous Porous Media in Natural Geologic Environments

We make here a distinction between a geologic "medium" and a "structure", mainly as a matter of convenience.

- A geologic "structure" might be an isolated "object" or set of "objects", like clay lens inclusions, fractures, cavities, fractures, or faults, all embedded in a geologic "medium";
- The geologic "medium" *per se* is viewed as a continuous porous matrix that may contain or imbed geologic structures. However, we do not exclude that this matrix be heterogeneous itself.

8.4.1 Subsurface Geologic Media and Structures

Let us list here a few examples of heterogeneous geologic media (soils, sedimentary aquifers, rocks, etc.) encountered in the subsurface.

8.4.1.1 Soils

Soils are a result of natural degradation of geologic formations. They are often heterogeneous and stratified; they can be fissured; and they contain plant roots, worm-holes and organic matter. Highly organic media like "peats" can also be included in this category. Note: porous media located at ground surface are commonly confused with "soils", while they may not be so "genetically".

8.4.1.2 Alluvial Aquifers

Alluvial aquifers are often layered, made up of a mixture of sand and gravel, and also contain clays, silts, nodules, etc.

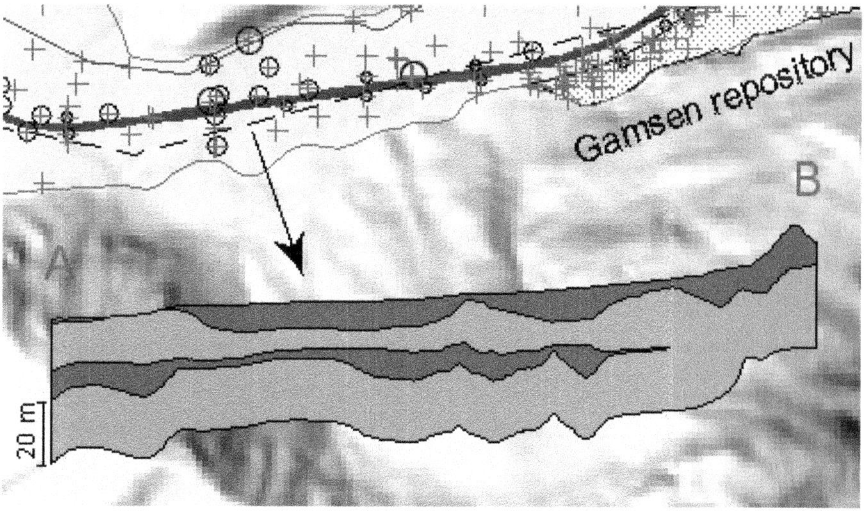

Fig. 8.7 Vertical axial cross-section of Rhône valley upper gravel alluvium aquifer(s) near the town of Visp, Valais, Switzerland. Groundwater flow is strongly conditioned by river streamflow (Rhône) and by canals (Brigerbad and Grossgrund). The valley deposits are quaternary sediments of different origins: glacial (moraine), lacustrine, and fluvial. Silts are shown in dark brown, and gravels light blue. The lithology is as follows, from top to bottom: (1) Shallow silts (10 m mean thickness); (2) Upper gravel aquifer (10—19 m thickness); (3) Middle silt unit (less than 20 m thick, and absent at places); and (4) Lower gravel aquifer (20 m average thickness) lying over an impervious substratum

a) An example of a multi-layered sedimentary structure is the alluvial Rhône valley aquifer in Switzerland, shown in Figure 8.7; the gravel aquifer layers appear to be "confined" at some places, but "unconfined" at other places, depending on silt thickness.
b) Sand-and-gravel sedimentary formations are often "imperfectly stratified" and "cross-bedded" (strata thickness can be *1 cm*).
c) In some cases the media appear to be "bimodal", given the presence of a large number of inclusions, such as: (i) clay lenses in sand; or (ii) shale inclusions in limestone.

8.4.1.3 Rocks

Altered rocks are heterogeneous and contain several transition zones (soil-rock transition, altered-to-intact rock transition, etc). For example, the "*saprolithe*" visible in the bottom part of Figure 8.8 results from alteration of a fractured gneiss rock.

Fig. 8.8 This photograph displays a saprolithe, i.e. an extremely altered fractured gneiss, beneath a layer of red clayey soil. This is an example of a sharp spatial variation of porous medium structure. The "structural" gradient occurs over a vertical scale of a fraction of a meter (the scale of the photograph here). The underlying rock (not shown here) is fractured gneiss. The photo was taken by the author at the instrumented basin site of Moole Hole, studied by the Indian–French Cell for Water Resources and the Indian Institute of Sciences of Bangalore. The basin is located in the Kerala/Karnataka region, near the city of Mysore, South India

Hard rocks (non altered) can still be fractured mechanically, and possibly also by other mechanisms. The fractures make them more permeable at large scale, due to the faster flow paths through the fracture network:

a) Individual fractures in hard rocks (like granite and other crystalline rocks) tend to have rough wall surfaces; the geometry of the aperture field has an important effect on its hydraulic behavior; fractured rock samples can be studied in the laboratory in order to analyse, for instance, the hydraulic transmissivity and tracer dispersion properties of a single rough planar fractures (Spiller, 2003).
b) Crystalline rocks, and other rocks, can be multiply-fractured. Other examples of commonly fractured rocks are *limestones* and *chalks* (see also below the case of highly fissured *karsts*).
c) In tectonically active regions, rocks can be both highly stratified and folded; the interfaces between folded strata are potential flow paths.

8.4.1.4 Highly Fissured Carbonate Aquifers and Basins (Karsts)

Karsts are highly fissured carbonate rocks, with networks of fissures, caves, galleries, underground streams, natural wells and springs. With all these structures, karsts form truly "3D hydrogeologic basins". A recent review on 3D karsts and related flow modeling issues can be found in Kiraly (2003).

8.4.2 Synthetic or Reconstructed Random Structures

For convenience, we will make a distinction between fully "synthetic" and partially "reconstructed" media (or structures).

- We label as "synthetic" a geologic medium (or structure) that is entirely generated by a mathematical model (statistical, genetic) without incorporating any detailed site-specific data (such as point measurements of texture, porosity, permeability);
- In contrast, we label as "reconstructed" those media and structures that are generated by a combination of "modeled" variables (e.g. random fields with given prior statistics) and site-specific information (e.g. point observations).

In the latter case, data integration can be performed *via* bayesian conditioning (kriging, conditional simulation), or *via* optimal calibration of the statistical generator (optimization).

8.4.2.1 Reconstructed Permeabilities in an Alluvial Aquifer

We will present here only one example of reconstruction, obtained via "kriging", a geostatistical technique based on the theory of regionalized variables (initiated by Matheron and others at the Paris School of Mines).

In Figure 8.9 the permeability field of the multi-layered sedimentary structure of the Rhône valley aquifer near Visp (Switzerland) was partially reconstructed (estimated) by kriging using the ISATIS3 package. Standard kriging with omni-directional variogram was implemented. There were 791 flow meter permeability measurements K(x,y,z) collected every meter along a number of vertical boreholes. The cross-sectional permeability map of Figure 8.9 is to be compared with the lithology represented earlier in Figure 8.7.

Note: this work was performed in collaboration with researchers at the University of Neuchâtel in Switzerland; most of the work on data estimation was accomplished by M. Bouzelboudjen and F. Kimmeieir in Neuchâtel. For more details, see Kimmeieir et al. (2001).

For completeness regarding the mathematical method, the reader interested in aquifer geostatistical estimation (kriging) should consult the pioneering paper by Delhomme (1979). A "bayesian" point of view on "kriging", estimation and simulation of random fields, can also be found in the introductory part of Ababou et al. (1994a). Theoretical aspects and further examples of statistical media, synthetic or reconstructed from data, will be presented later (Section 8.6).

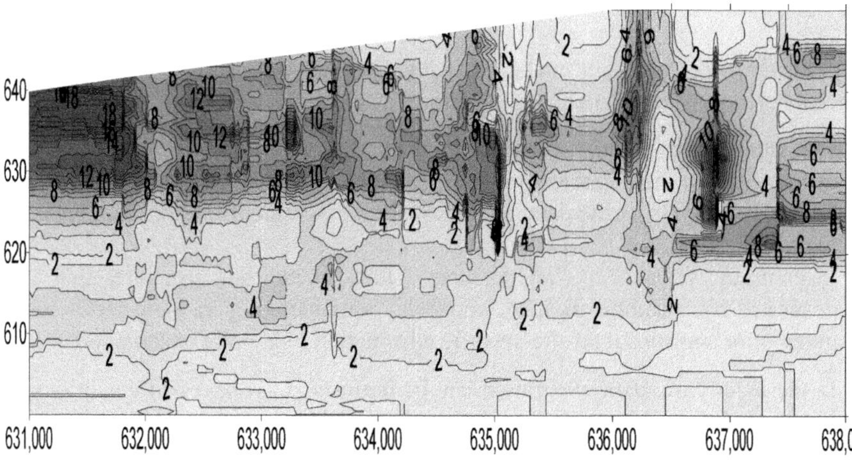

Fig. 8.9 Visualization of a 2-D vertical cross-section of permeability K(x,z), extracted and transformed from the kriged 3D log-permeability field ln K(x,y,z) in the Rhône valley aquifer near Visp (Switzerland). This view shows labeled contours of $K \times 10^4$ m/s. The range of values of K is approximately 1×10^{-4} to 20×10^{-4} m/s in this cross-section. The scale of the map is 40 m vertically (at left) and 7 km horizontally

8.5 Quantitative Hydrogeology: Porous Media Flow Equations

This section presents the fundamental bases for flow modeling in hydrogeology. We focus particularly on spatially distributed "equational" models that can capture the effects of heterogeneity on groundwater flow, in terms of pore pressure, hydraulic head, flux density and/or velocity vector.

8.5.1 Porous Media Concepts

There are many different possible views about what a porous medium really is. Whether a given material should be considered as a porous medium or not, depends much on the scale of interest. It can be largely a matter of point of view. However, several useful notions like porosity, permeability, and "Representative Elementary Volume" (REV), make it possible to have a more objective definition of the "porous medium" concept.

First of all, many types of materials, natural or man-made, are "porous":

- Unconsolidated granular media like sand on a beach (e.g. Golden Sands beach, Black Sea, Varna, Bulgaria);
- Consolidated granular media, like sandstone, with cemented grains;
- Volcanic rocks with alveoli or vesicles (e.g. gas bubbles in vesicular basalt and pummice);
- Bimodal geologic media made up of a porous matrix that contains inclusions of other materials like metallic nodules, or clay lenses (in sand), or oil shale layers (in sandstone);
- Fractured hard rocks, with networks of relatively tight planar joints;
- Carbonate rocks, with extremely open and variable dissolution patterns (fissures, holes, conduits, cavities);
- Extremely altered rocks, soils, peats, etc.;
- Biological media that can be treated as porous media from a hydrodynamic viewpoint (soil-plant-root systems, biological tissues and bones, bio-films, capillary and vascular networks);
- Man-made or natural materials composed of plates, lamellae (sandwich materials), fibers (textiles, wood, paper), and cells (cellular materials for isolation), etc.

Given this diversity, it is useful to state now some of the essential characteristics held in common by all these different types of "porous" media:

- A porous medium is a complex solid with many inter-connected pores;
- In a porous sample, flow and mass transport phenomena are limited by the small width of the pores, relative to sample scale.

As a consequence, we see that:

a) A proper porous medium should be not only "porous", but also "permeable", since the pores must be interconnected (see Darcy's law).
b) Also, there should be "many" (not a few) interconnected pores; this implies that a material that contains only a few pores, holes or conduits, does not constitute a true porous medium.
c) And finally, because the openings of the interconnected pores are small, the average fluid/solid contact area is large, and the velocity is limited by a large amount of viscous dissipation due to the solids.

All the properties listed above lead to using Darcy's law as a specialized hydrodynamic model for porous media. The coefficient of "permeability" in Darcy's law eliminates the need for explicitly representing the complex set of solid/fluid interfaces (grains and pores).

Figure 8.10 shows one example of a simplified granular medium in 2D. It should be viewed more properly as the cross-sectional view of a 3D random packing of cylindrical grains. The packing is "poly-disperse", meaning that there are grains of different diameters. The sub-square at bottom right suggests the possible scale of the "REV" for this medium. The correct REV size can be checked by measuring the average porosity over blocks of increasing sizes, and defining the REV as the block size for which a stable porosity is attained. This example leads to a definition of the REV.

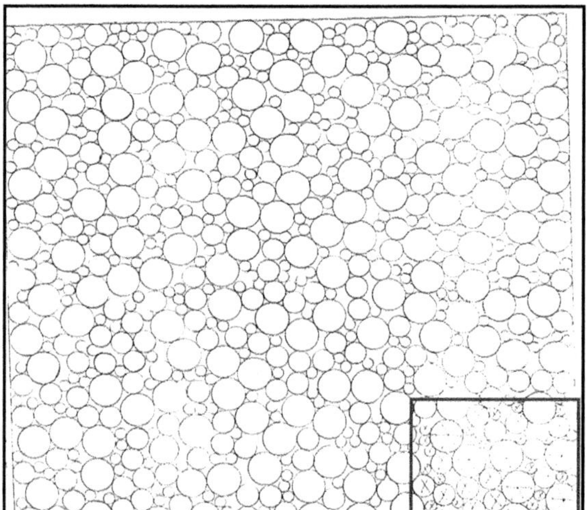

Fig. 8.10 A simplified porous medium and its possible "REV" indicated by the sub-square at bottom right. The "grains" are a poly-disperse random packing of cylindrical grains, viewed in cross-section. Note: there are roughly 100 grains in the "REV" sub-square (Courtesy L.W.G.)

8.5.1.1 Averaging and REV (Representative Elementary Volume)

By definition, the REV is the scale at which macroscopic porous medium quantities, obtained by spatial averaging over the REV, are meaningful and remain "stable" as the averaging volume increases further.

For instance, porosity, permeability, pore pressure and tracer velocity are all macroscopic quantities. The spatial resolution of these macroscopic quantities is equal to REV size (e.g. golf ball, tennis ball, or football size). Here "macroscopic" means averaged, as opposed to "pore-scale". A simple version of a phase averaging operator can be used to define, for instance, macroscopic porosity (more general filters can also be devised):

$$\langle * \rangle = \frac{1}{\mathrm{Vol}(\Omega)} \iiint_{\Omega} * \varphi(x, y, z)\, dxdydz \tag{6}$$

where "*" represents the quantity to be averaged, Ω is the region of space to be used for averaging (e.g. REV size), and $\varphi(x,y,z)$ is the characteristic function, or indicator function, of the subset of Ω comprised of pores ($\varphi=1$ in the pores, $\varphi=0$ in the solids). The phase average <*> is adequate for defining porosity as a ratio of two volumes, but it is not obvious that it is also correct, say, for defining the macroscopic pressure or the macroscopic velocity vector.

Unfortunately, there are some other difficulties with the REV concept and with the whole concept of a "porous medium". Here are two common problems:

- The REV size may be significantly different for different quantities like porosity and permeability, or pressure and velocity.
- The REV may be undetermined or inexistent given the scale of investigation (e.g., poorly connected or sparsely fissured medium).

These difficulties are intimately related to the so called "upscaling" problem. One way to get around the absence of a clearly defined REV is to develop porous media hydraulic models that depend explicitly on the averaging method and scale, i.e. to accept the idea that the properties of the medium, and of the flow itself, may be scale dependent (Ababou et al., 1990).

8.5.1.2 Porosity, Void Ratio, Specific Area

Let us now define more precisely two important macroscopic quantities, namely *porosity* and *specific area*, assuming that an REV exists for both quantities. First, porosity is defined as the volume of pores per unit volume of the medium; it is therefore obviously related to the water storage capacity of the medium:

Porosity:

$$\phi = \frac{Volume\ Pores}{Volume\ Medium} = \frac{V_P}{V_M} \tag{7}$$

The void ratio is another quantity related to porosity. It does not depend on the volume of the medium. This is particularly useful in geotechnical applications where the medium can be highly deformable (e.g. swelling clays):

Void ratio:
$$e = \frac{Volume\ Pores}{Volume\ Solids} = \frac{V_P}{V_S} \Rightarrow \phi = \frac{e}{1+e} \quad (8)$$

In order to evaluate porosity, it is also useful to have a fair *a priori* estimate of the average density of solids (e.g. minerals, grains) in a porous medium:

Solid density:
$$\rho_S = \frac{Mass\ Solids}{Volume\ Solids} = \frac{M_S}{V_S} \quad (9)$$

For most natural porous media (soils, rocks), we have: $\rho_S \approx 2.65\ g/cm^3$.

On the other hand, *specific area* is related to solid/fluid contact area, therefore, indirectly, to friction and viscous dissipation, and to permeability via Darcy's head loss law (more on this later). It can be defined as follows (although there are several other definitions in the literature):

Specific area:
$$s = \frac{Fluid\ /\ Solid\ Contact\ Area}{Total\ Volume\ of\ Medium} \quad [m^2/m^3] \quad (10)$$

Table 8.3 Specific areas of three types of porous materials (gravel, silt, and clay)

	Specific area "s" (m^2/m^3)
Gravel, spherical grains, diameter d = 1 cm	400 m^2/m^3
Silt, spherical grains, diameter d = 0.02 cm	1.6 ha/m^3
Montmorillonite clay, disc plates, interspacing e = 10 Å = 10^{-7} cm.	800 m^2/cm^3
Note: 1 Ångström = 1 Å = 10^{-10} m.	*Note*: 1 hectare = 1 ha = 10,000 m^2

Note that the specific area "s" has the units of an inverse length scale, which is roughly on the order of the characteristic radius of pores. Table 8.3 gives a rough evaluation of "s" for three types of geologic porous media: (i) gravel; (ii) silt; and (iii) clay, based on their characteristic pore sizes. It can be seen that a small sample of clay, contained in a small goblet or a thimble, has a total "area" as large as an entire soccer stadium! As a consequence, resistance to flow is expected to be very high, and permeability very low, in such a clay.

8.5.2 *From Navier-Stokes to Darcy's Law*

8.5.2.1 The Navier-Stokes Equations of Fluid Mechanics

In "open flow" systems (non-porous, classical fluid mechanics) the incompressible Navier-Stokes system of equations is given by:

$$div(\mathbf{V}) = 0 \tag{11}$$

$$\rho_0 \frac{\partial V_i}{\partial t} + \rho_0 (\mathbf{V} \bullet \nabla)\mathrm{V}_i = -(\nabla p + \rho_0 g \nabla z)_i + \mu \nabla^2 V \tag{12}$$

The first equation is the mass conservation PDE for an incompressible fluid. It states that the velocity divergence must be zero. The second equation is a system of three PDE's enforcing the conservation of momentum, a vector quantity (momentum is mass times velocity). The momentum equation was formulated here by taking into account the zero divergence of velocity inferred from mass conservation. In total, the 3D Navier-Stokes equations constitute a system of four equations (one for mass and three for momentum) with four unknown variables: pressure (p), and velocity components (V_x, V_y, V_z).

8.5.2.2 From Navier-Stokes to Darcy's Law for Saturated Porous Media

Figure 8.11 shows a simplified presentation of the derivation of Darcy's law from the conservation of momentum in Navier-Stokes' equations. Darcy's law is obtained by averaging the N-S equations over many pores (REV scale) and neglecting acceleration terms (eulerian and inertial).

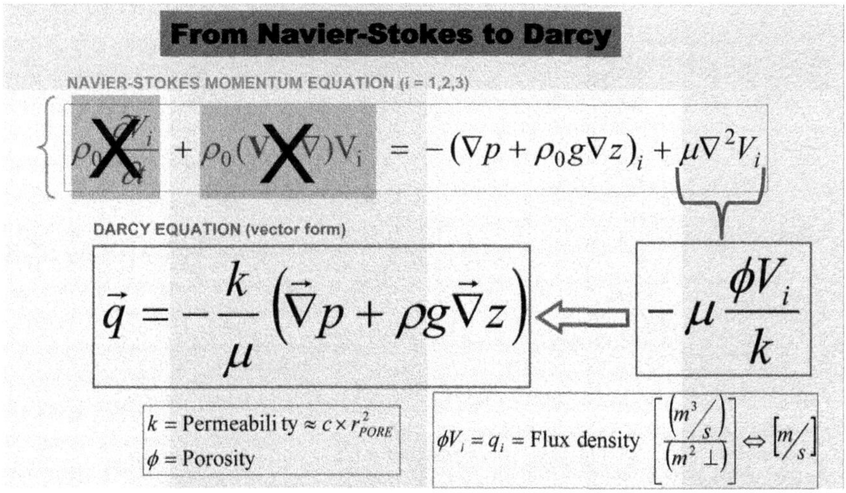

Fig. 8.11 A simplified presentation of the derivation of Darcy's law from Navier-Stokes

The viscous dissipation term ($\mu\nabla^2 V$) becomes proportional to V upon averaging over many pores. This finally leads to Darcy's linear flux-gradient law, usually expressed in terms of flux density "q" rather than velocity "V".

8.5.2.3 Darcy's Law and Permeability (Saturated Isotropic Media)

Figures 8.12 and 8.13 summarize Darcy's law and Darcy's permeability experiments with sand columns (water-saturated, isotropic porous media).

Darcy's law and permeability

Darcy's law (hydraulic formulation)

$$\mathbf{q} = -K \ \mathbf{grad}\ H \qquad H = \frac{p - p_{ATM}}{\rho g} + z(\uparrow)$$

$$K = \frac{g}{\nu} k$$

« k » = intrinsic permeability of P.M., in [m²] (*or in Darcy «Da»*)
« K » = hydraulic conductivity of P.M. for water, in [m/s]
« ν » = kinematic viscosity of water, in [m²/s] (or in *Centi-Stokes «cSt»*)

Relations between permeability and hydraulic conductivity

K (m/s) = 1.15 x 10^{+5} k (m²) | 1 Darcy = 0.9866 10^{-12} m²

Typical values of permeability

	Darcy's « k »	Hydraulic « K »
Gravel	*10^{+3} Darcy*	*10^{-2} m/s*
Sand	*1 Darcy*	*10^{-5} m/s*
Clay	*10^{-4} Darcy*	*10^{-9} m/s*

Fig. 8.12 Summary of Darcy's law for a saturated porous medium (hydrogeologic formulation, in terms of hydraulic head). The porous medium is assumed isotropic (scalar permeability)

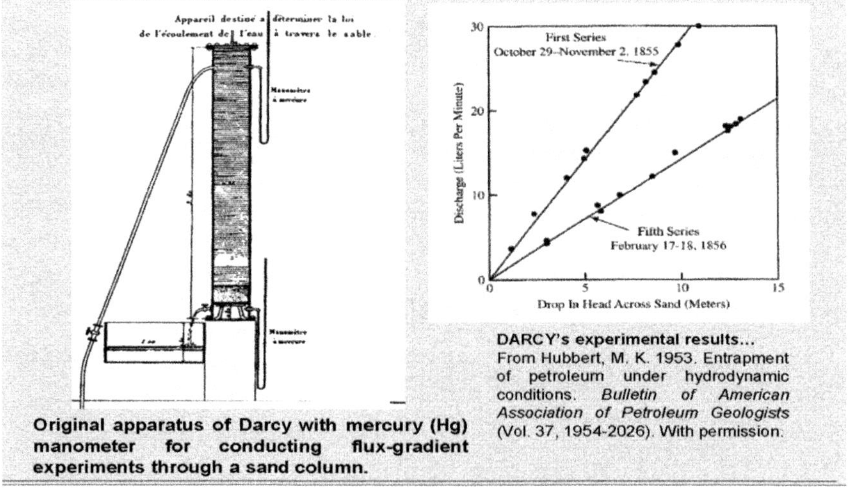

Fig. 8.13 Presentation of Darcy's 1856 permeameter experiments

In Figure 8.13, the left part displays the original Darcy permeameter, and the right part shows Darcy's 1855 and 1856 experimental results, as replotted by Hubbert (1953). In both series, it can be seen that the discharge rate of the sand column increases linearly with the drop of hydraulic head: this is Darcy's law.

8.5.2.4 Tensorial Darcy Law for Anisotropic Media

For a non-isotropic medium, Darcy's law can be expressed in tensorial form as:

$$\mathbf{q} = -\underline{\underline{\mathbf{K}}}\,\mathbf{grad}H \tag{13}$$

or equivalently, with Einstein's notation (implicit sum on repeated indices):

$$q_i = -K_{ij}\frac{\partial H}{\partial x_j} \tag{14}$$

where (K_{ij}) is a symmetric second rank tensor, which is usually positive definite, or at least positive ($\mathbf{K} \geq \mathbf{0}$). This tensor can also be written as a matrix:

$$\mathbf{K} = \begin{bmatrix} K_{XX} & K_{XY} & K_{XZ} \\ K_{YX} & K_{YY} & K_{YZ} \\ K_{ZX} & K_{ZY} & K_{ZZ} \end{bmatrix} \tag{15}$$

satisfying $K_{ij} = K_{ji}$, that is: $K_{YX} = K_{XY}$, etc. Now, what is the relation between permeability anisotropy and porous medium structure?

- Permeability is often anisotropic, because most geologic media have a layered or stratified or fractured structure (*the case of fracture networks will be treated further below*);
- Let (K//) be the permeability parallel to strata, and ($K\perp$) the permeability orthogonal to strata: in most cases it is found that (K//) is larger than ($K\perp$).

This last point can be demonstrated rigorously for a perfectly layered medium with parallel layers of infinite extent. In this case, regardless of the precise alternation of the layers, it is always true that:

- K// equals K_A, the arithmetic mean permeability of all layers;
- $K\perp$ equals K_H, the harmonic mean permeability of all layers.

On the other hand, it can be shown that the harmonic mean is always less or equal than the arithmetic mean, and therefore, we obtain the stated result:

- $K_H \leq K_A \Rightarrow K\perp \leq K//$.

8.5.2.5 Darcy Flow and Stokes Flow in Planar Fractures (Cubic Aperture Law)

In the case of steady state incompressible flow in a single infinite plane fracture made up of two smooth parallel walls: the Navier-Stokes equations boil down to the simple Poiseuille/Couette equation of viscous flow, which yields a linear relation between flux and total pressure gradient. This flux-gradient relation is similar to Darcy's law. It is known as the "*cubic law*" because the flux is proportional to the *cube* of fracture aperture (Figure 8.14).

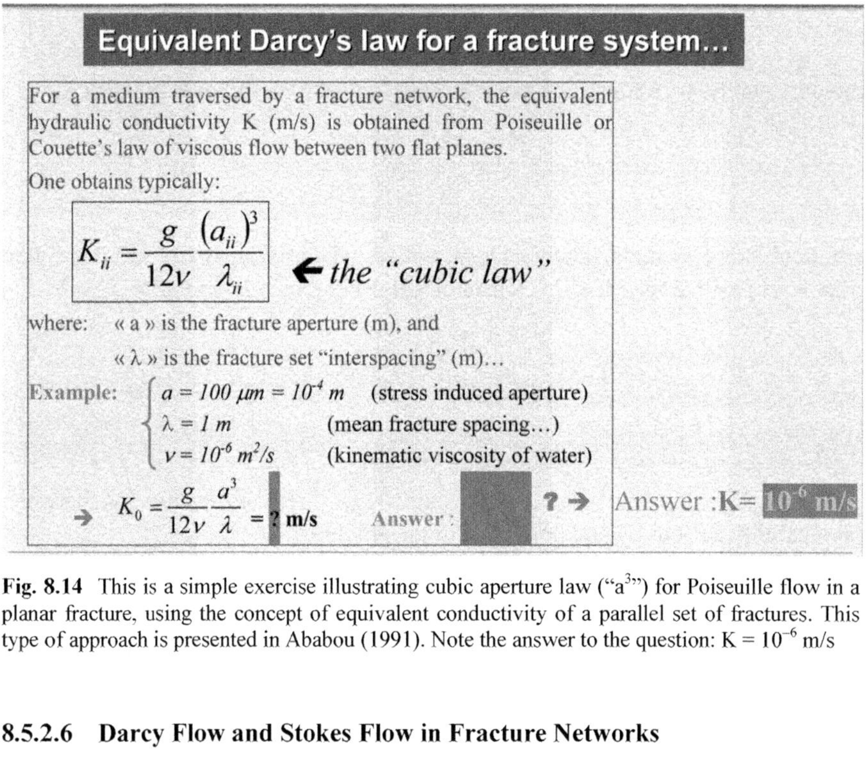

Fig. 8.14 This is a simple exercise illustrating cubic aperture law ("a^3") for Poiseuille flow in a planar fracture, using the concept of equivalent conductivity of a parallel set of fractures. This type of approach is presented in Ababou (1991). Note the answer to the question: $K = 10^{-6}$ m/s

8.5.2.6 Darcy Flow and Stokes Flow in Fracture Networks

(Concerning this theme, see also Section 8.0 on statistical fracture networks.) In the case of a fracture network, an *approximate* equivalent permeability can be obtained by a simple superposition approach. The approach is to sum the individual fluxes of all fractures, assuming these fluxes to be produced by the same constant mean (frozen) pressure gradient. The resulting equivalent law is a tensorial Darcy law. This approach was initially implemented by Snow (1969) and Kiraly (1969), and later by Oda (1986) and others, assuming the intact matrix to be impervious (as we do here). See Ababou (1991) for a review on fracture flow. The reader is also referred to Oda (1986), Ababou et al. (1994b) and Stietel et al.

(1996) for the case of coupled Thermo-Hydro-Mechanics in elastic fractured rocks, neglecting intact rock permeability.

Let us start with an example. For the orthogonal network of Figure 8.15 (left-bottom), the equivalent permeability tensor obtained by superposition is:

$$\begin{bmatrix} K_{XX} & K_{XY} \\ K_{YX} & K_{YY} \end{bmatrix} = \begin{bmatrix} K_{=} & 0 \\ 0 & K_{II} \end{bmatrix} = \begin{bmatrix} K_2 & 0 \\ 0 & K_1 \end{bmatrix} \tag{16}$$

where K_1 and K_2 are the fracture permeabilities of each of the two families:

$$K_1 = \frac{g}{12\nu}\frac{(a_1)^3}{\lambda_1} \qquad K_2 = \frac{g}{12\nu}\frac{(a_2)^3}{\lambda_2} \qquad \text{(17) and (18)}$$

It can be seen that the equivalent permeability K_{ij} of an orthogonal fracture network aligned with the (Ox,Oy) axes is a diagonal matrix.

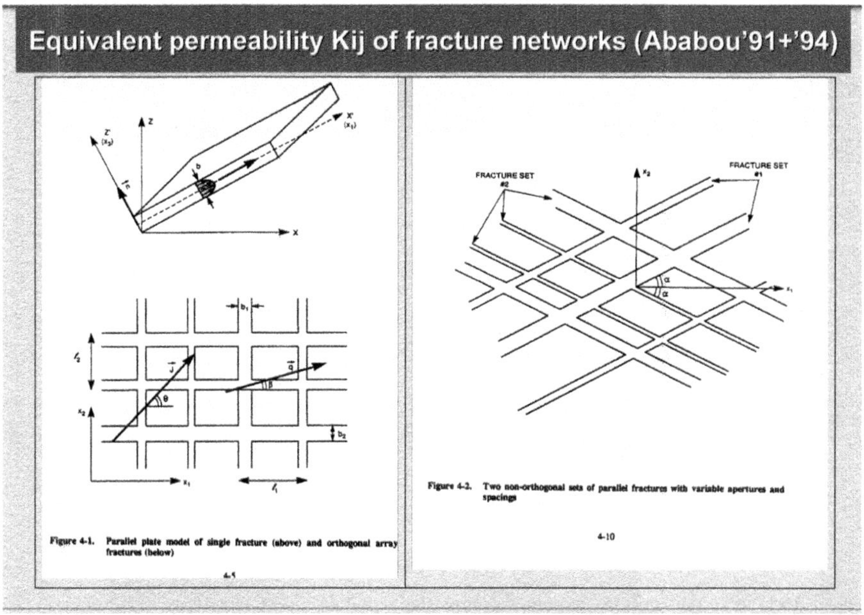

Fig. 8.15 Left-top: a single "fracture" idealized as a pair of parallel plates. **Left-bottom**: a cartesian orthogonal network comprising two orthogonal families of fractures with differing apertures (the hydraulic gradient vector J and the flux density vector q are not aligned because of the anisotropic geometry of the network). **Right**: a slightly more general, non-orthogonal network made up of two slanted families of fractures (the angle between them is 2α)

On the other hand, for the non-orthogonal network of Figure 8.15 (right), the equivalent permeability tensor obtained by superposition is:

$$\begin{bmatrix} K_{XX} & K_{XY} \\ K_{YX} & K_{YY} \end{bmatrix} = \begin{bmatrix} (K_1 + K_2)\cos^2\alpha & (K_1 - K_2)\cos\alpha\,\sin\alpha \\ (K_1 - K_2)\cos\alpha\,\sin\alpha & (K_1 + K_2)\sin^2\alpha \end{bmatrix} \quad (19)$$

where K_1 and K_2 are the same as before. It can be seen now that the permeability of this network is non-diagonal in the (Ox,Oy) system, provided that $K_1 \neq K_2$, that is, if $a_1 \neq a_2$ or if $\lambda_1 \neq \lambda_2$. The principal directions of K_{ij} have an inclination angle β with respect to (Ox,Oy), where β is obtained from:

$$\tan(2\beta) = \frac{K_1 - K_2}{K_1 + K_2}\tan(2\alpha) \quad (20)$$

Finally, Figure 8.16 (left) shows a more realistic (albeit 2-dimensional) network comprising over 7,000 randomly distributed fractures. The superposition approach outlined above was used to compute the equivalent permeability tensor at the scale of subdomains or blocks. The resulting block-scale permeabilities are second rank symmetric positive tensors. They are represented by ellipses in Figure 8.16 (right) (this classical type of representation is explained for instance in the textbook by Bear 1972, 1988).

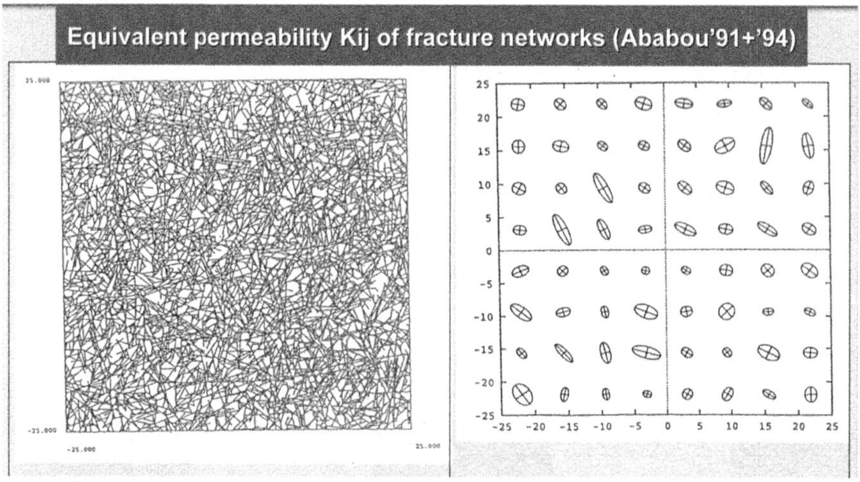

Fig. 8.16 Plane network of fractures comprising roughly 7,000 fractures. The domain is partitioned into 64 blocks or subdomains. The tensorial equivalent permeabilities K_{ij} are shown for each subdomain. Each tensor K_{ij} is represented by an ellipse (After Ababou et al., 1994b)

The resulting equivalent medium will be, by construction, a discontinuous composite, assembled from blocks with different permeabilities. However, it will be less highly heterogeneous than the initial fractured medium. Moreover, it is also possible to construct a fully continuous permeability tensor by using moving averages (or other filters) instead of block partitioning. In this way, the discontinuous fractured medium is replaced by a smoother porous medium "equivalent", with continuously varying anisotropic permeability.

Therefore, *upscaling* techniques tend to smooth out variability, and they can even transform a discrete medium into a continuous one. Moreover, this approach (upscaling and smoothing) can be extended in various ways. For instance, Kfoury et al. (2006) developed a two-step sequential upscaling for dual matrix/fracture media. The first step was to map the discrete medium into a continuous equivalent based on dual continuum flow equations, and the second step was to transform the heterogeneous continuum into a homogeneous equivalent. This two step approach was compared with single step upscaling, and was used to study the difficult problem of "REV" size for dual matrix/fracture media. See also Section 8.0 further below.

8.5.2.7 Darcy-Ward-Forchheimer: Fast Flows and Quadratic Head Loss Law

Figure 8.17 summarizes the so-called quadratic head loss law. The proposed law is in fact linear with head gradient, and linear-quadratic with velocity.

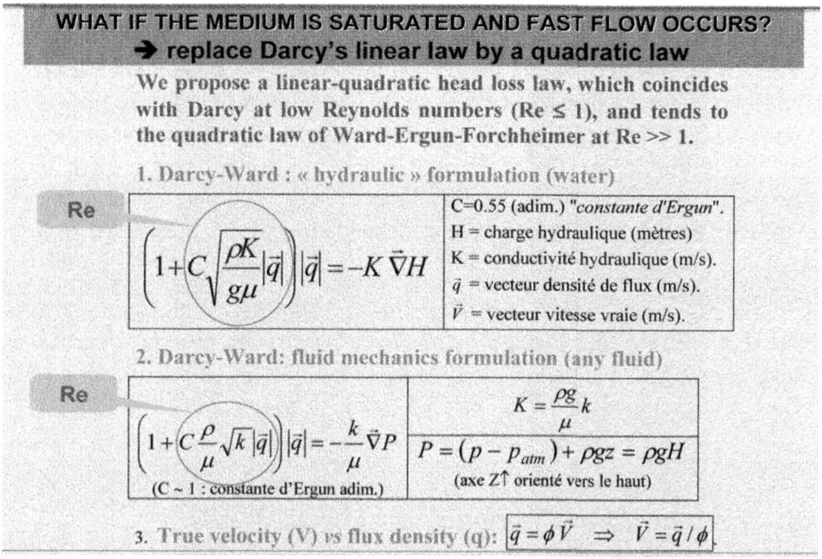

Fig. 8.17 Darcy-Forchheimer linear-quadratic head loss law (also named after Ergun and Ward), as implemented in the BIGFLOW code for partially saturated flow in 3D heterogeneous media

8.5.2.8 Darcy's Law Extension to Unsaturated Flows

To model flow in unsaturated or partially saturated regions, the following extension of Darcy's law must be considered:

$$\mathbf{q} = -K(h)\, grad(h+z),\ \forall \mathbf{x}\ ,\ \forall t \tag{21}$$

where $H = h+z$ is the hydraulic head (or total head), and h is pressure head, i.e. pressure expressed in equivalent water head; the axis (z) is vertical and oriented upwards, towards the sky. Note that it is assumed that air pressure remains constant. This is a simplified air/water flow model, where air pressure rapidly equilibrates as water pressure varies.

Historically, Darcy's law was first applied to unsaturated media (soils) by Buckingham (1907). For this reason, it is sometimes known as the Darcy-Buckingham equation. However the first modern formulation of the complete transient unsaturated flow equation, taking into account changes in water content, was produced by Richards (1931) (more on this later).

8.5.2.9 Other Extensions of Darcy's Law (Summary)

For completeness, we list below several extensions of Darcy's law that are frequently used in different areas of environmental sciences and engineering:

- Anisotropic media with tensorial permeability (K_{ij});
- Fractured rocks and fracture network (K_{ij});
- Unsaturated media and the Darcy-Buckingham law ($K(\theta)$ or $K(h)$);
- Quadratic extension of Darcy's head loss law (Ward-Ergun-Forchheimer);
- Fluids with variable density $\rho(x,y,z)$ due to salt solutions or thermal effects;
- Compressible barotropic fluids with a thermodynamic state relation $\rho=\rho(p)$;
- Immiscible multiphase flows (air-water, oil-water, oil-air-water, etc.): in this case the pressure-dependent permeabilities of each fluid phase intervene in the extended Darcy's law, in a way that depends mostly on pressure differences (capillary pressures) among the different fluids (this generalization is attributed to Muskat).

Note that the first four extensions of Darcy's law were presented earlier, and the full governing flow equations will be presented further below in Section 8.5.3 (3D flow, non averaged) and Section 8.5.4 (plane flow, vertically averaged).

8.5.3 The Local Flow Equation for Porous Media in 3D (Darcy's Law and Mass Conservation)

In this section, we formulate the "local" form of the water flow equations for 3D porous domains. Here "local" means "pointwise", as opposed for instance to "vertically averaged" models. The localized "3D equations" can also be used in 1D and 2D, given some assumed symmetries of the flow system.

The "localized" point of view is useful for dealing with flow systems that cannot be vertically averaged. The local equations may also occur in reduced dimensions, e.g. regional groundwater circulations with vertical flow components modeled in (x,z) cross-sections, or vertical unsaturated infiltration in a soil column (z). Now, the "local" flow equations combine three relations:

a) Darcy's head loss law (or another, generalized flux-gradient relation);
b) Mass conservation (the "continuity" or "flux divergence" equation);
c) Mass storage law (relating water storage capacity to pressure head).

In what follows, we present successively, first the general form of these equations, secondly, the special case of confined compressible aquifers (where water storage is characterized by a "specific storativity"), and thirdly, the case of unsaturated or partially saturated flow systems (where water storage is characterized by a "specific moisture capacity").

8.5.3.1 The 3D Saturated Flow Equation (General Form)

Figure 8.18 presents the general formulation of porous media flow equations based on Darcy's law in 3D space and time (x,y,z,t).

These equations are spatially localized, i.e. "point-wise" PDE's (Partial Differential Equations). The divergence (*div*) and gradient (*grad*) vector differential operators are used. The unknown variables (flux density vector **q**, pressure head *h*) depend on (x,y,z,t), and coefficients like permeability may depend on space, e.g. K(x,y,z) for saturated or K(h;x,y,z) for unsaturated media.

1. Mass conservation :

$$\frac{\partial(\rho\theta)}{\partial t} + div(\rho\mathbf{q}) = 0 \quad , \forall \mathbf{x} \, , \forall t$$

2. Darcy's law:

Saturated media (water)	Unsaturated media (air/water)
$\mathbf{q} = -K \; grad(H)$	$\mathbf{q} = -K(h) \; grad(h+z)$

with $H = h+z$ and $h=h(\theta)$

3. Combining these two equations gives :

$$\frac{\partial\{\rho\theta(h)\}}{\partial t} = -div\{-\rho K(h) \, grad(h+z)\} \quad , (\forall \mathbf{x} \, , \forall t)$$

Fig. 8.18 The general formulation of 3D porous media flow equations based on Darcy's law

However, these equations are not closed because the mass storage relation in "$\partial(\rho\theta)/\partial t$" has not yet been given. Two cases of interest are examined next:

- Compressible saturated media (confined aquifers, saturated clays in geotechnical applications, quasi-impervious saturated rocks serving as isolation layers in nuclear waste geological repositories);
- Unsaturated media with known water retention properties (unsaturated surficial soils, intermittently or partially saturated zones; desaturation problems in pumping and mining operations).

8.5.3.2 Saturated Flow Equation for 3D Confined Aquifers (Elastic Storativity)

We assume elastic behavior of the combined system is made up of:

- The solid porous matrix, with bulk stiffness coefficient α (Pa^{-1});
- Pore water, with volumetric compressibility or stiffness β (Pa^{-1}).

Assuming elastic deformations, and using the effective stress concept of Terzaghi (1936) relating pore pressure to stress at grain contacts (Bear 1972, Freeze and Cherry 1979), one obtains finally:

$$C_S \frac{\partial H}{\partial t} = div\left(K_S \; grad\left(H\right)\right) \quad , \forall(x,y,z,t) \tag{22}$$

where Cs (also denoted "Ss" in the literature) is specific elastic storativity $[m^{-1}]$:

$$C_S = \rho g \left(\alpha + \phi\beta\right) [m^{-1}] \tag{23}$$

Table 8.4 Compressibility coefficients of geologic porous media (gravel, silt, and clay)

Type of Geologic Material	Compressibility or Stiffness (Pa^{-1}).
Gravel aquifer	$\alpha \approx 10^{-9}$ Pa^{-1}
Sand aquifer	$\alpha \approx 10^{-8}$ Pa^{-1}
Clay	$\alpha \approx 10^{-7}$ Pa^{-1}
Intact granite	$\alpha \approx 10^{-10}$ Pa^{-1}
Water	$\beta = 4.4 \times 10^{-10}$ Pa^{-1}

The capacity coefficient "C_S" is often noted "S_S" for "specific storativity". It is in fact an elastic storage capacity, in m^3 of stored water per m^3 of medium and per meter of increase of hydraulic head (or pressure head). Its typical value for a confined aquifer is $C_S \approx 10^{-4}$ *to* 10^{-5} m^{-1}. "Compressibility" expresses the relative variation of volume $\delta m^3/m^3$ per variation of pressure (δPascal), whence the units (Pa^{-1}). Typical values of compressibility are listed in Table 8.4, however, only water compressibility is known precisely.

8.5.3.3 The 3D Nonlinear Unsaturated Flow Equation

In the case of unsaturated porous media (soils, rocks), the dynamic storage or drainage of water is not caused by soil-water compressibility, but by invasion of the free pore space by water, a process which depends on pore size distribution and the wetting properties of air/water/solids. The corresponding storage coefficient is called "specific capillary moisture capacity", to be defined below.

It should be noted that the unsaturated flow model presented here, after Buckingham (1907) and Richards (1931), does not take into account air compression effects, nor does it take into account the viscous dissipation occurring in air flow. These effects are taken into account only in the more complete two-phase Darcy-Muskat model. Nevertheless, in the "unsaturated" flow model, the presence of air is indirectly acknowledged via variations of water content, since air content is just porosity minus water content.

General Unsaturated Flow Equations

At any given time (t), the porous domain being modeled may be "strictly unsaturated" if water content is strictly less than porosity everywhere; otherwise the porous domain is either partially saturated, or possibly, fully saturated. In unsaturated regions, two nonlinear coefficients, both dependent on pressure head "h", emerge from the darcian approach:

a) *Unsaturated permeability*, or pressure-dependent unsaturated hydraulic conductivity *K(h),* a nonlinear coefficient in the unsaturated flux law of *Darcy-Buckingham* (see earlier);
b) *Water retention curve,* or pressure-dependent volumetric water content $\theta(h)$, leading to the specific moisture capacity $C = \partial\theta/\partial h$.

Assuming that the constitutive curves θ(h) and K(h) are known, and also, neglecting the compressibilities of air, water, and solid matrix, leads to:

$$\frac{\partial\{\theta(h)\}}{\partial t} = div\{K(h)\,grad(h+z)\},\ \forall(x,y,z,t) \tag{24}$$

Once this equation is solved in terms of pressure head h(x,y,z,t), the water content distribution θ(x,y,z,t) can be deduced from the water retention curve θ(h), and the Darcy flux density vector **q**(x,y,z,t) can be obtained by computing:

$$\mathbf{q} = -K(h)\,grad(h+z),\ \forall(x,y,z,t) \tag{25}$$

given the known permeability–pressure relation K(h).

Infiltration in Unsaturated Soils

The "theory of infiltration" essentially deals with analyzing the solutions of unsaturated flow equations for various infiltration processes taking place vertically or in 1D soil columns. The reader is referred to the original works of Green and Ampt (1911), Philip (1969), Parlange et al. (1982), and references therein. The reader is also referred to recent reviews on soil hydrodynamics and infiltration, such as: Espinoza (1999); Musy and Souter (1991).

In particular, a quasi-analytical solution of the nonlinear unsaturated flow equation (with unsaturated moisture content as the primary variable) was developed by J. R. Philip in 1957 to obtain the infiltration rate and moisture profiles in a homogeneous soil "column", under imposed surface water content (restricted to non positive pressure head, without ponding). This infiltration problem, and many others, are described in the monograph by Philip (1969).

In addition, see Parlange et al. (2002) for a recent presentation of so-called "explicit infiltration equations".

8.5.4 The 2D Plane Flow Equations (Vertically Integrated Dupuit-Boussinesq Model)

The "plane flow equations" of Dupuit-Boussinesq are useful for describing quasi-horizontal groundwater circulations. They are obtained by vertically averaging the previous "3D" equations between two surfaces: the impervious substratum at bottom (Zinf), and at the top, either the free surface (unconfined flow) or the superstratum (confined flow). The resulting plane flow equations are simpler. However, several points should be kept in mind:

- Vertical circulations are neglected (except possibly for exchange fluxes through leaky layers); the averaging procedure is based on a *vertically hydrostatic* approximation of the real 3D flow system;
- For this reason, in unconfined flow, the plane flow approximation requires that both substratum and water table be quasi-horizontal. More generally, the velocity field needs to be parallel (not convergent as it happens near a pumping well or a discharge point).
- The unsaturated zone located above the water table is excluded from the Dupuit-Boussinesq model. Retarded flow to or from the unsaturated zone is neglected. In unconfined aquifers, it is assumed that a known "effective" fraction of total porosity instantaneously fills up (or drains out) as the water table moves up (or down).

For the sake of brevity, we now present the Dupuit-Boussinesq plane flow equations *only* for the case of unconfined aquifers. (*The equations for confined aquifers are of a similar form except that C becomes the storage coefficient due to aquifer compressibility, and T is fixed rather than head-dependent*).

- Darcy's law in (x,y), in terms of vertically integrated flux density:

$$\mathbf{Q} = -T(h)\mathbf{grad}(H) \tag{26}$$

- Mass conservation in (x,y), vertically integrated at each point (x,y):

$$C\frac{\partial Z_{SUP}}{\partial t} = -div(\mathbf{Q}) + I_{SUP} + I_{INF} \tag{27}$$

Combining these two equations, and assimilating as before the water table height Zsup to the vertically averaged hydraulic head H, we obtain:

$$C\frac{\partial H}{\partial t} = div\big(T(h)\mathbf{grad}(H)\big) + I_{SUP} + I_{INF} \tag{28}$$

where the "I" terms on the right-hand side are sink/source terms standing for incoming and outgoing vertical fluxes (infiltration, evaporation, and leakage):

- I_{SUP} is a sink/source term representing the flux through the roof of the aquifer (Zsup); $I_{SUP} > 0$ if the flux is in-going (e.g. infiltration);
- I_{INF} is a sink/source term representing the flux through the floor of the aquifer (Z_{inf}); $I_{INF} > 0$ if the flux is in-going (e.g. upwards leakage through the floor).

Coefficient C [m^3/m^3] represents the storage capacity of the aquifer due to storage/drainage of water during fluctuations of the water table: thus it is a fraction of total porosity, called "efficient" or "effective" porosity (typically about ½ to ¾ of total porosity in classical aquifers).

Coefficient T [m^2/s] represents the hydraulic transmissivity, product of hydraulic conductivity K (m/s), and saturated thickness "h" (m):

$$T(h) \;=\; K \times h \tag{29}$$

Note that h(x,y,t) is water depth, from water table to substratum, so:

$$h(x,y,t) \;=\; Z_{SUP}(x,y,t) - Z_{INF}(x,y) \approx H(x,y,t) - Z_{INF}(x,y) \tag{30}$$

The quantity Z_{SUP} is the unknown elevation of the water table with respect to the frame of reference (e.g. sea level); but in order to obtain a closed system of equations, we have again assimilated Z_{SUP} to the hydraulic head H. Finally:

$$T(H) \;\approx\; K \times \big(H(x,y,t) - Z_{INF}(x,y)\big) \tag{31}$$

8.5.5 Summary of Hydraulic Properties of Aquifers

The following table summarizes the hydraulic properties of aquifers for:

- Vertically integrated 2D plane flow *vs.* non-integrated 3D flow;
- Confined *vs.* unconfined aquifers (in 2D plane flow models).

Note that two main types of hydraulic properties are distinguished:

- Conductive (flux transmission): K, T;
- Capacitive (water storage): C, Φ.

Table 8.5 Typical values of aquifer hydraulic properties

Capacitive Terms (Storativity, Storage or Eff. Porosity)		Conductive Terms (Conductivity, Transmissivity)	
Confined aquifer $C_S \approx 10^{-4}$–10^{-5} m^{-1} [m^3/m^3/m] $C \approx 10^{-3}$–10^{-4} [m^3/m^3].	Unconfined aquifer C_S [m^{-1}] negligible. $C=\Phi_{EFF} \approx 0.10$; $0.01 \leq \Phi_{EFF} \leq 0.30$ m^3/m^3].	Confined aquifer $K \approx 10^{-2}$–10^{-5} m/s [(m^3/s)/m^2] T = K e [m^2/s] [ou: (m^3/s)/m], $e = Z_{SUP}$-Z_{INF}.	Unconfined aquifer $K \approx 10^{-2}$–10^{-5} m/s [(m^3/s)/m^2] T=K.(H-Z_{INF}) m^2/s H = Z_{SUP}: watertable elev. Z_{INF}: substratum elevation

Finally, we explain how specific storativity (C_s) was estimated in Table 8.5:

Specific weight of water: $\rho g \approx 10^{+4}$ N/m^3 = 10^{+4} Pa/m,

Compressibility of solid matrix: $\alpha \approx 10^{-8}$–10^{-9} Pa^{-1},

Compressibility of pore water: $\Phi\beta \approx 0.25 \times (4.4 \times 10^{-10}$ Pa$^{-1}) \approx 10^{-10}$ Pa^{-1},

whence:

Specific storativity: C_S [(m^3/m^3)/m] = $\rho g(\alpha + \Phi\beta)$ [m^{-1}] $\approx 10^{-4}$–10^{-5} m^{-1}.

8.6 Stochastic Hydrogeology: 2D/3D Flow in Random Geologic Media

This section is devoted to flow modeling in randomly heterogeneous geologic media, such as heterogeneous soils and aquifers. We present methods for generating the geologic media using stochastic approaches, and we also present several simulated flow systems in such "random geologic media".

8.6.1 Numerical Tool (BIGFLOW 2D/3D Code)

A numerical tool, the BIGFLOW 2D/3D code, was used for flow simulations in this section, except where mentioned otherwise. The BIGFLOW code is based on implicit finite volumes), with sparse Preconditioned Conjugate Gradient matrix solver, and modified Picard iterations for nonlinear problems. The code was developed for high resolution simulation of 3D groundwater and unsaturated flow in heterogeneous media, and was tested for high-performance computations, e.g. on parallel Cray machines (Ababou et al., 1992).

The equational model being solved is a generalized Darcy-type equation, with a mixed formulation of mass conservation, capable of simulating various types of flows within the same domain. For more, see in this *Book*, the *Chapter* entitled "Coupled Surface/Subsurface Flow Systems: Numerical Modeling". BIGFLOW's generic equation, for 3D as well as 2D plane flow, is of the form:

$$\begin{aligned} &(1): \quad \frac{\partial \Theta(h,\vec{x})}{\partial t} = -\vec{\nabla} \bullet \vec{q} \\ &(2): \quad \vec{q} = -\vec{\vec{K}}\left(h, \vec{\nabla} H, \vec{x}\right) \vec{\nabla} H \\ &(3): \quad H = h + \vec{g}(\vec{x}) \bullet \vec{x} \end{aligned} \tag{32}$$

where only the first equation is actually solved, once the second and third equations have been inserted. The first equation expresses mass conservation in a partially saturated medium with known water retention or storage law Θ(h); the second equation is the generalized nonlinear flux-gradient head loss law with tensorial hydraulic conductivity/transmissivity "K"; and the third equation is the relation between total head or elevation (H) and pressure head or water depth (h) via a normalized gravitational vector (**g**).

A natural extension of this section would concern the study of tracers and contaminants migrating in heterogeneous flow systems. The relevant transport mechanisms are advection, diffusion, and subgrid hydrodynamic dispersion. However, due to lack of space, this topic will not be presented here. The reader is referred to the forementioned literature. See also Lam et al. (2006a,b) and references therein, concerning 3D particle tracking simulations in the case of randomly heterogeneous and discontinuous matrix-fracture media.

8.6.2 Literature Review on Stochastic Hydrogeology (Summary)

We start with a cursory, non exhaustive review on stochastic hydrogeologic modeling in the literature. The goal here is to point to appropriate bibliographic references, rather than to present the concepts and methods themselves. Note:

some of the methods will be presented and illustrated in the remainder of this chapter, further below, based on examples developed by the author.

The interested reader will find below a list of several general reviews or monographs on stochastic approaches in hydrogeology, including discussions on scale issues (upscaling, spatial resolution, domain size), stochastic solution techniques (spectral representations and Green's function perturbations), and in some cases, a review of evidence from field data:

- Ababou, 1988; Ababou, 1991; Dagan, 1989 ("Flow and Transport in Porous Formations"); Gelhar, 1986 (in French); Gelhar, 1993 ("Stochastic Subsurface Hydrology"); Marsily (1993, 1994, 1986a, in French); Marsily, 1986b ("Quantitative Hydrogeology"); Rubin 2003 ("Applied Stochastic Hydrogeology"); Zhang, 2002 ("Stochastic Methods for Flow in Porous Media"); among others.

One of the earliest analytical studies of stochastic groundwater flow with randomly distributed permeabilities was presented by Freeze (1975). The frequently cited article by Gelhar and Axness (1983) contains an analytical study of groundwater flow, effective permeability, and macro-dispersion in anisotropic aquifers, based on the spectral/Fourier approach to spatial randomness. Gelhar's book (1993) presents the latter theory and its applications to both saturated and unsaturated media. The book by Dagan (1989) summarizes a number of stochastic results obtained with the Green's function approach, and with conditional probability concepts.

A new formulation of stochastic flow based on groundwater velocity as the dependent variable, was developed in Ababou (1988), and was later described and used in Zijl and Nawalany (1993) ("Natural Groundwater Flow"; Section 3.5 and Appendix B). The velocity formulation is useful for analysing stochastic media, but also, for the classical case as well. It shows, in particular, how the gradient of log-permeability produces vorticity in groundwater flow.

Randomly heterogeneous geologic media can be generated by using the theory of spatially correlated random fields, i.e. random functions of space. Random fields F(x,y,z) are multidimensional generalizations of random processes F(t). If data or observations are available, random fields can be "conditioned" to honor the observed values at data points. This leads to spatial interpolation (or extrapolation) methods known as "geostatistics". There are two ways to obtain an interpolated field on a finer grid than the observation grid:

- A single mean field interpolation (bayesian estimation, or kriging);
- Multiple random replicates of the field (conditional simulation).

Some of the relevant concepts on random functions and estimation theory are developed in the following texts (with applications to electronics, communications, civil and geotechnical engineering, mechanics, hydrology, and earth sciences):

- Bras and Rodriguez-Iturbe, 1985 ("Random Functions in Hydrology"); Journel and Huijbregts, 1978 ("Mining Geostatistics"); Kitanidis, 1997 ("Introduction to Geostatistics"); Matheron, (1973); Papoulis (1964); or Papoulis and Pillai, 2002 ("Probability, Random Variables, and Stochastic Processes"); Tarantola 1987, 2005; Vanmarcke, 1983 ("Random Fields").

Particular algorithms to generate unconditional and conditional random fields in N-dimensional space (N = 1, 2, or 3) are described in:

- Ababou et al. (1994a); Delhomme (1979): on bayesian estimation, kriging, and conditional simulations.
- Mantoglou (1987) and Tompson et al. (1989): on unconditional random field generation by the Turning Band method in 2D and 3D.

The statistical generation of random sets of discrete fracture networks and their applications in hydrogeology, are addressed in:

- Chiles and Marsily (1993); Long and Billaux (1987);

and also in the following review article and book by Sahimi:

- Sahimi (1993, 1995) (Flow and Transport in Porous Media and Fractured Rock).

Scale dependence of water flow, multiphase fluid displacement, and solute transport in fractal porous media is also reviewed by Sahimi (1993, 1995). Groundwater flow in sedimentary aquifers with self-affine (fractal) distributions of permeabilities, was studied theoretically in terms of scale-dependent macro-permeability and macro-dispersion, in Ababou and Gelhar (1990).

Effective macro-scale permeability models were reviewed extensively by Renard and Marsily (1997). An earlier review can be found in Ababou (1988). A heuristic conjecture on the effective permeability tensor in anisotropic media was given in Ababou (1996), considering both the case of gaussian and non-gaussian log-permeability fields (the conjecture satisfies all known bounds and identities). Many early results on effective permeability can be found in the mathematically oriented monographs of Matheron (1967, in French) and Shvidler (1985, in Russian), and in the works by Dagan, Gelhar, and others.

Concerning various applications and reviews of scale problems for flow and transport in hetererogeneous soil-aquifer systems, the reader is referred to a recent special issue of Advances in Water Resources edited by Miller et al. (2002). The problems examined concern hydraulic permeability and tracer dispersion, but also, microbial and chemically reactive transport, rainfall processes, infiltration processes, and ecohydrology.

Finally, the interested reader may look up the (pre-)proceedings of a recent international symposium devoted to "GroundWater Hydraulics in Complex Environments", held in Toulouse (France): see reference "IAHR-GW 2006".

8.6.3 *Flow in Continuous Random Media with Spatially Correlated Random Fields*

Stochastic hydrogeology is built upon a stochastic or statistical representation of the spatial distribution of naturally heterogeneous geologic media in 3D space, in 2D plane views and cross-sections, or even, along 1D vertical profiles ("logs"). As a first step, the geologic medium needs to be "*parameterized*", e.g. described quantitatively in terms of its texture, porosity, permeability, etc. The remaining task is then to represent, reconstruct, or synthesize the spatial patterns of these parameters by statistical means, which is the subject of this section. See Section 8.0 concerning the different types of geologic media and structures encountered in hydrogeologic applications.

8.6.3.1 Generation of Synthetic Media with Continuous, Spatially Correlated Random Field Properties F(x,y,z)

Continuous Medium

We define loosely a "*continuous medium*" as one that can be represented by a set of spatially correlated random field parameters. In this way, a distinction is made between a continuous medium and a composite, discrete, or boolean medium. However, it should be kept in mind that some of the random fields F(x,y,z) discussed in this section are not necessarily pointwise continuous in the root-mean-square sense. Here are two examples:

- A log-permeability field F=ln K(x,y,z) with exponential correlation function is not differentiable; the variance of grad(F) diverges;
- A more extreme example is the white noise field, where every point in space is uncorrelated to any other point; the point field F(x,y,z) is not continuous; its variance diverges unless it is pre-filtered.

The relationship between random field differentiability and order of accuracy was examined based on Fourier analysis for a finite difference discretization scheme of the 3D saturated flow equation in a random medium (Ababou 1988).

Spatially Correlated Random Field

A *random field* is a function of space $F(\mathbf{x})$ such that $F(\mathbf{x_0})$ is a random variable for each fixed position ($\mathbf{x_0}$). The random field is spatially correlated if there exists some spatial correlation between distinct points. A single replicate of $F(\mathbf{x})$ is like a classical deterministic function of space, albeit highly irregular and possibly not differentiable.

A random field $F(\mathbf{x})$ can be partially characterized, through probability space, by its single-point moments (mean, variance) and two-point moments (spatial covariance $C(\mathbf{x_1},\mathbf{x_2})$). In fact, it is *entirely* characterized by these moments if it is *gaussian* (such as log-permeability).

A random field $F(\mathbf{x})$ is *statistically homogeneous* (stationary) if its moments are invariant by translation. The mean and variance are then constant, and the two-point covariance depends only on the separation vector between the points: $C \equiv C(\mathbf{x_2}-\mathbf{x_1})$. Note also that $C(\mathbf{0}) = Var(F(\mathbf{x}))$ and $C(\mathbf{x_2}-\mathbf{x_1}) \to 0$ as $|\mathbf{x_2}-\mathbf{x_1}| \to \infty$.

Random Field Generation

There are several ways to generate statistically homogeneous N-dimensional random fields:

i) Covariance matrix factorization;
ii) N-dimensional Fourier representation (Wiener-Kinchine in IR^N);
iii) N-dimensional Turning Band method (combined with 1D Fourier).

The Turning Band method is based on a line projection theorem (Matheron 1973). It is applicable for isotropic random fields, and for anisotropic fields with ellipsoidal covariance functions that can be transformed into isotropic ones. A number of authors studied the method (Journel, Deutsch, Mantoglou, and others). In particular, the Turning Band was implemented and analyzed in detail by Tompson et al. (1989) for the 3D case; the line processes of the TB method can be generated by 1D Fourier decomposition.

Figure 8.19 shows a single replicate of ln K(x,y,z) generated with the 3D Turning Band method on a $64 \times 64 \times 64$ cubic grid with mesh size $\Delta x=1$. The ln K field is gaussian; its two-point covariance structure is exponential isotropic with same correlation scale $\lambda=4$ in all three directions. These ensemble properties were satisfactorily tested in terms of sample statistics of the single replicate.

Fig. 8.19 Single replicate of log-permeability field with exponential isotropic covariance. It was generated on a 64-cube grid using the Turning Band module "CTURN" with a 64-bit random number generator. CTURN is part of a conditional simulation code, XIMUL 123D, available from the author upon request (Credits: P. Renard, 2003 helped process and analyze this test)

8.6.3.2 3D Flow in a Saturated Confined Aquifer with Random Permeability

The BIGFLOW code, with finite volume discretization and Preconditioned Conjugate Gradients solver, was used to conduct saturated flow simulations through 3D random isotropic and anisotropic media, generally with ellipsoidal exponential correlation function. The degree of heterogeneity, in terms of ln K standard deviation (σ), ranged from 1.00 to 2.30 or more. The 3D grid size was from one to ten million cells (Ababou et al., 1992; Ababou 1996).

8.6.3.3 3D Flow in a Random Imperfectly Stratified Soil (Moisture Plume)

For the case presented here, detailed flow simulations with randomly heterogeneous unsaturated soil parameters were conducted in 3D by Ababou (1988); comparisons with mean flow from stochastic unsaturated theory were presented in Polmann et al. (1991) using cross-sections of the detailed flow; and full 3D colour views of the detailed flow were shown in Ababou et al. (1992).

Figure 8.20 displays a 3D moisture plume θ(x,y,z,t) after t=10 days of constant flux, strip source infiltration in a randomly heterogeneous, statistically anisotropic soil. The soil is discretized into 300,000 finite volume cells (domain size 15m × 15m × 5m). The colors indicate moisture contents: unsaturated «wet» zones are blue-violet, drier zones are yellow-brownish. The volumetric representation is truncated in order to better visualize the 3D moisture field.

Figure 8.21 displays the 3D distribution of saturated permeability $K_S(x,y,z)$, for the imperfectly stratified random soil used in the moisture plume simulation. The other random parameter was the slope of ln(K(h)), the unsaturated permeability curve. The mean properties of the random soil are approximately those of Las Cruces' Jornada experiment (Polmann et al., 1991).

8.6.3.4 Plane Flow: Seawater Intrusion in a Random Aquifer (Sharp Interface)

Steady-state simulations of seawater intrusion in a freshwater aquifer were conducted based on the quasi-static sharp interface approach implemented with SWIM2D (a plane flow module of the BIGFLOW 2D/3D code).

The basic equations are summarized in Figure 8.22. The vertically averaged permeability distribution is stochastic in the (x,y) plane, with σ(ln K) as high as 2.30 or more. Single replicate numerical simulations were performed and analyzed statistically, on finite volume grids from 300 × 300 to 1000 × 1000 cells and more. Grid resolution was kept at $\Delta x/\lambda = 1/10$. For more details, see Ababou et al. (2004) and Al-Bitar et al. (2005).

Examples of single replicate simulations of 2D seawater intrusion are shown in Figures 8.23 and 8.24 in plane view and perspective view, respectively. The log-permeability structure ln K(x,y) is smooth (gauss-shaped covariance) but the degree of heterogeneity is fairly large, with σ(ln K)=2.30 here. The grid size was only 300×300 cells here, but much finer grids were used for statistical interpretations of salt wedge heterogeneity (*not shown here*).

Fig. 8.20 Spatial distribution of soil moisture: 3D moisture plume simulation (details in text)

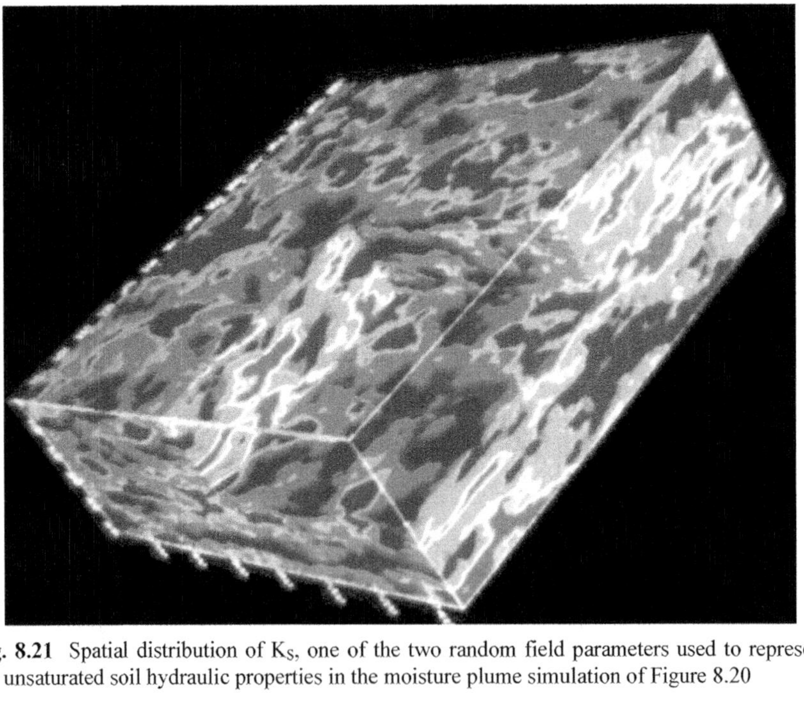

Fig. 8.21 Spatial distribution of K_S, one of the two random field parameters used to represent the unsaturated soil hydraulic properties in the moisture plume simulation of Figure 8.20

(1) **mass conservation in fresh water :**

$$\frac{\partial \theta}{\partial t} = -div(\mathbf{Q})$$

(2) **Darcy's Law + Boussinesq approximation:**

$$\mathbf{Q} = -T(H, Z_{salt}, Z_{inf}, x)\,\mathbf{grad}(H)$$

$$T = \begin{cases} K(x,y) \times (H - Z_{inf}(x,y)) & if\ Z_{salt} < Z_{inf} \\ K(x,y) \times (H - Z_{salt}(x,y)) & if\ Z_{salt} \geq Z_{inf} \end{cases}$$

(3) **Pressure continuity at the interface:**

$$Z_{salt} = Z_{sea} - \frac{(H - Z_{sea})}{\varepsilon} - \delta Z$$

Fig. 8.22 Plane flow equations of seawater intrusion with sharp interface in a heterogeneous aquifer (BIGFLOW 2D/3D: SWIM2D). Freshwater transmissivity "T" is highly nonlinear; it depends on both H (fresh water table elevation) and "Zsalt" (fresh/saltwater interface)

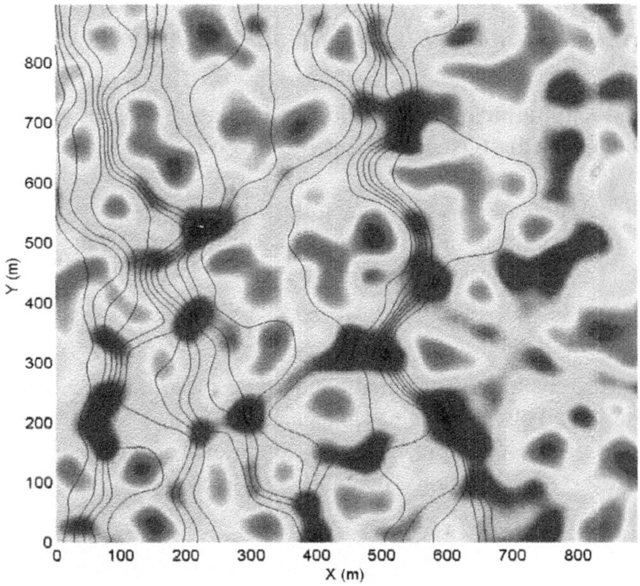

Fig. 8.23 Plane view of Zsalt(x,y) (contours) and log-K(x,y) (color map). The sea line is the left boundary (x = 0). The last Zsalt contour at right corresponds to level z = 0; it is the contour of the toe of the salt wedge on the impervious substratum. Low K values (in dark blue) range from 1E-2 to 1E-3, and high K values (in dark red) range from 1E-2 to 1E-3, relative to geometric mean Kg

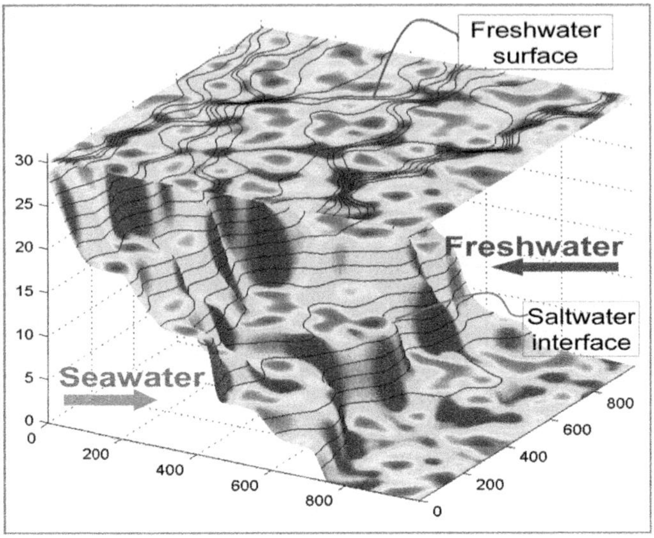

Fig. 8.24 Perspective view of seawater intrusion, same case as Figure 8.23. Both the fresh water table (top) and salt/fresh interface are shown. Observe the irregular contour of the salt wedge toe

8.6.4 Flow in Randomly Heterogeneous Aquifers Based on Bayesian Conditioning or Kriging from Point Data

8.6.4.1 Optimal Estimation and Conditional Simulation of Geologic Media Based on Point Data (Bayes, Krige)

The question to be addressed now is the following: how does one generate a "random" geologic medium in multi-dimensional space, while at the same time, taking into account a set of measurements at data points?

Here is a qualitative procedure that answers this question, based on statistical conditioning:

- First, estimate the spatially distributed conditional mean of the medium on a fine grid, honoring all observed values at data points;
- Secondly, on the same fine grid, generate conditional simulations (single or multiple replicates) of the geologic medium, also honoring all observed values at data points.

The "fine grid" is an interpolation grid. The suggested procedure is a geostatistical interpolation, applicable in any number of dimensions, e.g.: 1D for logs and boreholes, 2D for plane flow, and 3D as well.

Bayesian Approach

Let us outline here the *bayesian* approach to achieve the above purpose, i.e. spatial interpolation on statistical grounds. Without loss of generality, assume that one can create a regular interpolation grid containing, as a subset, all the “data points” (other grid points are “unobserved” points).

On this grid, as shown in Figure 8.25, let us name the discrete field ($\underline{Y}$), and distinguish observed ($\underline{Z}$) *versus* unobserved values ($\underline{X}$):

$$Y(\vec{x}) \rightarrow \begin{bmatrix} Y(\vec{x}_1) \\ \dots \\ \dots \\ \dots \\ \dots \\ \dots \\ \dots \\ Y(\vec{x}_N) \end{bmatrix} \rightarrow \begin{bmatrix} X_1 \\ \dots \\ \dots \\ X_M \\ \equiv \\ Z_1 \\ \dots \\ Z_K \end{bmatrix} = \begin{bmatrix} \underline{X} \\ \equiv \\ \underline{Z} \end{bmatrix} = \underline{Y} \qquad (33)$$

In this equation, the arrows “→” indicate transformations (projection on a grid, then re-ordering). The “≡” symbol is a separator between sub-vectors. The continuous field Y(**x**) is transformed into a state vector $\underline{Y}$ (N grid points), with subvector $\underline{X}$ containing all unobserved values (at M interpolation points) and subvector $\underline{Z}$ containing all observed values (at K data points). We have $M + K = N$, and usually, $K << N$.

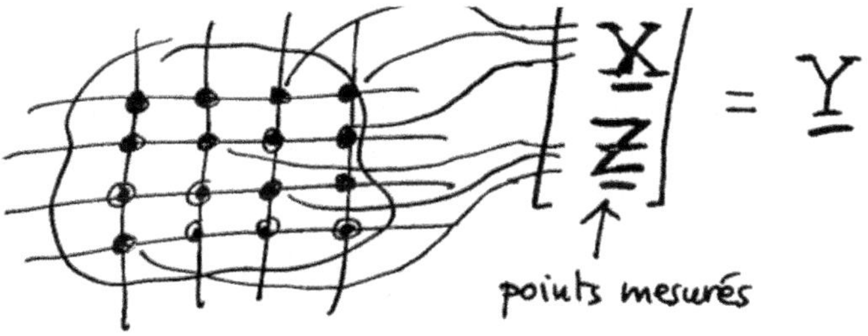

Fig. 8.25 Schematic of geostatistical interpolation grid in 2D, depicting the different grid locations of observed ($\underline{Z}$) and unobserved ($\underline{X}$) values of the field ($\underline{Y}$). For fine grid interpolation, the length of vector ($\underline{X}$) is typically chosen much larger than that of the data vector ($\underline{Z}$)

Optimal Bayesian Estimation

The procedure for obtaining an optimal mean field estimate of Y(x) on the fine grid is based on Bayes theorem of conditional probability in terms of PDF’s (Probability Density Functions):

$$\text{(Bayes)}\ p_{X|Z}\left(\underline{X}|\underline{Z}\right) = \frac{p_{X,Z}\left(\underline{X},\underline{Z}\right)}{p_Z\left(\underline{Z}\right)} = \frac{p_Y\left(\underline{Y}\right)}{\int p_Y\left(\underline{Y}\right) d\underline{X}} \tag{34}$$

This is a random field generalization of Bayes theorem. For a similar formulation in the simpler case of random processes, see Papoulis (1964) or Papoulis et al. (1985).

The left hand side, above, is the conditional PDF of (X) given (Z). In other words, it is the posterior distribution of (X) once the observations (Z) are known. Therefore, in p(X|Z), (X) is random state vector but (Z) is a set of fixed, known, deterministic parameters (the "data").

On the right hand side, the denominator of the last term is a multiple integral over all components of the "unobserved" state vector (X). Integrating the PDF of (Y) over all components of (X) yields $p_Z(Z)$, the marginal PDF of (Z). It should be understood that pZ(Z) is the prior (a priori) distribution of the random field Y at data points before knowing their measured values.

We now seek an optimal estimate ($\underline{Y}$*) of the unknown ($\underline{Y}$) given the measurements ($\underline{Z}$). Minimizing the variance of error ($\underline{Y}$*-$\underline{Y}$) (Bayes criterion), ensuring that the estimator is unbiased (zero mean error), and using Bayes' theorem, leads to the following results (Papoulis, 1964; Bras et al., 1985).

- *First result*: the best estimator ($\underline{Y}$*) of the unknown state vector ($\underline{Y}$), given the data vector ($\underline{Z}$), is the conditional mean of ($\underline{Y}$);
- *Second result*: if ($\underline{Y}$) is *a priori* gaussian, the conditional mean estimator ($\underline{Y}$*) is linear in ($\underline{Z}$), as given below:

$$\underline{Y}^* = \langle \underline{Y} \rangle + \underline{\underline{G}} \,.\left(\underline{Z} - \underline{\underline{A}} . \langle \underline{Y} \rangle\right) \tag{35}$$

This equation gives explicitly the optimal estimator. Matrix $\underline{\underline{G}}$ is a "vertical" rectangular *gain matrix* (N×K), and $\underline{\underline{A}}$ is a "horizontal" rectangular *measurement matrix* (K×N). The $\underline{\underline{A}}$ matrix indicates with 0's and 1's where data points are located on the grid. The brackets <•> indicate the prior mean, or unconditional mean (mathematical expectation).

The gain matrix $\underline{\underline{G}}$ is solution of:

$$\underline{\underline{G}} \,.\left(\underline{\underline{A}} . \underline{\underline{C}} . \underline{\underline{A}}^T\right) = \underline{\underline{C}} . \underline{\underline{A}}^T \tag{36}$$

where $\underline{\underline{C}}$ is the (N×N) prior covariance matrix of vector ($\underline{Y}$), known from the prior two-point covariance of random field Y(**x**). The computational difficulty of the estimation procedure resides in inverting the square system above, where (ACA^T) is a dense square matrix (Ababou et al. 1994a).

Figure 8.26 shows a geostatistical interpolation of soil permeability measurements in a costal watershed, assuming stationary, constant mean permeability field. In this case, our bayesian approach is equivalent to ordinary kriging. The calculations were done with XIMUL 123D, a code which performs both estimation (as shown) and multiple conditional simulations (not shown).

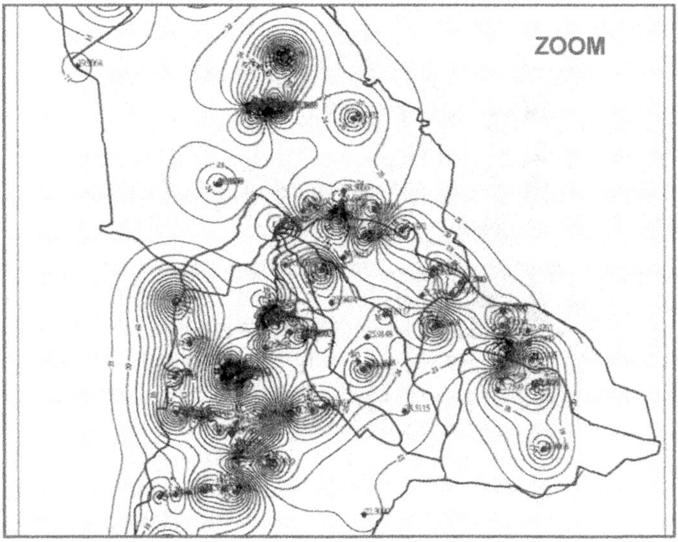

Fig. 8.26 Zoomed view of a kriged map of surficial permeability K(x,y) in the Leyre river basin, Landes de Gascogne, near Arcachon on the french atlantic coast. Collaboration with T. Lê and J. C. Chossat (Cemagref), who also provided the data

A few more remarks are in order. First, it is interesting to note that the gain G(i,j) measures the amount of information provided by data point Z(j) for estimating Y(i). The product $G.(Z-A\langle Y\rangle)$ measures the "*innovations*". For instance, if a "*datum*" happens to be equal to the prior mean, its innovation is zero. Secondly, if the prior distribution of Y is *non gaussian*, the linear estimator Y* is still the *Best Unbiased Linear Estimator (BLUE)*, but not the best of all estimators. And finally, the BLUE estimator (Y*) is identical with the "kriged" field in the theory of regionalized variables (in its *ordinary* form).

Conditional Simulation

The previous bayesian estimator Y* is a deterministic interpolator of Y. However, it can also be used to construct any number of random replicates of Y conditioned on Z by using the following orthogonality theorem: $\langle (\underline{Y}^*-\underline{Y}).\underline{Y}^{*T} \rangle = 0$, where the expectation $\langle ... \rangle$ is similar to a sample mean, taken over the ensemble of random values of the field at data points (these values are random prior to measurements).

This orthogonality property leads to the construction of conditional random fields (Y^C) by a linear combination "trick", such that:

- Each replicate ($\underline{Y}^C$) honors the data;
- The ensemble mean of all replicates ($\underline{Y}^C$) is the best estimator ($\underline{Y}^*$).

Each conditional replicate (Y^C) is then obtained as follows:

$$\underline{Y}^C = \underline{Y}(Z)^* + \left(\underline{Y}^U - \underline{Y}(Z^U)^* \right) \tag{37}$$

- $Y(Z)^*$ is the best estimator of Y given the data values Z;
- Y^U is an unconditional (prior) replicate of Y, ignoring the data;
- Z^U is the subset of Y^U on data points (values of Y^U at data points);
- $Y(Z^U)^*$ is the best estimator of Y given the fictitious data values Z^U.

This procedure was implemented by Delhomme (1979) in the framework of regionalized variables and kriging. See Ababou et al. (1994a) for a bayesian presentation, and for an algebraic analysis of the effects of measurement grid on the conditioning of the system. See also Chirlin and Wood (1982) concerning the equivalence between kriging and state-space linear estimation theory.

Finally, there exist other methods for estimating, conditioning and simulating random fields. One such method is the *NNM* (*Nearest Neighbor Markov*) of Bagtzoglou and Ababou (1997), where the spatial distribution of the random field is driven by a stochastic PDE (Klein-Gordon). This method can be viewed as analogous to "genetic" models (*to be discussed later*). The NNM implementation tested so far does not handle discontinuities or boolean structures. But its advantage is that bayesian conditioning is easy to handle, by imposing boundary/internal conditions on the governing PDE at data points.

8.6.4.2 Plane Flow in a Phreatic Aquifer with Kriged Transmissivities

Once the required hydraulic parameters have been reconstituted on a fine grid based on the conditional mean estimator and/or conditional simulations, the hydraulic behavior of the system remains to be examined. The computational task is heavier with conditional simulations, because of their random variability, compared to the smoother variability of a conditional mean or kriged field.

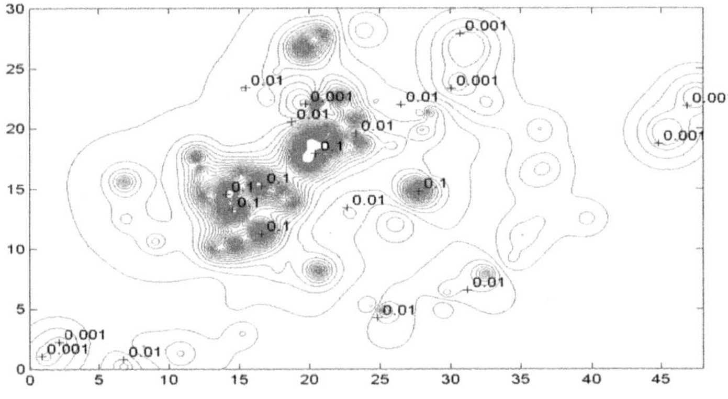

Fig. 8.27 Subset of transmissivity data and kriged transmissivity map contours for the Tadla aquifer, using a stationary exponential isotropic variogram with range 1 km (scales are in km's)

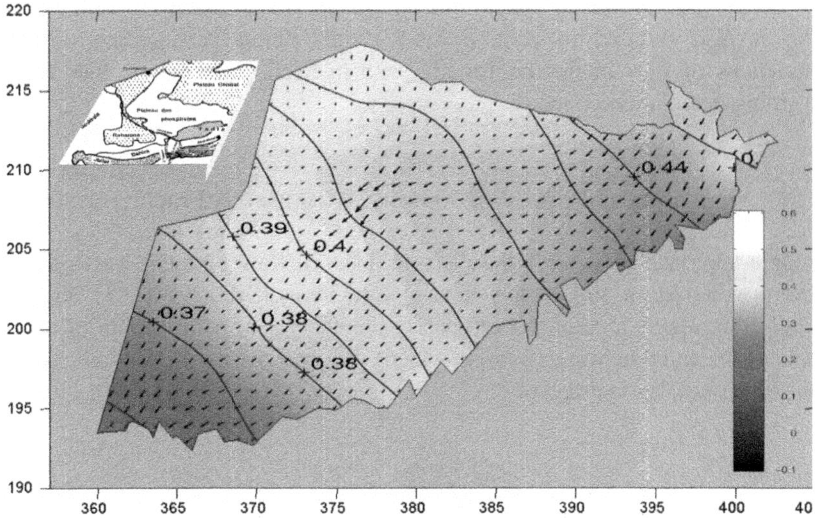

Fig. 8.28 Steady flow in a phreatic aquifer with kriged transmissivities (Tadla, Morocco). The domain is about 50 km (east-west) by 25 km (north-south). The south-east boundary is a river

Figure 8.27 displays data points and the resulting kriged transmissivity map for a phreatic aquifer located in an intensively irrigated plain in Tadla (Morocco).

Figure 8.28 shows the steady groundwater flow system obtained by solving the Dupuit-Boussinesq plane flow equations for the Tadla phreatic aquifer with kriged transmissivities (Trégarot, 2000). The simulated piezometric maps helped study groundwater exfiltration and salt accumulation at soil surface.

8.6.5 Hydrogeologic Modeling with Discrete Boolean Media

8.6.5.1 Flow in 2D Statistical Fracture Networks with Poiseuille's Law

Discrete fracture networks constitute an especially important type of discontinuous structure, e.g. in fissured soils, carbonate rocks (karsts), and/or in mechanically fractured hard rocks. If the intact porous matrix can be assumed impervious, then the flow takes place in the network, not in the matrix. In this case, the appropriate discrete flow model is based on Poiseuille's head loss law in the linear or planar fractures (similar to Darcy), and mass conservation at fracture intersections. See Sections 8.5.2.5 and 8.5.2.6 concerning the Poiseuille flow model and the related "cubic law".

The resulting discrete flow equations are relatively easy to formulate and solve in the case of 2D networks of linear fractures. Detailed head and velocity fields can be obtained by solving the discrete equations on statistical networks on the order of hundreds of thousands of links or nodes, using compact Conjugate Gradient type solvers (*due to lack of space, this topic will not be discussed further)*. The reader is also referred to Section 8.6.5.2 (just below) concerning upscaled formulations of flow in fractured media, and to Section 8.6.5.3 (further below) concerning percolation effects.

8.6.5.2 Simulating and Upscaling Flow in 2D/3D Matrix/Fracture Systems

A simple approach to *upscale* the discrete network flow problem was indicated earlier: it is based on Stokes flow for single fractures (Section 8.5.2.5), and on a superposition approximation for estimating the equivalent permeability of a large sample of the entire network (Section 8.5.2.6). We will discuss below some related issues (non percolation effects).

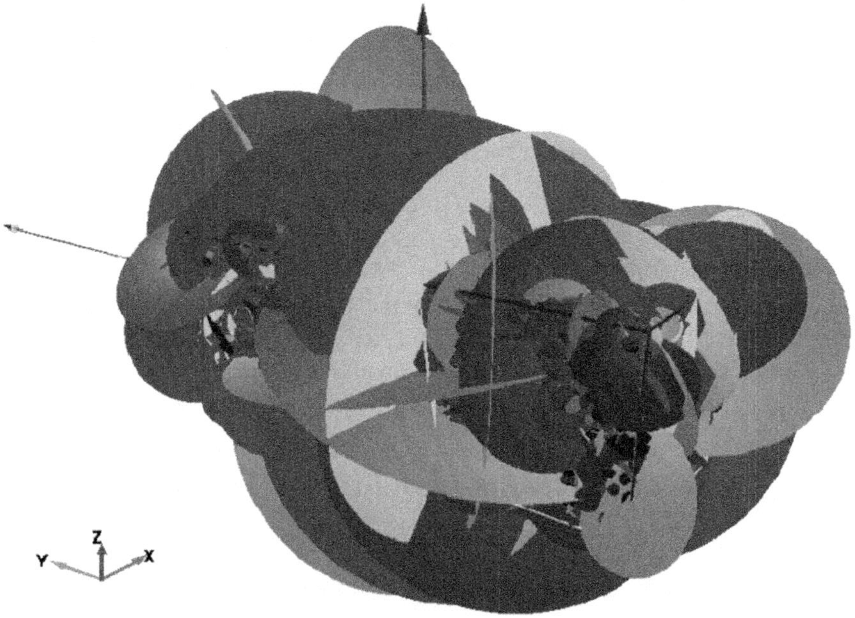

Fig. 8.29 Single replicate of a reconstructed 3D random network of planar disc fractures for the FEBEX in-situ experiment, located at the granitic Grimsel Test Site (Swiss Alps). Fracture traces on the walls of the FEBEX tunnel were used for reconstruction (the tunnel axis is visible). Length scales are 70 × 200 × 70 meters. In this view, the largest fractures hide many other smaller ones inside the domain. In total, about three million fractures of many different sizes were generated

Here, let us briefly indicate how fracture network approaches can be extended to deal with matrix/fracture flow taking into account a *permeable* matrix in either 2D with linear fractures, or 3D with planar fractures.

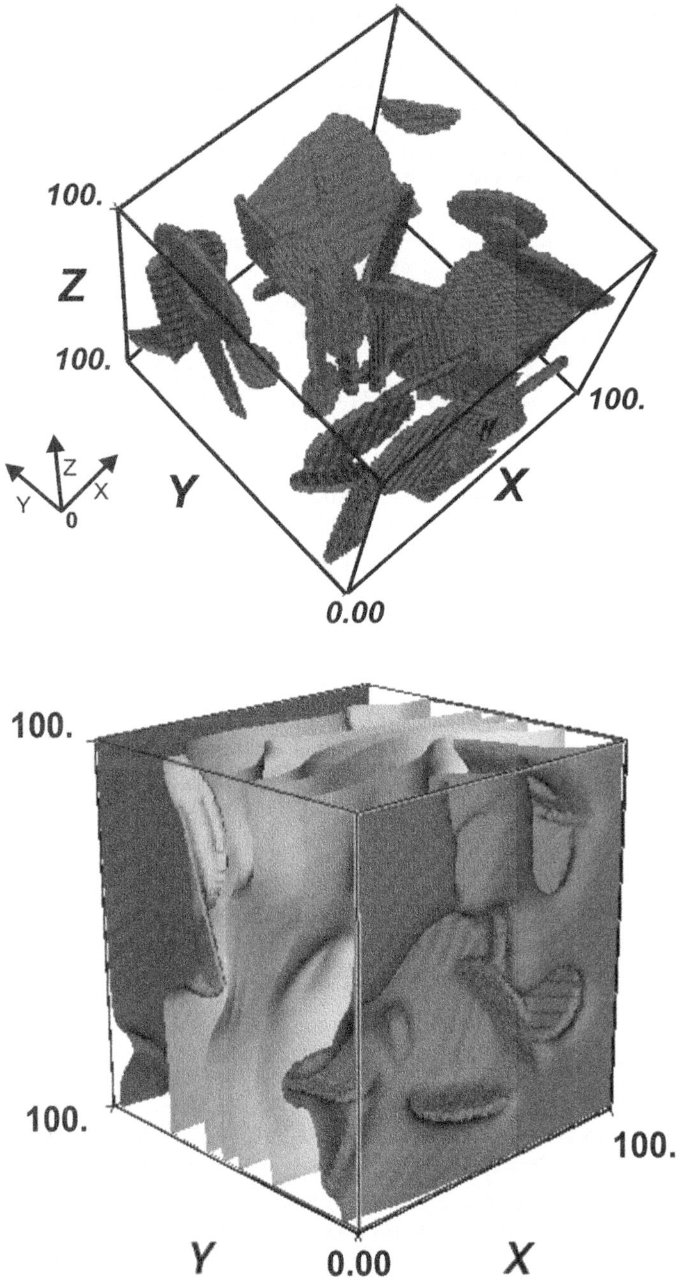

Fig. 8.30 A detailed groundwater flow simulation with Darcy's law for a 3D matrix/fracture system on a 1 million node grid. **Top**: 3D network of planar disc fractures with Poisson centers and isotropic orientations. **Bottom:** 3D hydraulic head field (piezometric isosurfaces)

See Canamon et al. (2006, 2007) for 3D model development and application to coupled Thermo-Hydro-Mechanical processes in fractured rock. See also Lam et al. (2006a,b) for further developments and applications to particle transport modeling in 3D fractured porous rock. These references present recent works conducted by the author and collaborators in the context of high-level radioactive waste isolation in deep geologic repositories.

To give an example from research on radioactive waste isolation, we show in Figure 8.29 one replicate of a reconstructed, 3D random fracture network in a granitic rock (after Canamon et al., 2006).

A numerically simulated flow field is shown in Figure 8.30 for another 3D matrix/fracture system, resolved on a one million node cubic grid. The finite volume BIGFLOW code was used with the following options: 3D steady flow, fully saturated, with a linear head loss law (Darcy). Other simulations were performed on the same system, with matrix/fracture permeability contrast $\varepsilon = K_M/K_F$ varying from 1/1 to 1/1,000,000 or even less (Bailly et al. 2007).

It is interesting to note that the percolation cluster can be approximately obtained in this way, i.e. through hydraulic calculations rather than topological search, by letting the permeability ratio $\varepsilon = K_M/K_F$ decrease to 1E-6.

8.6.5.3 Percolation Analyses in Fracture Networks

Here we analyze more specifically "percolation" effects, particularly for 2D discrete fracture networks (impervious matrix).

In the case of impervious matrix, a *non percolating* network causes the algebraic system of discrete flow equations to become singular. Also, if a subset of the network does not participate to flow, this is not taken into account when upscaling permeability with the simple superposition approach.

In such cases, a separate analysis of the *percolation cluster* may be necessary. Theoretical results on bond percolation can be found in the literature (*not detailed here for lack of space*). In addition, for 2D fracture networks, a direct topological search on a planar graph is possible. Some results of this approach are illustrated just below.

Figure 8.31 depicts the "oriented" percolation cluster in a 2D fracture network. The network was first analyzed in terms of nodes and links, and the percolation cluster was then obtained by a topological search on the planar graph of the network, based on a classification of different types of clusters (isolated links, dead-end links, isolated clusters, dead-end clusters, etc.).

Although percolation analyses of 2D fracture networks are commonly found in the literature, there are two main points worth noting in our procedure here:

- Percolation was analyzed as a function of orientation, and it was found to vary significantly with orientation. Orientation was imposed by partitioning the boundary of the sample into two subsets, "inlet" and "outlet", separated by a straight line. The direction of percolation (or non percolation) is orthogonal to the separation line. The percolation cluster shown in Figure 8.31 is for the particular orientation indicated in the figure.

- We applied "thresholding" on fracture apertures in order to analyze percolation effects *versus* minimum allowed aperture. Thus, in Figure 8.31, all fractures with apertures smaller than 4 microns were eliminated. The original network comprised 6,580 fractures, while the thresholded network comprises only 2,762 fractures, of which 1,287 did not percolate, and the remaining 1,475 formed the percolation cluster. In contrast, the original network with 6,580 fractures was almost totally percolating in all directions.

Now, Figure 8.32 shows an original result on "critical" percolation apertures a_{CRIT} (in microns) versus direction angle for a subdomain sample of the fracture network (BMT3 test sample). The aperture thresholding operation described above was used. In short, the graph is a polar histogram of the critical aperture above which the network will always percolate along the given direction. If the full

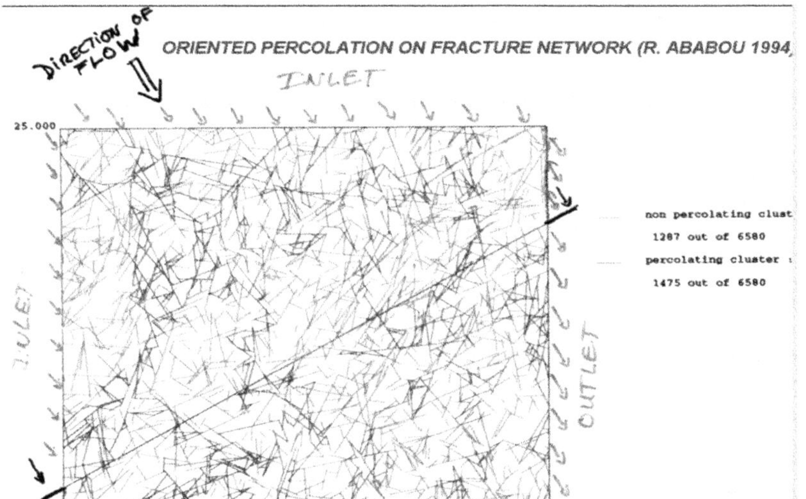

Fig. 8.31 Oriented percolation effects in a 2D fracture network. The percolation cluster (dark blue) was obtained by a topological search on the planar graph of the fracture network. The non percolating fractures are in light blue. The search algorithm was devised by the author, R. Ababou, in collaboration with E. Treille (circa 1994)

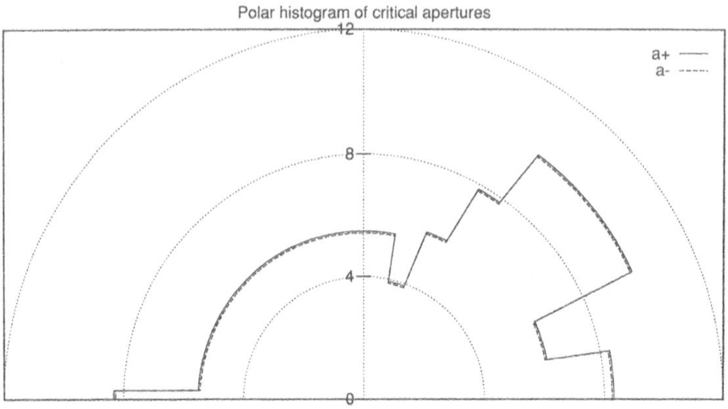

Fig. 8.32 Polar histogram of "critical" percolation apertures (microns) versus direction angle. It can be seen that the network "strongly percolates" for direction $\theta \approx 45°$

network does not percolate at all in a certain direction, this is indicated by $a_{CRIT} = 0$. If, on the contrary, it is very "strongly" percolating, then a_{CRIT} is large.

8.6.5.4 Matrix/Fracture Exchanges and Dual Continuum Approaches for Fractured Porous Media

When the porous matrix of the fractured rock is permeable, the hydraulic model needs to take into account, at a minimum, the flow pathways from fractures to porous matrix and vice-versa (even if permeability is negligible, solutes can be trapped into, or release from, the porous matrix).

One way to deal with the permeable matrix as well as the 2D/3D fracture network is to develop upscaled flow equations that take into account both matrix and fracture flow. The simplest method of upscaling is achieved by seeking an equivalent "*single medium*", with macroscale permeability tensor that captures both the permeability of the isotropic matrix (K_M) and the tensorial permeability ($\underline{\underline{K}}_F$) of the fracture system.

However, a more complete upscaling method will take into account the transient (non equilibrium) water exchanges between fractures and matrix, which leads to an equivalent "*dual continuum*" system involving a water exchange coefficient (α) in addition to matrix and fracture permeabilities. The exchange coefficient (α) typically depends on the *specific area* of fractures. Along these lines, the interested reader is referred to recent work by Kfoury et al. (2006) with 2D cartesian networks, and by Canamon et al. (2006, 2007) with 3D random fractures partially constrained/optimized from field observations. See also Sections 8.5.2.6 and 8.6.5.2.

8.6.5.5 Generation of Composite, Boolean, Karstic Media with 3D Objects (Fissures, Conduits, Cavities)

Synthetic modeling of 3D karstic media, and other highly fissured media, can be done by embedding boolean (discrete) objects in an embedding porous matrix. The spatial distribution of the "matrix/objects" system can be represented numerically on a regular, voxel-based grid. To begin with, very simple examples of purely statistical boolean media on 2D/3D cartesian grids are shown in Figure 8.33.

In the case of karsts, both the porous matrix and (at least) a subset of the discrete objects should be generated based on statistical distributions. However, the coarser part of the discrete structure (main conduits, etc.) can be considered deterministic, e.g. conditioned on observations.

The random boolean part of the "karst" model should include several classes of discrete objects, such as: points (0-D); lines (1-D); circular and ellipsoïdal plane fractures (2-D); and a variety of fully 3-D objects like cylindrical conduits, spheroïdal, and ovoïdal cavities. See Bailly et al. (2007).

An example of a boolean "karstic" structure and its percolation cluster is shown in Figure 8.34. The 3D grid has about one million cubic voxels, representing a cubic domain of size 99 m in each direction.

The synthetic "karstic medium" in Figure 8.34 was conceptualized using two karstic subdomains and three classes of geomorphological objects. The two sub-domains are:

- An epikarst, encompassing the top 25 m layer;
- A fully developed karst, below the top 25 m layer;
- The three classes of objects imbedded in the porous matrix are:
- A horizontal cylindrical conduit near the center of the domain (50 m length, 40 m diameter);
- A set of 100 planar disc fractures of moderate sizes (12.5 m mean diameter) in the epikarst;
- A set of 50 larger plane disc fractures or faults (the diameter follows a Pareto distribution between 0 and 100 m, with a power of 2.5).

All fracture distributions are Poissonian (with uniformly distributed centroids in 3D space) and their orientations are statistically isotropic. The distribution of a statistically isotropic vector in 3D space can be found in Tompson et al. (1989).

In Figure 8.34, observe that the 3D percolation cluster includes the largest conduit. This percolation cluster was obtained by a 3D morphological search (*not described here*). Morphological analyses like these, complemented with numerical flow experiments, are being used to further "constrain" the synthetic fissured medium in order to mimic the 3D structure of karst aquifers.

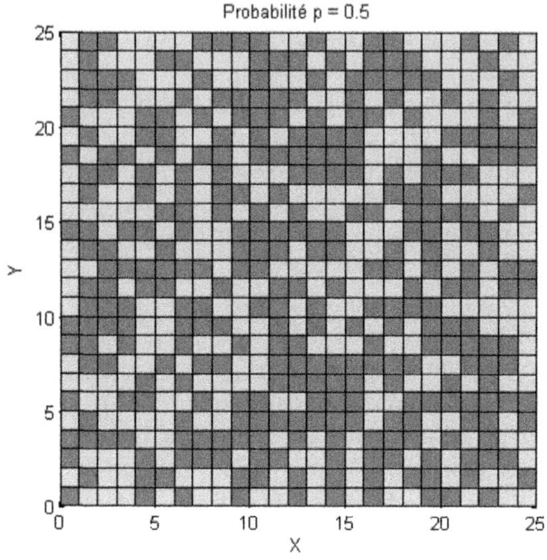

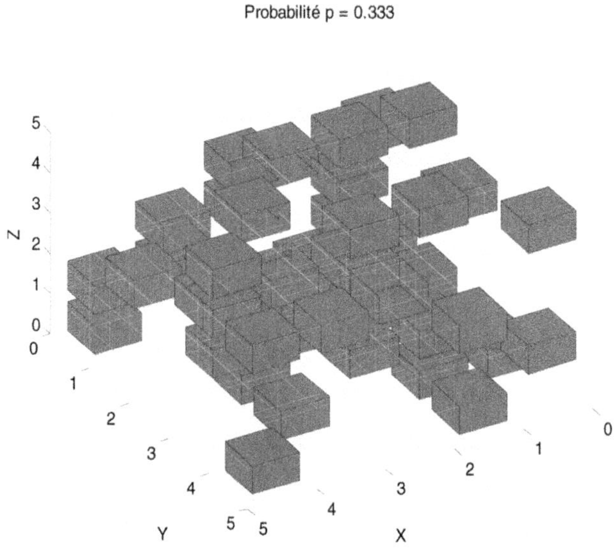

Fig. 8.33 A simple statistical boolean model, containing only one type of objects (cubic cells) embedded in a homogenous matrix. The dark cells are distributed in a purely random fashion with probability "P" on the voxel grid. **Top**: 2D model with P = 1/2. **Bottom**: 3D model with P = 1/3

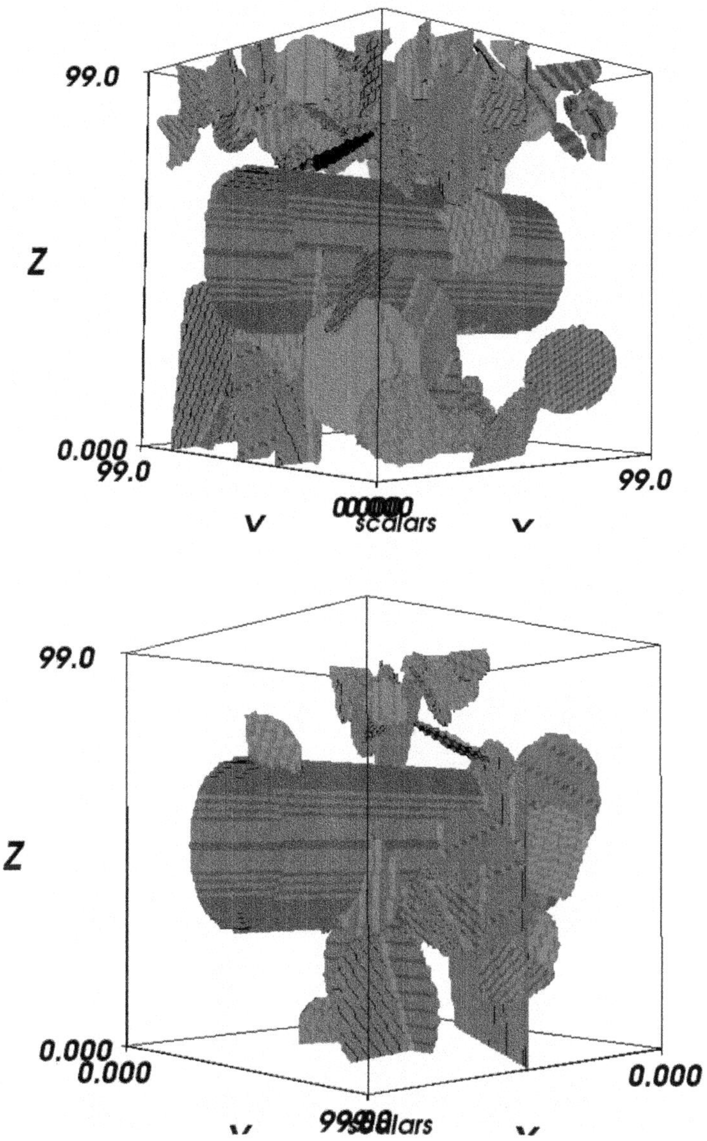

Fig. 8.34 **Top:** view of the complete set of objects ("fissures") described in the text, not showing the embedding porous matrix (transparent). **Bottom:** a view of the percolation cluster for flow oriented in the vertical direction (from top to bottom boundary faces); this view shows all "objects" participating to flow (compare to the complete set of objects shown in the left view)

8.6.5.6 Genetic Models and Cellular Automata (CA's)

Genetic models have been used in the literature to generate many types of geologic media: sedimentary basins and aquifers; facies of river deltas and dejection cones; and fissured structures in karsts.

Some genetic models are based on Markov chain transition probabilities. Others are governed by simplified models of hydrologic flow, solid transport, dissolution, and other geological processes. For example, a small sedimentary basin (Alameda Creek, San Francisco bay) was entirely reconstructed based on a combination of hydrologic flood hydrographs and erosion/deposit models, over a time scale of 0.6 Ma (600,000 years): Kolterman and Gorelick (1992, 1996). For a review, see also Marsily et al. (2005) and references therein.

In the case of karsts, the genesis of the highly fissured medium is governed by the dissolution of carbonate rocks by acidic waters. With this idea in mind, a simplified *Cellular Automaton (CA)* genetic model was implemented by *Jacquet (1998),* to model karst genesis by dissolution, as well as the flow itself.

More generally, Cellular Automata (CA's) can be very useful for generating synthetic media in 3D, and possibly, for reconstruction algorithms incorporating in-situ data. However, CA's come under many different guises, e.g. lattice gas CA's *versus* totalistic CA's (cf. Wolfram's classification).

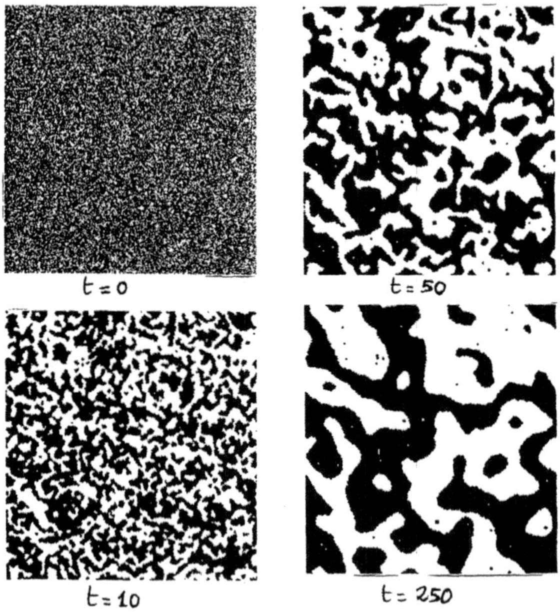

Fig 8.35 Evolution of a 2D Cellular Automaton, the twisted voting rule, shown at time steps t = 0, t = 10, t = 50, t = 250. The initial distribution of black/white cells is purely random and equilibrated, 50%-50%. This CA was programmed on a Silicone Graphics machine by the author

To give one example, we show in Figure 8.35 the evolution of a 2D cellular automaton known as a totalistic CA, with a simple "voting rule" (a "twisted" voting rule here). The black and white cells might represent fluid and/or solid phases. We started with a purely random 50–50% initial distribution. Note the appearance of a slow meta-stable coalescence process in each of the "phases".

Acknowledgments The contributions of several collaborators to models and results described in this work, under various forms, directly or indirectly, are gratefully acknowledged. The long list includes: doctoral students A. Al-Bitar, D. Bailly, M-P. Lam; former doctoral students I. Canamon and G. Trégarot; fellow researchers, professors, and engineers (J. C. Chossat, J. Elorza, P. Renard, E. Treille) and many others. The author has benefited from research programs and grants along the years, too numerous to name. The author's research is currently supported in part by the French program GdR 2990 "PNRH" INSU/CNRS (Centre National de Recherche Scientifique), among others.

References

Ababou, R., 1996, Random porous media flow on large 3-D grids: Numerics, performance, and application to homogenization, Chap.1, pp.1–25, in: IMA Vol. N°.79 *Mathematics and its Applications: Environmental Studies—Mathematical, Computational and Statistical Analysis.* M. F. Wheeler, ed., Springer, New-York, 410 p.

Ababou, R., and Al-Bitar, A., 2004, Salt water intrusion with heterogeneity and uncertainty: Mathematical modeling and analyses, in: Proceedings of *Computational Methods in Water Resources (CMWR04)*, Elsevier, Chapel Hill, USA, 13–17 June 2004, 12 p.

Ababou, R., and Bagtzoglou, A. C., 1993, *BIGFLOW: A Numerical Code for Simulating Flow in Variably Saturated, Heterogeneous Geologic Media (Theory and User's Manual, Version 1.1)*, Report NUREG/CR-6028, U.S. Nuclear Regulatory Commission, Gov. Printing Office, Washington D.C., USA, 139 pp., (available from U.S.DOE web site http://www.osti.gov).

Ababou, R., and Gelhar, L. W., 1990, Self-similar randomness and spectral conditioning: Analysis of scale effects in subsurface hydrology, Chapter XIV, in: *Dynamics of Fluids in Hierarchical Porous Media*, J. Cushman, ed., Academic Press, New York, pp. 393–428.

Ababou, R., Gelhar, L. W., and Hempel, C., 1992, Serial and parallel performance on large matrix systems, *Cray Channels*, **14**:22–25.

Ababou, R., Bagtzoglou, A. C., and Wood, E. F., 1994a, On the condition number of covariance matrices arising in kriging, estimation, and simulation of random fields, *Mathematical Geology*, **26**(1):99–133.

Ababou, R., Millard, A., Treille, E., Durin, M., and Plas, F., 1994b, Continuum modeling of coupled thermo-hydro-mechanical processes in fractured rock, in: *Computational Methods in Water Resources*, A. Peters et al., eds., Kluwer Acad. Publishers, Vol. 1, Chap. 6, pp. 651–658.

Al-Bitar, A., and Ababou, R., 2005, Random field approach to seawater intrusion in heterogeneous coastal aquifers: Unconditional simulations and statistical analysis, in: *GeoENV: Geostatistics for Environmental Applications,* P. Renard, H. Demougeot-Renard, and R. Froidevaux, eds., ISBN: 3-540-26533-3, Springer.

Bagtzoglou, A. C., and Ababou, R., 1997, Conditional simulation and estimation of Gauss-Markov random fields using the bayesian nearest neighbor method, in: *GeoENV I-Geostatistics for Environmental Applications,* (Lisbon 1996), A. Soares et al., eds., Kluwer Academic Publishers, The Netherlands, pp. 477–491.

Bailly, D., Ababou, R., and Quintard, M., 2007, Geometric characterization, hydraulic behavior and upscaling of highly fissured 3D geologic media (karsts), Proceedings of *International Conference on Approximation Methods and Numerical Modeling in Environment and Natural Resources*. Granada (Spain), 11–13 July 2007, 4 p.

Bear, J., 1972, 1988, *Dynamics of Fluids in Porous Media*, American Elsevier, N.Y., 1972, republished by Dover Publications Inc., N.Y., 764 p.

Bertoldi, G. L., Johnston, R. H., and Evenson, K. D., 1991, Groundwater in the Central Valley, California—A summary report, *U.S. Geological Survey Professional Paper* 1401-A, 44 p.

Bras, R., and Rodriguez-Iturbe, I., 1985, *Random Functions in Hydrology*, Dover, New York, re-edited in 1993.

Buckingham, E., 1907, *Studies on the Movement of Soil Moisture*, U.S. Dept. Agric., Bureau of Soils. Bulletin No. 8.

Canamon, I., Elorza, F. J., Ababou, R., 2006, 3D Fracture networks: Optimal identification and reconstruction, in: Proc. *IAMG'06: Internat. Assoc. Math. Geol.* (XIth Internat. Congress, 3–8 Sept. 2006, Liège,Belgium), 5 p.

Canamon, I., Elorza, F. J., and Ababou, R., 2007, Thermo-hydro-mechanical simulation of a 3D fractured porous rock: Preliminary study of coupled matrix-fracture hydraulics, in: Proc. *Reactive Heat Transfer in Porous Media—EUROTHERM Seminar No.81*, Albi, France (4–6 June 2007), Paper ET81-11, 8 p.

Chiles, J. P., and de Marsily, G., 1993, Stochastic models of fracture systems and their use in flow and transport modeling, in: *Flow and Contaminant Transport in Fractured Rock*, J. Bear, Chin-Fu Tsang, and G. de Marsily, eds., Springer-Verlag, pp. 169–231.

Chirlin, G. R., and Wood, E. F., 1982, On the relationship between kriging and state estimation, *Water Resources Research*, **18**(2):432–438.

Dagan, G., 1989, *Flow and Transport in Porous Formations* Springer-Verlag, 465 p.

Darcy, H. P. G., 1856, *Les Fontaines Publiques de la Ville de Dijon: Exposition et Application des Principes à Suivre et des Formules à Employer dans les Questions de Distribution d'Eau*, Victor Dalmont (publisher), Paris, (see Appendix D: "*Annexe D*").

Delhomme, J. P., 1979, Spatial variability and uncertainty in groundwater flow parameters: A geostatistical approach, *Water Resources Research*, **15**(2):269–280.

Espinoza, R. D., 1999, Infiltration, Chapter 6, in: *The Handbook of Groundwater Engineering*, J.W. Delleur, ed., CRC Press LLC, Boca Raton.

Freeze, R. A., and Cherry, J. A., 1979, *Groundwater*, Prentice-Hall, Englewood Cliffs, NJ 07632, USA.

Freeze, R. A., 1975, A stochastic-conceptual analysis of one-dimensional groundwater flow in nonuniform homogeneous media, *Water Resources Research*, **11**:725–741.

Green, W. H., and Ampt, G. A., 1911, Studies on soil physics: Flow of air and water through soils, *Journal of Agricultural Research*, 4:1–24.

Gelhar, L. W., 1993, *Stochastic Subsurface Hydrology*. Prentice Hall, Englewood Cliffs, New Jersey, 390 p.

Gelhar, L. W., and Axness, C. L., 1983, Three dimensional stochastic analysis of macrodispersion in aquifers, *Water Resources Research*, **19**(1):161–180.

Grenet, B., Haeffner, H., Samson, A., Suzanne, P., 1996, Stratégie d'exploitation des nappes de l'Albien et du Néocomien en région parisienne, in: *La Houille Blanche - Revue Internationale de l'Eau*, Dossier "Les nappes Profondes", Part. 2, No. 4, pp. 20–26, (in French).

Hubbert, M. K., 1953, Entrapment of petroleum under hydrodynamic conditions, *Bulletin of American Association of Petroleum Geologists*, **37**:1954–2026.

IAHR-GW 2006, 2006, International Symposium on "*Groundwater Hydraulics in Complex Environments*", Sponsored by IAHR: International Association of Hydraulics Research. Toulouse, France, 12–14 June 2006: (September 26, 2007); http://www.iahr-gw2006.cict.fr.

Jacquet, O., 1998, *Modèle stochastique de la géométrie des réseaux karstiques*, Doctoral thesis, Lausanne University, Switzerland, (in French).

Journel, A. G., and Huijbregts C. J., 1978, *Mining Geostatistics,* Academic Press, New York, 600 p.

Kfoury, M., Ababou, R., Noetinger, B., and Quintard, M., 2006, Upscaling fractured heterogeneous media: Permeability and mass exchange coefficient, *Journal of Applied Mechanics, Transactions ASME,* **73**(1):41–46.

Kimmeier, F., Bouzelboudjen, M., Ababou, R., and Ribeiro, L., 2001, Travel path uncertainty: A case study combining stochastic and deterministic hydraulic models in the Rhône valley of Switzerland, Proceedings of *3rd Internat. Conf. on Future Groundwater Resour. at Risk (FGR'01),* 25–27 June 2001, Lisbon, Portugal, 8 p.

Kiraly, L., 1969, Anisotropie et hétérogénéité de la perméabilité dans les calcaires fissurés, *Eclogae Geol. Helv.*, **62**(2):613–619.

Kiraly, L., 2003, Karstification and Groundwater Flow, in: *Speleogenesis and Evolution of Karst Aquifers* (*The Virtual Scientific Journal*), 1(3), Sept. 2003, 26 pp., (September 22, 2007); www.speleogenesis.info.

Kitanidis, P. K., 1997, *Introduction to Geostatistics*, Cambridge University Press, 249 p.

Kolterman, C. E., and Gorelick, S. M., 1992, Paleoclimatic signatures in terrestrial flood deposits, *Science*, **256**:1775–1782.

Kolterman, C. E., and Gorelick, S. M., 1996, Heterogeneity in sedimentary deposits: A review of structure-imitating, process-imitating, and descriptive approaches, *Water Resources Research*, **32**:2617–2658.

Lam, M. -P., Nebauer, R., and Ababou, R., 2006a, Asynchronous particle tracking for transport in 3D fractured porous media, Proc. *7th International Conference on Hydro Informatics* (HIC 2006), Nice, France, 8 p.

Lam, M. -P., Nebauer, R., and Ababou, R., 2006b, Asynchronous particle tracking for contaminant migration in heterogeneous 3D media with unstructured tetrahedral mesh, Proc. *IAHR-GW2006*: International Symposium on "*Groundwater Hydraulics in Complex Environments*" (Toulouse, France, 12–14 June 2006), IAHR, ed., *(to be published).*

Long, J. C. S., and Billaux, D., 1987, From field data to fracture network modeling: an example incorporating spatial structure, *Water Resources Research*, **23**(7):1201–1216.

Mantoglou, A., 1987, Digital simulation of multivariate two- and three-dimensional stochastic processes with a spectral turning bands method, *Mathematical Geology*, **19**(2):129–149.

Marsily, G. (de), 1986, *Quantitative Hydrogeology (Groundwater Hydrology for Engineers,* Academic Press, New York, 440 p.

Marsily, G. (de), Delay, F., Gonçalvès, J., Renard, P., Teles, V., and Violette, S., 2005, Dealing with spatial heterogeneity, *Hydrogeology Journal*, **13**:161–183, DOI 10.1007/s10040-004-0432-3.

Matheron, G., 1967, *Eléments pour une théorie des milieux poreux*, Masson et Cie, Paris, (in French).

Matheron, G., 1973, The intrinsic random functions and their applications, *Advances in Applied Probability*, **5**:438–468.

Miller, C. T., Parlange, M. B., and Hassanizadeh, S. M., eds., 2002, Special 25th Anniv. Issue, *Advances in Water Resources*, **25**(8–12):859–1396.

Musy, A., and Soutter, M., 1991, *Physique du sol.* Collection "Gérer l'Environnement" (Vol.6). Presses Polytech. et Univ. Romandes, Lausanne, 335 pp., (in French).

Oda, M., 1986, An equivalent continuum model for coupled stress and fluid flow analysis in jointed rock masses, *Water Resources Research*, **22**(13):1845–1856.

Papoulis, A., 1964, *Probability, Random Variables, and Stochastic Processes*, McGraw Hill, 583 p.

Papoulis, A., and Unnikrishna Pillai, S., 2002, *Probability, Random Variables, and Stochastic Processes*, Mc-Graw Hill, 4th edition, 852 pp.

Parlange, J. -Y., Barry, D. A., and Haverkamp, R., 2002, Explicit infiltration equations and the Lambert *W*-function, pp. 1119–1124, in: *Advances in Water Resources*, 25(8–12):859–1396. Special 25th Anniv. Issue, C. T. Miller, M. B. Parlange, and S. M. Hassanizadeh (eds.).

Parlange, J.-Y., Lisle, I., Braddock, R. D., and Smith, R. E., 1982, The three-parameter infiltration equation, *Soil Science Society of American Proceedings*, **5**:133, **6**:337–341.

Philip, J. R., 1969, Theory of infiltration, *Advances in Hydroscience,* Volume 5, New York, NY: Academic Press,

Polmann, D. J., McLaughlin, D., Gelhar, L.W., and Ababou, R., 1991, Stochastic modeling of large-scale flow in heterogeneous unsaturated soils, *Water Resources Research*, **27**(7):1447–1458.

Qahman, K., 2004, *Optimal management of coastal aquifers hindered by seawater intrusion: regional model for the Gaza aquifer, Palestine*, Doctoral Thesis, Water and Envir. Research Unit, Civil Engg. Dept., Mohammadia School of Engineers (EMI), University Mohammed V-Agdal, Rabat, Morocco.

Qahman, K, and Larabi, L., 2005, Evaluation and numerical modeling of seawater intrusion in the Gaza aquifer (Palestine), *Hydrogeology Journal* DOI 10.1007/s10040-005-003-2.

Renard, Ph., and Marsily, G. (de), 1997, Calculating equivalent permeability: a review, *Advances in Water Resources*, **20**(5–6):253–278.

Richards, L. A., 1931, Capillary conduction of liquids through porous medium, *Physics*, **1**:318–333.

Rubin, Y., 2003, *Applied Stochastic Hydrogeology*, Oxford University Press.

Sahimi, M., 1993, Flow phenomena in rocks: From continuum models to fractals, percolation, cellular automata, and simulated annealing, *Review of Modern Physics,* 65(4):1393–1533.

Sahimi, M., 1995, *Flow and Transport in Porous Media and Fractured Rock (from Classical Methods to Modern Approaches)*, VCH, Weinheim and New York.

Shvidler, M. I., 1985, *Statistical Hydrodynamics of Porous Media*. Nedra, Moscow (USSR), 288 pp., (in Russian).

Snow, D. T., 1969, Anisotropic permeability of fractured media, *Water Resources Research*, **5**(6):1273–1289.

Stietel, A., Millard, A., Treille, E., Vuillod, E., Thoraval, A., and Ababou, R., 1996, Continuum representation of coupled hydro-mechanical processes of fractured media: Homogenisation and parameter identification, developments, in: *Geotechnical Engineering: Coupled Thermo-Hydro-Mechanical processes (DECOVALEX project for radioactive waste repositories),* O. Stephansson, L. Jing, and C-F.Tsang, eds., Elsevier, **79**:135–164

Tarantola A., 1987, *Inverse Problem Theory, Methods for Data Fitting and Model Parameter Estimation*, Elsevier, Amsterdam.

Tarantola A., 2005, *Inverse Problem Theory and Methods for Model Parameter Estimation,* SIAM, Philadelphia, 342 p.

Terzaghi, Von K., 1936, The shearing resistance of saturated soils and the angle between the planes of shear, *First Int. Conf. Soil Mech.*, Harvard University, Vol. 1, pp. 54–56.

Tompson, A. F. B., Ababou, R., and Gelhar, L. W., 1989, Implementation of the three-dimensional turning bands random field generator, *Water Resources Research*, **25**(10): 2227–2243.

Trégarot, G., 2000, *Modélisation Couplée des Ecoulements à Saturation Variable avec Hétérogénéités, Forçages, et Interfaces Hydrologiques*, Thèse de Doctorat de l'Institut National Polytechnique de Toulouse: No. 1663 (dir.: R. Ababou), Institut de Mécanique des Fluides de Toulouse, France, May 2000, 496 pp. (in French).

Vanmarcke, E., 1983, *Random Fields (Analysis and Synthesis)*, Massachusetts Institute of Technology Press, Cambridge, 382 p.

Zijl, W., and Nawalany, M., 1993, *Natural Groundwater Flow*. Lewis Pub., Boca Raton, 321 p.

Part IV
Sustainable Measures to Palliate the Overexploitation of Groundwater and Decontaminate Aquifers

Chapter 9
Coastal Aquifers and Saltwater Intrusion

Christophe J. G. Darnault*, Itzel G. Godinez

Abstract As a large portion of the world's population (e.g. about 70%) dwells in coastal zones, sustainable and optimal coastal aquifer exploitation that controls saltwater intrusion is necessary. Especially in coastal zones, groundwater overexploitation has disturbed the naturally established equilibrium between saltwater and freshwater which has resulted in uncontrolled saltwater encroachment into coastal aquifers as well as in groundwater uses impairment and environmental degradation. This chapter presents the essential elements of the saltwater intrusion phenomenon related to coastal aquifer exploitation, such as hydrodynamics, hydrogeochemistry, geophysics, as well as optimization models to achieve a sustainable and conservative management of coastal aquifers.

Keywords Coastal aquifer, saltwater intrusion, hydrodynamics, hydrochemistry, geophysics, aquifer management, optimization methods

9.1 Introduction

Approximately 78% of the total freshwater on Earth is located and stored in the subsurface. Nevertheless, despite the abundance of groundwater resources, uncontrolled and unregulated extraction of the resource can contribute to changes in the aquifer systems. In coastal areas for instance, overexploitation has disturbed the naturally established equilibrium between seawater and freshwater resulting in uncontrolled saltwater encroachment into coastal aquifers. As a large portion of the world's population (e.g. about 70%) dwells in coastal zones, it is therefore important to develop optimal exploitation of groundwater and to control saltwater intrusion (Bear et al., 1999).

Christophe J. G. Darnault, Itel G. Godinez
University of Illinois at Chicago, Department of Civil and Materials Engineering, 842 W. Taylor St., Chicago, IL 60607, U.S.A.
* To whom correspondence should be addressed. E-mail: darnault@uic.edu

C.J.G. Darnault (ed.), *Overexploitation and Contamination of Shared Groundwater Resources.*

9.2 Hydrodynamics

Understanding of the saltwater intrusion process is essential for sustainable management and conservation of coastal aquifer resources. It is therefore important to characterize groundwater flow to understand the interactions existing between saltwater and freshwater, especially in the transition zone generated by these two fluids when coming into contact with each other. Brackish and saline groundwater can be found in coastal aquifers, but also further inland (Oude Essink, 2001). The main actors that contribute to saltwater intrusion and affect coastal aquifer systems are

- Saltwater encroachment in coastal aquifers;
- Characteristics of aquifer formations;
- Human activities producing saline waste;
- Tidal effects;
- Freshwater head fluctuations.

Numerous coastal aquifers in the world are severely impacted or threatened by saltwater intrusion. Coastal aquifers located within the zone of influence of mean sea level (MSL) are also threatened by an accelerated rise in global mean sea level (Oude Essink, 2001). The Intergovernmental Panel on Climate Change (IPCC) predicts that by the end of this century, an increase in the global mean sea level of approximately 50 cm will be attained. The later situation places coastal aquifers located within the mean sea level influence zone at greater risk since the process of saltwater intrusion will be accelerated.

9.2.1 Saltwater–Freshwater Interactions

W. Badon Ghijben (1888/1889) and A. Herzberg (1901) discovered that the saltwater occurrence underground was not at sea level but at a depth beneath sea level of approximately 40 times the height of freshwater above sea level. The hydrostatic equilibrium existing between saltwater and freshwater contributes to this phenomenon. The resulting expression derived to explain this phenomenon is known as the Ghijben-Herzberg principle.

The balanced pressures between the two liquids are defined as:

$$\rho_s gz = \rho_f g(z + h_f) \tag{1}$$

where ρ_s and ρ_f are saline and freshwater densities, respectively, g represents the acceleration due to gravity, h_f and z are as depicted in the figure above. Solving Eq. (1) for z provides the Ghijben-Herzberg principle. The saltwater–freshwater intrusion process in an unconfined aquifer is depicted in Figure 9.1.

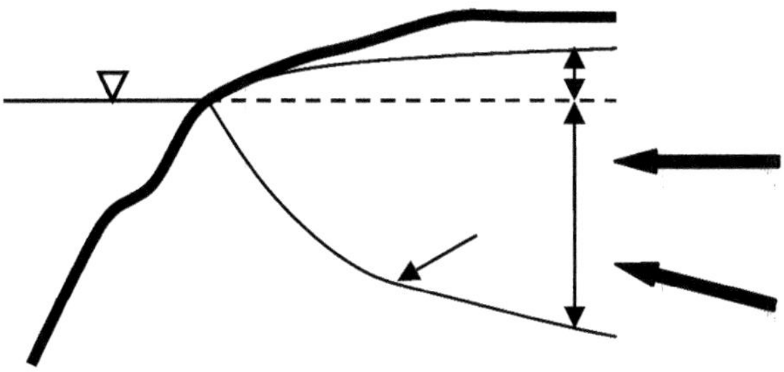

Fig. 9.1 Saltwater–freshwater intrusion in an unconfined aquifer

The Ghijben-Herzberg principle states that the depth to the saltwater–freshwater interface z, as measured from MSL, is given by

$$z = \frac{\rho_f}{\rho_s - \rho_f} h_f \approx 40 h_f \tag{2}$$

where h_f is the water table elevation above MSL. Equation (2) is valid under the existence of horizontal flow in the freshwater zone and the lack of progressive movement of saline water. Equation (2) can be applied in the case of confined aquifers by replacing water table by piezometric surface. From the Ghijben-Herzberg relation, it can be noted that the fresh–salt water equilibrium requires two conditions with respect to the water table or piezometric surface: it must lie above sea level and have a slope downward toward the ocean (Todd and Mays, 2005). If these conditions are not established, saltwater will move inland.

In the case of non-equilibrium conditions, where underlying seawater is in motion with heads above and below MSL, h_f and h_s respectively (Figure 9.2), Lusczynski (1961) generalized the Ghijben-Herzberg relation as:

$$z = \frac{\rho_f}{\rho_s - \rho_f} h_f - \frac{\rho_f}{\rho_s - \rho_f} h_S \tag{3}$$

Equation (3) determines the saline–freshwater interface depth from water level measurements in observation wells. These measurements require that observation wells be placed close to each other where a first well penetrates the saltwater region and a second well enters the freshwater region (Figure 9.2).

Analytical solutions of the flow corresponding to saltwater intrusion have been developed. Glover derived a solution where the aquifer is homogeneous, confined at the top, and unbounded at the bottom, and the saltwater is stagnant (Cheng and Ouazar, 1999). The depth to the interface for this situation is determined by:

$$z^2 = \frac{2\rho_f q x}{\Delta\rho K} + \left(\frac{\rho_f q}{\Delta\rho K}\right)^2 \tag{4}$$

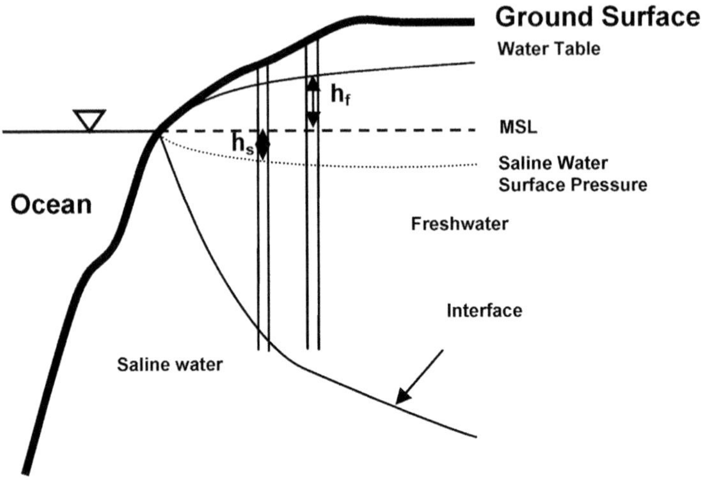

Fig. 9.2 Unconfined coastal aquifer with non equilibrium conditions between saline and freshwater

where q represents the freshwater flow per unit length of shore line, $\Delta\rho = \rho_s - \rho_f$ and K is the hydraulic conductivity. The width x_o of the submarine zone through which freshwater discharges into the sea can be obtained from Eq. (4) by setting z equal to 0; the latter results in Eq. (5). The depth of the interface below the shoreline z_o takes place where x is zero and is defined by Eq. (6).

$$x_o = \frac{\rho_f q}{2\Delta\rho K} \tag{5}$$

$$z_o = \frac{\rho_f q}{\Delta\rho K} \tag{6}$$

9.2.2 Structure of Saltwater–Freshwater Transition Zone

The sharp interface boundary between saltwater and freshwater does not occur under ordinary field conditions. Instead, a brackish transition zone of finite thickness separates the two fluids (Todd and Mays, 2005). The different factors that contribute to the increase in thickness of the transition zone are freshwater dispersion by flow, tides effect, groundwater recharge and wells pumping.

Groundwater abstraction from a coastal aquifer induces groundwater head decreases in both freshwater and saltwater bodies, which result not only in drawdown head depressions and submarine groundwater discharge reductions into coastal zones, but also in increased saltwater inflow and modified isocone contour surfaces which progressively increase the mixing zone thickness (Figure 9.3) (Custodio, 2002). The salinity distribution established over time depends on coastal aquifer characteristics and location of groundwater extraction.

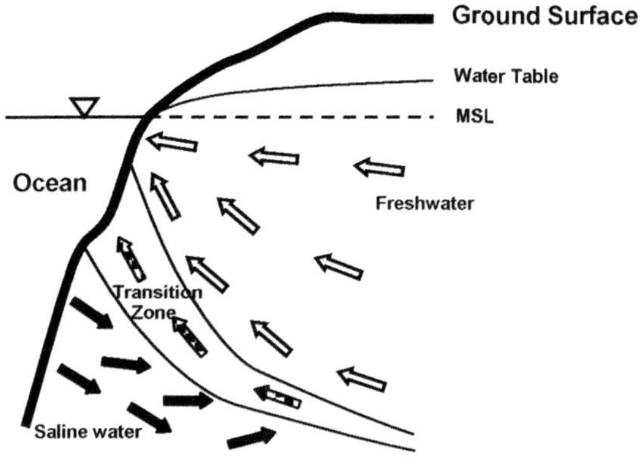

Fig. 9.3 Saline–freshwater transition zone in unconfined coastal aquifer

9.2.3 *Upconing of Saline Groundwater*

In areas where saline groundwater is present below fresh groundwater, the interface between fresh and saline groundwater may rise when piezometric heads are lowered due to well extraction and lead to an upconing process (Oude Essink, 2001). Continued abstraction of fresh groundwater draws the interface up to higher levels until coming into contact with the bottom of the well. In this case, brackish water will be pumped out of the well forcing it to close due to the degradation of the abstracted groundwater. In order to control the upconing process, the discharge rate has to be limited.

An analytical solution for ultimate rise upcoming below a pumping well can be derived based on the Dupuit assumptions and the Ghijben-Herzberg principle. The ultimate rise of the interface is expressed as:

$$z = \frac{Q}{2\pi dK\left(\frac{\Delta\rho}{\rho_f}\right)} \quad (7)$$

where d is the distance between the initial interface and the bottom of the well, Q is the pumping rate and $\Delta\rho = \rho_s - \rho_f$. The critical rise must have a z to d ratio between 0.3 and 0.5. Adopting the upper limit of $z/d = 0.5$, the maximum permissible pumping rate, Q_{max}, without salt entering the well is:

$$Q_{\max} \le \pi d^2 K\left(\frac{\Delta\rho}{\rho_f}\right) \tag{8}$$

9.2.4 *Effects of Mean Sea Level Rise*

The possible impacts of mean sea level rise on coastal water resources systems are listed in Table 9.1 and illustrated in Figure 9.4 (Oude Essink, 2001). The effects of mean sea level rise should be analyzed in relation to human activities. The mean sea level is a slow process occurring over a period of decade while groundwater overexploitation induces saltwater intrusion in a couple of years.

Table 9.1 Impacts of mean sea level (MSL) rise on coastal water resources (Adapted from Oude Essink, 2001)

Causes	Possible Effects or Consequences
Storm surges	Inundation of populated coastal zones.
Coastal erosion	Sand dunes reduction along aquifers coastline.
River bed elevation and sea level rise	Saltwater intrusion in aquifer adjacent to rivers.
	Saltwater intrusion impacts on beneficial uses.
Accelerated saltwater intrusion	Decrease in fresh groundwater resources.
	Transition zone between seawater and freshwater moves further inland where extraction wells may be located.

9.3 Hydrogeochemistry

Saltwater intrusion involves mixing between saline and freshwater systems. Because of its significant salt content, a small fraction of saltwater dominates the chemical composition of groundwater mixture (Jones et al., 1999). Groundwater quality deteriorates when fresh groundwater is mixed with 2% saltwater (Custodio, 2002). If saltwater fraction is 4%, significant groundwater use impairments and environmental impacts occur. If saltwater mixing is 6%, freshwater groundwater is almost unusable for human consumption and may only be used for industrial purposes.

Early detection of the sources that contribute to salinity increase in coastal aquifers is important. Saltwater movement inland can be the main reason for salinity increase, however, other sources may be responsible for brackish water (Custodio, 2002):

- Unflushed old marine water;
- Intense evapo-concentration of surface and phreatic water in dry climates;

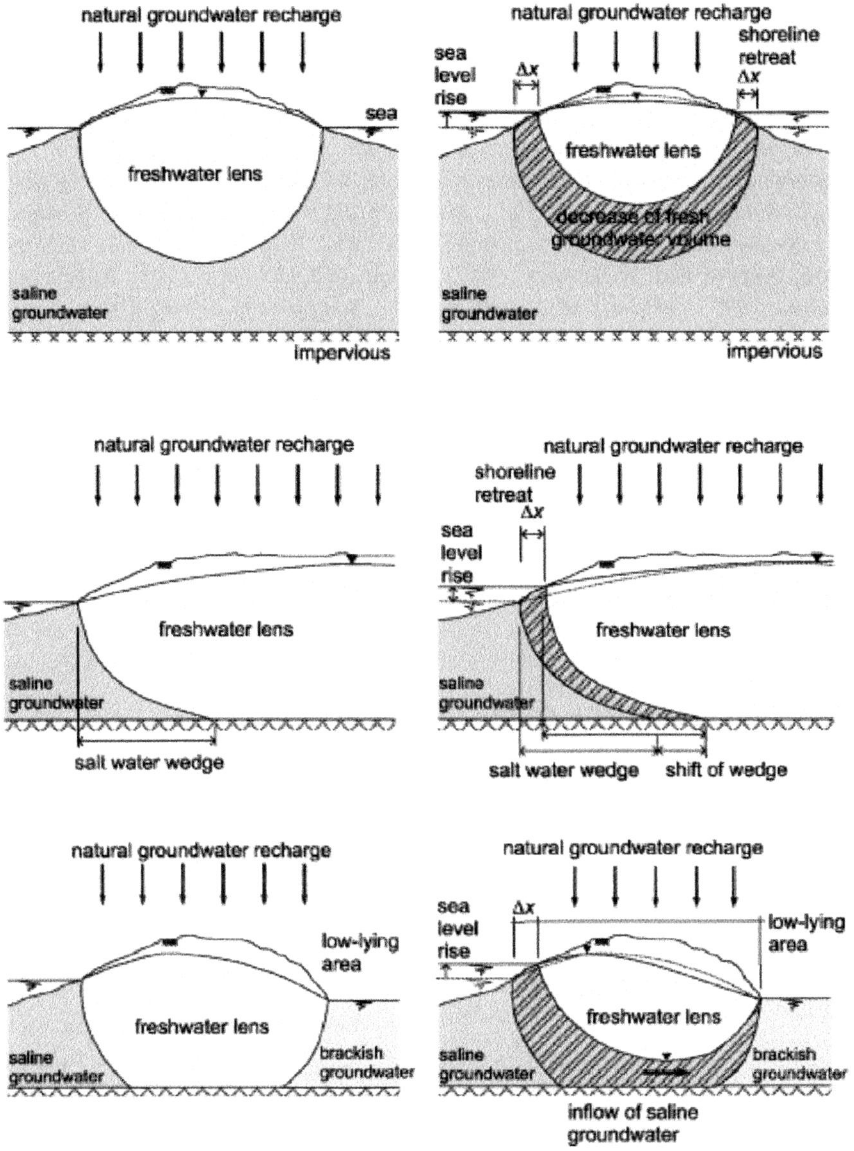

Fig. 9.4 Effects of a sea level rise on coastal aquifers (From Oude Essink, 2001. Reprinted from Ocean & Coastal Management with permission of Elsevier. Copyright 2001)

- Dissolution of evaporite salts existing in geologic formations;
- Seawater spray in windy coastal strips;
- Evaporation of outflowing groundwater in discharge areas and wetlands;
- Displacement of saline groundwater contained;

- Pollution by saline water resulting from drainage mining, leakage from industrial site using brackish water, effluent from softening, de-ionization and desalination plants or dissolution of de-icing road salts.

Jones et al. (1999) performed a critical review on the chemical composition of saline groundwater. The chemical composition of saline groundwater deviates from simple conservative saltwater–freshwater mixing to water-rock interactions, ion exchange with clay materials, carbonate dissolution-precipitation processes and contamination by subsurface brines (Mercado, 1985; Fidelibus and Tulipano, 1986; Appelo and Willemsen, 1987; Appelo and Geirnart, 1991; Appelo and Postma, 1993; Vengosh and Rosenthal, 1994; Sukhija et al., 1996). Chemical and isotopic parameters which behave conservatively (e.g. Cl, Br, deuterium) can be used to estimate the contribution of the different sources, while others give information on the extent of the interactions with the solid matrix (e.g. B and Sr isotopes) (Jones et al., 1999).

Four basic reaction processes are associated with the hydrologic environmental characteristics of saltwater intrusion. These include: mixing of groundwaters (including fluids associated with evaporites) and saltwater, carbonate precipitation and/or diagenesis (e.g. dolomitization), ion exchange and silicate (largely clay) diagenesis and redox reactions (Jones et al., 1999).

9.3.1 Mixing of Groundwater and Seawater

Mixing of normal dilute groundwater and saltwater may be achieved and traced using halogens. However, the presence of evaporitic conditions near the edge of the sea or the outflow of basinal brines from depth, present difficulties with hypersaline fluids and chloride possible non-conservancy (Jones et al., 1999).

9.3.2 Ionic Exchange Reactions

Cation exchangers in aquifers such as clay minerals, organic matter, oxyhydroxides or fine-grained rock materials have mostly Ca^{2+} adsorbed on their surfaces. In contrast, sediments in contact with saltwater have Na^{+} as the most prevalent sorbed cation (Sayles and Mangelsdorf, 1977; Jones et al., 1999). The ion exchange reactions between Ca^{2+} and Na^{+} takes place when saltwater enters a coastal freshwater aquifer, where Na^{+} substitutes partially Ca^{2+} in the aquifer. This reaction is described as:

$$Na^{+} + \frac{1}{2}\text{Ca-X}_2 \rightarrow \text{Na-X} + \frac{1}{2}Ca^{2+} \tag{9}$$

where X represents the natural exchanger. In such reactions, Na^+ is taken up by the solid phase, Ca^{2+} is released, and the solute composition changes from NaCl to $CaCl_2$ type water (Custodio, 1987; Appelo and Postma, 1993; Jones et al., 1999). The chloride ion concentration is taken as a reference parameter. Therefore, as saltwater intrudes coastal freshwater aquifers, Na/Cl ratio decreases and (Ca + Mg)/Cl ratio increases (Jones et al., 1999). If there is an inflow of freshwater, a reverse process of the ion exchange reaction above occurs:

$$\frac{1}{2}Ca^{2+} + NaX \rightarrow \frac{1}{2}CaX_2 + Na^+ \tag{10}$$

The mixing zone flushing by freshwater will result in Ca^{2+} and Mg^{2+} uptake by exchangers with Na^+ release. This is observed in the increase of the Na/Cl ratio and a decrease of the (Ca + Mg)/Cl ratio value, and the formation of $NaHCO_3$-type fluids (Jones et al., 1999). Groundwater characteristics may indicate oscillations of the saltwater–freshwater mixing while the change in ionic ratios may reflect ion exchange reactions.

Appelo and Postma (1993) suggest that the water type changes in the saltwater–freshwater transition zone are linked to the amounts and concentrations of exchangeable cations in solution. When diffusion is the primary process by which saltwater intrusion takes place and the cations exchange capacity is low, the effects of ion exchange are reduced leading to a simple mixture of brackish and freshwater as water composition.

9.3.3 Carbon Precipitation and/or Diagenesis

Coastal aquifers are generally composed of carbonate materials. The mixing of diluted groundwater in equilibrium with calcite in saltwater that is supersaturated with respect to calcite can produce a solution of intermediate composition and undersaturated with respect to calcite (Plummer et al., 1975; Wigley and Plummer, 1976). The degree of calcite undersaturation and dolomite supersaturation resulting from this mixing depends on temperature, chemical equilibria, ionic strength, and partial pressure of CO_2 of fresh groundwater during the early evolution in the vadose zone.

The dolomitization process, in which calcite and dolomite are in equilibrium can be described as:

$$2CaCO_3 + Mg^{2+} \rightarrow CaMg(CO_3)_2 + Ca^{2+} \tag{11}$$

It results in a progressive enrichment of Ca over Mg in solution (Jones et al., 1999).

9.3.4 Redox Reactions

Redox reactions can contribute to geochemical changes with saltwater intrusion through early diagenetic reactions involving organic matter and sulfur. The interpretation of salinization process should include geological and hydrochemical criteria. The geochemical criteria that may be selected to determine the origin of saltwater intrusion are salinity, Cl/Br ratio, Na/Cl ratio, Ca/Mg ratio, Ca/(HCO_3 + SO_4) ratio, O and H isotopes and Boron isotopes (Jones et al., 1999).

9.4 Geophysics

For coastal and island environments, the physical properties of interests are bulk conductivity and seismic velocity (Stewart, 1999). Bulk conductivity is a macroscopic property of the fluid/matrix system. Seismic velocity is related to the mechanical properties of materials such as elastic constants (Telford et al., 1990; Stewart, 1999). The coastal aquifer can be characterized by integrating surface and borehole geophysics with additional site data (Paillet, 2004). Paillet (2004) considers a five-step conceptual tool for coastal aquifer geophysics characterization (Table 9.2).

Table 9.2 Conceptual tool used in the integration of surface and borehole geophysics (Adapted from Paillet, 2004)

Conceptual Tool	Geophysical Measure	Example of Application
Scale of investigation	Vertical distribution of electrical Conductivity	Comparison of number and thickness of layers used in inversion of electromagnetic soundings with local structure given by induction log.
Regression of data	Electrical conductivity of formation around borehole	Comparison of induction log to water sample data for screened intervals to calibrate electric surveys in water-quality units.
Multivariate interpretation	Electrical conductivity of water-bearing sediments	Use of logs to identify the lithology to differentiate the effects of mineral matrix conductivity from the effects of pore water salinity on formation electrical conductivity.
Inversion methods	Inversion model structure and minimizing residuals	Comparison of logs to results of inversion models to define optimum size and shape of model layers or cells.
Verification boreholes	Subsurface structure from surface soundings	Use of log rotary-drilled boreholes to determine local depth and thickness of geo-electric layers as inferred from electromagnetic sounding interpretations.

9.5 Coastal Aquifer Management

The coastal aquifer management approaches are similar to those used by inland aquifer management. In both coastal and inland systems, the aquifer management is implemented within a framework establishing objectives, policies, and constraints (Bear, 2004). The main difference between coastal and inland aquifer managements is that in the former saltwater movement inland must be considered. Excessive extraction of fresh groundwater in coastal aquifers can result in saltwater intrusion, and affect not only the aquifer exploitation, but also the quality of the water extracted.

9.5.1 Efective Management

Coastal aquifer management aims at developing a strategy to maximize exploitation of a coastal aquifer to satisfy the water needs of different stakeholders and controlling the saltwater intrusion process and effects. Spatial distribution of the extraction wells and a pumping program need to be established to control aquifer salinity at locations closest and adjacent to the sea/ocean boundary. Constraints such as volume of water extracted, water quality, economics, operational restrictions, environmental and social-economic costs, combination of using surface and subsurface waters need also to be assessed. Effective management should include (Custodio, 2002; Bear, 2004):

- Incorporation of land use management;
- Integration of subsurface water resources, if present;
- Well established and properly equipped water management institutions;
- Governmental and society's will to achieve sustainability;
- Stakeholders' effective participation in management;
- Education, training and dissemination of knowledge and data;
- Correct understanding of aquifer behavior;
- Proper monitoring systems with early diction signals.

Most coastal aquifers suffer from intense population. This implies that large surface areas are covered by concrete and asphalt, resulting in a significant reduction in natural recharge and in severe groundwater quality deterioration due to contaminants present at the ground surface (Bear, 2004). It is evident that what occurs above the ground surface affects the underlying aquifer. Therefore, management of land-use needs to be incorporated as part of an effective coastal aquifer management.

In case a coastal region features surface water resources, these need to be incorporated along with groundwater resources to create a comprehensive and integrated management approach with equitable and suitable water resources allocation (Bear, 2004). Water allocation should establish a recycling process that can minimize the waste of the resource. Reclaimed sewage water, pumped saline groundwater, and desalinized seawater should be considered as potential water sources during water resources allocation planning in coastal systems.

Management institutions must include qualified and experienced professionals of diverse disciplines as many of the aquifer management challenges arise from conflicts resulting from multiple operation objectives. Therefore, a multi-disciplinary group can target the different aspects involving management as well as formulate coherent ideas to resolve conflicts. This implementation creates a water development program capable of protecting the subsistence of coastal aquifer resources.

Coastal aquifer management is not only a technical issue, but also an administrative, social and legal issue. Aquifer water users have important roles once common goals and individual rights are established. Intensive cooperation between governmental institutions and stakeholders is crucial to restrain the water extraction per capita. All parties must be involved in the management process to discuss their needs and concerns as well as to contribute with feasible problem solutions. Educating, training and exchange of data and information in regards to the designated program to administer the coastal aquifer can result beneficial for all parts implicated. Public institutions and stakeholders should participate in a collaborative approach to achieve a sustainable coastal aquifer management (Bear, 2004).

9.5.2 *Optimization Models*

The exploitation of coastal aquifers hydraulically connected with saltwater bodies may result in saltwater intrusion. Efficient management strategies are needed for optimal and sustainable water withdrawals from coastal aquifers, while maintaining salt concentration under specific limits (Maimone et al., 2004). To achieve optimal and sustainable extraction of groundwater, Cheng et al. (2004) consider the following issues for coastal aquifer management:

- For existing wells: how should pumping rate be apportioned and regulated to achieve total maximum extraction?
- For new wells: where should they be located and how much can they pump?
- How can recharge wells and canals be used to protect pumping wells, and where should they be located?
- If recycled water is used in the injection, how recovery percentage could be maximized?

Physical processes governing groundwater flow and transport in a coastal aquifer should be simulated with accuracy and reliability to obtain meaningful optimal management strategies. Therefore, a saltwater intrusion simulation model is incorporated within the management model to ensure the feasibility of the obtained optimal strategies (Bhattacharjya and Datta, 2005). The flow and transport simulation processes involved in coastal aquifers are based on the density-dependent flow and transport processes. The incorporation of this simulation model into an optimization-based management model is complex (Das and Datta, 2000; Bhattacharjya and Datta, 2005).

The physical processes governing saltwater intrusion in coastal aquifers are well understood and governing equations of flow and salinity transport in coastal aquifers have been developed and solved numerically using different methods (Huyakorn et al., 1987; Andersen et al., 1988; Essaid, 1990; Bear et al., 1999; Gambolati et al., 1999; Oude Essink, 2001). A common approach for saltwater intrusion simulation in coastal aquifers is based on the sharp interface approximation and the Ghyben–Herzberg relation (Bear, 1979; Essaid, 1990; Bear et al., 1999; Essaid, 1999). Emch and Yeh (1998), Cheng and Ouazar (1999), Cheng et al. (2000) and Mantoglou (2003) presented models of coastal aquifer management based on the sharp interface approximation. Strack's (1976) flow potential is used to trace the toe of saltwater lens (Cheng and Ouazar, 1999; Mantoglou, 2003). More complex saltwater intrusion models integrate the transport processes that occur in the mixing zone (Das and Datta, 1999a,b; Gordon et al., 2000). Model to describe saltwater intrusion should be three-dimensional, transient, and account for varied densities and for dispersion (van Dam, 1999).

The saltwater intrusion mechanism in coastal aquifer involves generally an outflow of freshwater zone, overlaying an invasion of saltwater zone (Qahman et al., 2005). The region in between these two areas is referred to as the transition zone. According to the relative thickness of the transition zone, two methods can be employed to model the process. When the transition zone is abrupt and narrow in thickness, the saltwater intrusion phenomena is modeled as a two-phase fluid flow separated by a sharp interface (Qahman et al., 2005). On the other hand, when the transition zone is gradual and spans the thickness of the aquifer, the saltwater intrusion is modeled using the density-dependent miscible flow and transport approach (Qahman et al., 2005).

Optimization aims at maximizing an objective function subject to constraints that protect and conserve the aquifer resources in terms of quality and quantity (Mantoglou et al., 2004). The differential flow and transport equations are solved using analytical solutions or numerical simulation to represent the aquifer response to various pumping scenarios. The resulting optimization problem may be linear (Ahlfeld and Heidari, 1994; Hallaji and Yazicigil, 1996; Mantoglou, 2003) or nonlinear (Gorelick et al., 1984; Shamir et al., 1984; Wang and Ahlfeld, 1994; Hallaji and Yazicigil, 1996; Emch and Yeh, 1998; Mantoglou, 2003). The linearity of the optimization problem, the continuity of the objective function, the availability of information for the derivatives and the existence of local minima determine the optimization method selected (Cheng et al., 2000).

Depending on the problem, various objective functions and constraints have been applied. The objective of optimization is usually to maximize the total pumping rate from wells while controlling saltwater intrusion into the aquifer. Shamir et al. (1984), Hallaji and Yazicigil (1996), Cheng et al. (2000) and Mantoglou (2003) aim at maximizing the total pumping rate, whereas Das and Datta (1999a) aim at minimizing of the salinity of pumped water. Emch and Yeh (1998) and Gordon et al. (2000) include the pumping cost in the objective function. It is also possible to consider multiple objectives, constituting a multi-objective optimization problem (Shamir et al., 1984; Emch and Yeh, 1998; Das and Datta, 1999b). The constraints control the well pumping rate between a minimum and a maximum value (Hallaji and Yazicigil, 1996; Emch and Yeh, 1998; Das and Datta, 1999b; Cheng et al., 2000). Constraints may also include the control of the location of the toe as well as maintaining water levels, flow potential or salt concentration of the pumped water at desired levels (Cheng et al., 2000; Mantoglou, 2003; Mantoglou et al., 2004).

Optimization models have been developed to assist management decision-makers. These optimizations incorporate physically based groundwater models and include the embedding, the response matrix, and the linked simulation–optimization approaches (Qahman et al., 2005, Bhattacharjya and Datta, 2005). Embedding technique and response matrix approaches are two methods that integrate the simulation model within the management model (Gorelick, 1983). Embedded optimization models use finite difference or finite element approximation of flow and transport equations as equality constraints within the management model, along with additional physical and managerial constraints. The embedding technique has several limitations for large-scale aquifer systems and is numerically inefficient when applied to large heterogeneous aquifer systems (Willis and Finney, 1988; Das and Datta 1999a,b; Bhattacharjya and Datta, 2005). The response matrix approach relies on the principle of superposition and linearity. However, this method is unsatisfactory for highly nonlinear systems (Rosenwald and Green, 1974). As an alternative to these approaches, a linked simulation–optimization approach may be applied for saltwater intrusion management model (Gorelick, 1983; Emch and Yeh, 1998). In the linked simulation–optimization approach, the management model is repetitively called to compute state variable values (groundwater head and concentration) and their gradients subject to the perturbation of control variables (decision). The performance of the linked simulation–optimization approach is highly dependent on the performance of the saltwater intrusion simulation model, as repetitive simulations are required to achieve an optimal management strategy.

References

Ahlfeld, D. P., and Heidari, M., 1994, Applications of optimal hydraulic control to ground-water systems, *J. Water Resour. Plan. Manag.* **120**(3):350–365.

Andersen, P. F., White, H. O. Jr., and Mercer, J. W., 1988, Numerical modeling of saltwater intrusion at Hallandale, Florida, *Ground Water* **26**(5):619–630.

Appelo, C. A. J., and Willemsen, A., 1987, Geochemical calculations and observations on salt water intrusions, a combined geochemical/mixing cell model, *J. Hydrol.* **94**:313–330.

Appelo, C. A .J., and Geirnart, W., 1991, Processes accompanying the intrusion of salt water, in: *Hydrogeology of Salt Water Intrusion—A Selection of SWIM Papers*, II:291–304.

Appelo, C. A. J., and Postma, D., 1993, *Geochemistry, Groundwater and Pollution*, Balkema, Rotterdam, pp. 536.

Badon Ghyben, W., 1888/1889, Nota in verband met de voorgenomen putboring nabij Amsterdam (Notes on the probable results of well drilling near Amsterdam), *Tijdschrift van het Koninklijk Instituut van Ingenieurs*, The Hague, **9**:8–22.

Bear, J., 1979, *Hydraulics of Groundwater*, Mcgraw-Hill, New York, U.S.A.

Bear, J., Cheng, A. H.-D., Sorek, S., Ouazar, D., and Herrera, I., 1999, *Seawater Intrusion in Coastal Aquifers: Concepts, Methods and Practices*, Kluwer, Dordrecht, pp. 1–625.

Bear, J., 2004, Management of a coastal aquifer, *Ground Water* **42**(3):317.

Bhattacharjya, R. K., and Datta, B., 2005, Optimal management of coastal aquifers using linked simulation optimization approach, *Water Resour. Manag.* **19**:295–320

Cheng, A. H. -D., and Ouazar, D., 1999. Analytical solutions, in: *Seawater Intrusion in Coastal Aquifers—Concepts, Methods and Practices,* J. Bear, A. H.-D., Cheng, S. Sorek, D. Ouazar, and I. Herrera, eds., Kluwer Academic Publishers, Dordrecht, The Netherlands, pp. 163–191.

Cheng, A. H.-D., Halhal, D., Naji, A., and Ouazar, D., 2000, Pumping optimization in saltwater-intruded coastal aquifers, *Water Resour. Res.* **36**(8):2155–2165.

Cheng, A. H. -D., Benhachmi, M. K., Halhal, D., Ouazar, D., Naji, A., and El Harrouni, K., 2004, Pumping optimization in saltwater-intruded aquifers, in: *Coastal Aquifer Management Monitoring, Modeling, and Case Studies*, A. H.-D. Cheng, and D. Ouazar, eds., Lewis Publishers, Boca Raton, Florida, U.S.A., pp. 233–256.

Custodio, E., 1987, Salt fresh water interrelationship under natural conditions, in: *Groundwater Problems in Coastal Aquifers*, UNECCO-IHP, pp. 14.

Custodio, E., 2002, Coastal aquifers as important natural hydrogeological structures, in: *Groundwater and Human Development*, M. Bocanegra and M. Massone, eds., Barcelone, pp. 1905–1918.

Dam, J. C. van., 1999, Exploitation, restoration and management, in: *Coastal Aquifers—Concepts, Methods and Practices*, J. Bear, A. H.-D. Cheng, S. Sorek, D. Ouazar, and I. Herrera, eds., Kluwer Academic Publishers, Dordrecht, The Netherlands, pp. 73–125.

Das, A., and Datta, B., 1999a, Development of management models for sustainable use of coastal aquifers, *J. Irrg. Drainage Eng.* **125**(3):112–121.

Das, A., and Datta, B., 1999b, Development of multiobjective management models for coastal aquifers, *J. Water Resour. Plan. Manag.* **125**(2):76–87.

Das, A., and Datta, B., 2000, Optimization based solution of density-dependent seawater intrusion in coastal aquifers, *J. Hydrol. Eng.* **5**(1):82–89.

Emch, P. G., and Yeh, W. W. -G., 1998, Management model for conjunctive use of coastal surface water and ground water, *J. Water Resour. Plan. Manag.* **124**(3):129–139.

Essaid, H. I., 1990, A multilayered sharp interface model of coupled freshwater and saltwater flow in coastal systems: model development and application, *Water Resour. Res.* **26**(7):1431–1454.

Essaid, H. I., 1999, USGS SHARP model, in: *Seawater Intrusion in Coastal Aquifers—Concepts, Methods and Practices,* J. Bear, A. H.-D., Cheng, S. Sorek, D. Ouazar, and I. Herrera, eds., Kluwer Academic Publishers, Dordrecht, The Netherlands, pp. 213–247.

Fidelibus, M. D., and Tulipano, L., 1986, Mixing phenomena owing to sea water intrusion for the interpretation of chemical and isotopic data of discharge waters in the Apulian coastal carbonate aquifer (southern Italy), *Proc. 9th Salt Water Intrusion Meeting*, Delft, pp. 591–600.

Gambolati, G., Putti, M., and Paniconi, C., 1999, Three-dimensional model of coupled flow and miscible salt transport, in: *Seawater Intrusion in Coastal Aquifers—Concepts, Methods and*

Practices, J. Bear, A. H.-D. Cheng, S. Sorek, D. Ouazar, and I. Herrera, eds., Kluwer Academic Publishers, Dordrecht, The Netherlands, pp. 315–362.

Gordon, E. E., Shamir, U., and Bensabat, J., 2000, Optimal management of a regional aquifer under salinization conditions, *Water Resour. Res.* **36**(11):3193–3203.

Gorelick, S. M., 1983, A review of distributed parameter groundwater management modeling methods, *Water Resour. Res.* **19**(2):305–319.

Gorelick, S. M., Voss, C. I., Gill, P. E., Murray, W., Saunders, M. A., and Wright, M. H., 1984, Aquifer reclamation design: The use of contaminant transport simulation combined with nonlinear programming, *Water Resour. Res.* **20**(4):415–427.

Hallaji, K., and Yazicigil, H., 1996, Optimal management of a coastal aquifer in southern Turkey, *J. Water Resour. Plan. Manag.* **122**(4):233–244.

Herzberg, A., 1901, Die Wasserversorgung einiger Nordseebder (The water supply of parts of the North Sea coast in Germany), *Z. Gasbeleucht. Wasserversorg* **44**:815–819 and **45**:842–844.

Huyakorn, P. S., Andersen, P. F., Mercer, J. W., and White, H. O., 1987, Saltwater intrusion in aquifers—development and testing of a 3-dimensional finite-element model, *Water Resour. Res.* **23**(2):293–312.

Jones, B. F., Vengosh, A., Rosenthal, E., and Yechieli, Y., 1999, Geochemical investigations, in: *Seawater Intrusion in Coastal Aquifers—Concepts, Methods and Practices*, J. Bear, A. H.-D. Cheng, S. Sorek, D. Ouazar, and I. Herrera, eds., Kluwer Academic Publishers, Dordrecht, The Netherlands, pp. 51–71.

Lusczynski, N. J., 1961, Head and flow of groundwater of variable density, *J. Geophys. Res.* **66**:4247–4256.

Maimone, M., Harley, B., Fitzgerald, R., Moe, H., Hossain, R., and Heywood, B., 2004, Coastal aquifer planning elements, in: *Coastal Aquifer Management Monitoring, Modeling, and Case Studies*, A. H.-D. Cheng, and D. Ouazar, eds., Lewis Publishers, Boca Raton, Florida, U.S.A., pp. 1–27.

Mantoglou, A., 2003, Pumping management of coastal aquifers using analytical models of saltwater intrusion, *Water Resour. Res.* **39**(12):1335, doi:10.1029/2002WR001891.

Mantoglou, A., Papantoniou, M., and Giannoulopoulos, P., 2004, Management of coastal aquifers based on nonlinear optimization and evolutionary algorithms, *J. Hydrol.* **297**:209–228.

Mercado, A., 1985, The use of hydrogeochemical patterns in carbonate sand and sandstone aquifers to identify intrusion and flushing of old saline water, *Ground Water* **23**:635–644.

Oude Essink, G. H. P., 2001, Improving fresh groundwater supply—problems and solutions, *Ocean Coast. Manag.* **44**(5):429–449.

Paillet, F. L., 2004, Integrating surface and borehole geophysics in the characterization of salinity in a coastal aquifer, in: *Coastal Aquifer Management Monitoring, Modeling, and Case Studies*, A. H.-D. Cheng and D. Ouazar, eds., Lewis Publishers, Boca Raton, Florida, U.S.A., pp. 167–181.

Plummer, L. N., Parkhurst, D. L., and Kosiur, D. R., 1975, MIX-2, A computer program for modeling chemical reactions in natural waters, *U.S. Geol. Surv. Water-Res. Inv. Rep.* **61–75**:1–68.

Qahman, K., Larabi, A., Ouazar, Naji, A., and Cheng, A. H.-D., 2005, Optimal and sustainable extraction of groundwater in coastal aquifers, *Stocheometry Environ. Res. Risk Assess.* **19**:99–110.

Rosenwald, G. W., and Green, D. W., 1974, A method for determining the optimal location of wells in a reservoir using mixed-integer programming, *Soc. Res. Eng. J.* **14**:44–54.

Sayles, F. L., and Mangelsdorf, P. C., 1977, The equilibrium of clay minerals with sea water; exchange reactions, *Geochim. et Cosmochim. Acta* **41**:951–960.

Shamir, U., Bear, J., and Gamliel, A., 1984, Optimal annual operation of a coastal aquifer, *Water Resour. Res.* **20**(4):435–444.

Stewart, M. T., 1999, Geophysical investigations, in: *Seawater Intrusion in Coastal Aquifers—Concepts, Methods and Practices,* J. Bear, A. H.-D., Cheng, S. Sorek, D. Ouazar, and I. Herrera, eds., Kluwer Academic Publishers, Dordrecht, The Netherlands, pp. 9–50.

Strack, O. D. L., 1976, A single-potential solution for regional interface problems in coastal aquifers. *Water Resour. Res.* **12**(6):1165–1174.

Sukhija, B. S., Varma, V. N., Nagabhushaman, P., and Reddy, D. V., 1996, Differentiation of paleomarine and modern intruded salinities in coastal groundwaters (of Karaikal and Tanjavur, India) based on inorganic chemistry, organic biomarker fingerprints and radiocarborn dating, *J. Hydrol.* **174**:173–201.

Telford, W. M., Geldart, L. P., and Sheriff, R. E., 1990, *Applied Geophysics*, Cambridge University Press, Cambridge, pp. 1–700.

Todd, D. K., and May, L. W., 2005, *Groundwater Hydrology*, Wiley, Hoboken, NJ, U.S.A., pp. 1–656.

Vengosh, A., and Rosenthal, A., 1994, Saline groundwater in Israel: its bearing on the water crisis in the country, *J. Hydrol.* **156**:389–430.

Wang, W., and Ahlfeld, D. P., 1994, Optimal groundwater remediation with well location as a decision variable: model development, *Water Resour. Res.* **30**(5):1605–1618.

Wigley, T. M. L., and Plummer, L. N., 1976, Mixing of carbonate waters, *Geochem. Cosmochem. Acta* **40**(9):989–995.

Willis, R., and Finney, B. A., 1988, Planning model for optimal control of saltwater intrusion, *J. Water Resour. Plan. Manag.* **114**(2):163–178.

Chapter 10
Karst Aquifers: Hydrogeology and Exploitation

Christophe J. G. Darnault*

Abstract Karstic terrains cover approximately 12% of the Earth's continental surface and 25% of the world's population is supplied partially or entirely by karst water resources. In the Mediterranean and Southeast Asia, karst aquifers are the primary water resources. Karst terrains results from karstification, a phenomenon of dissolution of carbonate rocks under the action of water followed by physico-chemical processes. Karst processes use the original discontinuities, fractures, joints, bedding planes or macroporosity, developing a hydraulic continuum from surface to spring. The exploitation of karst aquifers integrates the concept of the close relationship between the surface water and groundwater with the evolution of karst aquifer as the dissolution process progresses in order to manage and protect these water resources from overexploitation and pollution. This chapter presents the key elements related to the hydrogeology of karst aquifers and the karst water resources exploitation including methods of exploitation, impacts of exploitation such as overexploitation, salinization and sinkhole, and risk assessment and mapping for karst aquifer protection.

Keywords Karst aquifer, karst terrains, karstification, carbonate rocks dissolution, permeability, matrix flow, fracture flow, conduit flow, hydrodynamic functioning, spring hydrograph, Mangin's equation, exploitation methods, overexploitation, salinization, sinkhole, risk assessment and mapping

10.1 Introduction

Karstic terrains cover approximately 12% of the Earth's continental surface and 25% of the world's population is supplied partially or entirely by karst water resources (Ford and Williams, 2007; Williams, 1993). In the Mediterranean and

---

Christophe J. G. Darnault
University of Illinois at Chicago, Department of Civil and Materials Engineering, 842 W. Taylor St., ERF# 2071, Chicago, IL 60607, U.S.A.

* To whom correspondence should be addressed. E-mail: darnault@uic.edu

C.J.G. Darnault (ed.), *Overexploitation and Contamination of Shared Groundwater Resources.*

Southeast Asia, karst aquifers are the primary water resources. Karst features occur generally in carbonates rocks, limestones and dolomites. Most carbonate rocks outcrops and covered carbonate rocks undergo the karstification process during geological time. Carbonate rocks that contain groundwater can be considered as karst aquifers which can be exploited for water supply (Bakalowicz, 2005).

Karst terrains are classified as holokarst or fluviokarst. In holokarst terrain, closed depressions occur at various scales and precipitation is drained underground with little or no channelized flow (Cvijic, 1918; Jennings, 1985; White, 1988). Fluviokarst terrains are characterized by channelized surface water flow and karst landforms have generally been superimposed on a former fluvial landscape. Karst drainage networks are disrupted and few rivers can traverse these terrains in a continuous and non-segmented manner because of surface flow diversion into underground system. Rivers also lose water when river flow descends into swallow holes or swallet.

Karst processes use the original discontinuities, fractures, joints, bedding planes or macroporosity, developing a hydraulic continuum from surface to spring. Conduits range from several meters wide to kilometers longs in which flow conditions encountered may be similar to open channel flow characterized by free surface flow, high velocity and flow rate or to pipe flow occurring in phreatic conduits during flood (Bakalowicz, 2005). The conduit network pattern is determined by the type of porosity, the type of recharge and the hydraulic gradient (Palmer, 2000; Bakalowicz, 2005).

The exploitation of karst aquifers integrates the concept of the close relationship between the surface water and groundwater with the evolution of karst aquifer as the dissolution process progresses in order to manage and protect these water resources from overexploitation and pollution. Especially, the main engineering and management issues in karst aquifers exploitation are the successful drilling of high pumping rate wells and the high vulnerability of karst medium to pollution.

10.2 Karstification Process

10.2.1 Karstification

Karst terrains results from karstification. Karstification is the phenomenon of dissolution of rocks under the action of water followed by physicochemical processes (Bakalowicz, 1975). The dissolution phenomenon develops existing interstices into voids and induces an organized drainage structure (Bogli, 1980; Dreybrodt, 1988; Ford and Williams, 2007).

Karstification is mediated by water both in term of quality and quantity. The kinetics of chemical reaction is related to water quality and water flow velocity. Carbonate reactions involving carbon dioxide present in water and leading to rock dissolution are not instantaneous. Therefore, if the flow velocity is too high, the rock dissolution can not be completed over the short period of time; while if the flow velocity is too low, the rock dissolution physicochemical reaction stops as the water become saturated. An optimal flow velocity exists to create karst system. More rock dissolution will be achieved with more abundant water, however, more water will be drained in the large developing voids. These qualitative and quantitative conditions create an interrelation between the voids and the flow that results in karstic system hierarchically organized from upstream to downstream. The karst system will be defined by chemical and hydraulic mechanisms that occur according to available energy, the "karstification potential" with respect to water quantity CO_2 contents and strength of hydraulic potential (Mangin, 1994). The hierarchical structure of the karst system is determined by the flow mechanism while the chemical mechanism is responsible for the creation of voids.

10.2.2 Chemistry of Carbonate Rocks Dissolution

Karstic weathering results from the dissolution of carbonate rocks. The dissolution of limestone and dolomite in carbon dioxide water is described by the following reactions:

$$CaCO_3 + CO_2 + H_2O \rightarrow Ca_{2+} + 2HCO^{3-} \tag{1}$$

$$CaMg(CO_3)_2 + 2CO_2 + 2H_2O \rightarrow Ca^{2+} + Mg^{2+} + 4HCO_3^- \tag{2}$$

The development of karst aquifers is mostly based on the dissolution reaction than the final equilibrium state of these reactions. Plummer et al. (1978) defined the different chemical reactions as well as the individual rates to provide an overall dissolution of calcite. These reactions are:

$$CaCO_3 + H^+ \Rightarrow Ca^{2+} + HCO_3^- \tag{3}$$

$$CaCO_3 + H_2CO_3 \Rightarrow Ca^{2+} + 2HCO_3^- \tag{4}$$

$$CaCO_3 + H_2O \Rightarrow Ca^{2+} + HCO_3^- + OH^- \tag{5}$$

The Eqs. (3)–(5) are reactions described for a forward reaction term in the rate equation (Eq. 6).

$$Rate = k_1 a_{H^+} + k_2 a_{H2CO3} + k_3 a_{H2O} - k_4 a_{Ca^{2+}} a_{HCO_3^-} \tag{6}$$

The first term is mass-transfer controlled while the second and third terms are reaction rate-controlled. Therefore, in the pH range of karst groundwater, the dissolution rate is dependent on the flow regime. A generic rate equation has been developed to investigate the carbonate dissolution kinetics (Palmer, 1991; Dreybrodt and Buhmann, 1991):

$$Rate = \frac{A}{V}\frac{dC}{dt} = k\left(1 - \frac{C}{C_s}\right)^n \quad (7)$$

where A = area, V = volume of solution, k = reaction rate constant, C = concentration of dissolved carbonate, C_s = equilibrium saturation concentration for dissolved carbonate and n = reaction order.

The dissolution kinetics under turbulent flow and near equilibrium conditions demonstrate additional controls (Dreybrodt and Buhmann, 1991; Svensson and Dreybrodt, 1992; Dreybrodt et al., 1996; Liu and Dreybrodt, 1997):

- Hydration reaction of aqueous CO_2 to H_2CO_3 is rate-controlling when A/V ratio is high and under turbulent flow conditions;
- Adsorption of ions on reactive surface becomes rate-controlling under near-saturation conditions;
- Diffusion boundary layer becomes important under turbulent flow conditions.

10.3 Karst Hydrogeology

10.3.1 Karst Aquifers Classification, Characteristics and Development

The classification of karst aquifers is based on recharge types, flow media, flow types, conduit network topology, stores and storage capacity, and outflow response to recharge (Ford and Williams, 2007). The infiltration that will recharge the karst aquifers can be either diffuse or concentrated. Similarly, the flow types can be in the form of diffuse of concentrated circulation (Burdon and Papakis, 1963). A conceptual classification using a three-dimensional field model that incorporate recharge, storage and transmission in karst aquifers was developed by Hoobs and Smart (1986). By analyzing the spring hydrograph response to recharge which translate the structure and transfer functions of a karst drainage system, Bakalowicz and Mangin (1980) were able to provide a measure of the aquifer karstification, ranging from well-developed speleological networks to carbonate terrains.

A karst aquifer system can be described as an open system with boundary defined by the catchment basin limits and with input, throughput, and output flows, mechanisms and controls (Ford and Williams, 2007). When karst terrains are found within the catchment basin and the recharge is only resulting from precipitation falling directly on them, then autogenic recharge occurs. If runoff from neighboring or overlying non-karst terrains drains into the kart aquifer, allogenic recharge takes place. Autogenic recharge is usually diffuse across karst outcrop while allogenic recharge is normally concentrated as point-inputs of sinking streams. Karst springs represent the termination of underground river systems and signal the point where surface fluvial processes become dominant.

The water table elevation at the output of the aquifer is controlled by the vertical position of the spring. The water table slope upstream and its variation under different discharge scenarios is determined by the hydraulic conductivity and the throughout discharge. At the basin scale, the flow direction is determined by the hydraulic gradient direction, however, at the local scale, pathways made available by interconnected fissures and pores control the flow direction. Kiraly (2003) stated that groundwater flow depends on hydraulic parameters and boundary conditions, and that other factors such as geology, geomorphology and climate will exert their influence on groundwater flow solely through the hydraulic parameters and the boundary conditions.

Several methods were proposed for classifying karst aquifers, including hydrodynamic functioning (Mangin, 1975a,b,c; Smith and Atkinson, 1976; Bakalowicz and Mangin, 1980) and water geochemistry (Shuster and White, 1971; Smith and Atkinson, 1976; Bakalowicz, 1977). The classification method proposed by Mangin (1975a,b,c) and detailed by Marsaud (1997) is the most complete because it integrates a quantitative characterization of the infiltration and the storage in the karst phreatic zone through the analysis of the recession curve of spring flood hydrographs.

10.3.2 Permeability

The movement of groundwater in karst terrains is influenced by the heterogeneity of its milieu, especially the variations in porosity and permeability. Karst aquifers are classified in three types based on the voids characteristics in which water is stored and transmitted through: intergranular pores within the unfractured bedrock (matrix permeability), joints and bedding planes imparted to the strata following deposition and lithification (fracture permeability), and conduits that have been enlarged by dissolution processes (conduit permeability) (White, 1999). As the dimensions of fractures and conduits may change over time, the resulting aquifer permeability will also evolve. The components and characteristics of the triple permeability model for karstic aquifers are presented in Table 10.1 (White, 2007).

Table 10.1 Components and characteristics of triple permeability model for karstic aquifers (Adapted from White, 2007)

Permeability	Dimension	Travel Time	Flow Mechanism	Distribution
Matrix	µm to mm	Long	Darcy law, laminar flow	Continuous medium
Fracture	10 µm to 10 mm	Intermediate	Cube law, usually laminar flow	Localized
Conduit	> 10 mm	Short	Darcy-Weisbach, open channel and pipe flow, turbulent flow	Localized

10.3.3 Groundwater Flow

10.3.3.1 Matrix Flow

The groundwater flow in limestone or dolomite bedrock is similar to matrix flow described through Darcy's law:

$$Q = -AK\frac{\Delta h}{\Delta L} = -\frac{A\rho g N d^2}{\eta}\frac{\Delta h}{\Delta L} \tag{8}$$

where Q is flow discharge, A is cross-sectional area of aquifer, K is hydraulic conductivity, ρ is water density, Nd^2 is permeability and η is water viscosity.

10.3.3.2 Fracture Flow

The fracture flow occurring in a facture with parallel walls and uniform aperture is described by the cubic law derived from Navier-Stokes equation:

$$Q = -\frac{W\rho g b^3}{12\eta}\frac{\Delta h}{\Delta L} \tag{9}$$

where W is total width of fractures and b is full fracture aperture.

An empirical approach developed by Witherspoon et al. (1980) incorporate the notion of non-uniform rock fracture by adding an empirical fraction factor f_f and integrating the constants of Eq. (9) into a single constant C:

$$Q = \frac{C}{f_f} b^3 \frac{\Delta h}{\Delta L} \tag{10}$$

where laboratory measurements of f_f range from 1.04 to 1.64.

10.3.3.3 Conduit Flow

As turbulent flow conditions are frequently encountered in pipes and fissures in karst and may be considered predominant in cave systems, groundwater flow in conduits can be treated as pipes flow (White, 2007). When increasing velocity, sinuosity and roughness may result in turbulent pipe flow, the specific discharge may be calculated using the Darcy-Weisbach equation (Thrailkill, 1968):

$$Q = \left(\frac{2dga^2}{f}\right)^{1/2}\left(\frac{dh}{dl}\right) \tag{11}$$

where f is a friction factor, g is gravitational acceleration, d is pipe diameter, and dh/dl is hydraulic gradient.

10.4 Karst Drainage Systems: Hydrological Analysis and Modeling

The characterization of the karst aquifers structure and properties is essential for water resources estimation, planning, protection and management. Although each karst aquifer is unique because of its geological context, storage and flow conditions, some structural linkages within karst drainage systems are widely found. The conceptualization of karst aquifer was first developed by White (1969) by focusing on geologic settings and their controlling influence on groundwater flow patterns. Further classification and integration of parameters have resulted in a conceptual three-dimensional field model of recharge, storage and transmission in karst aquifers (Hobbs and Smart, 1986). Karst drainage analysis involves the determination of the following elements (Ford and Williams, 2007):

- Areal and vertical extent of the system;
- Boundary conditions;
- Input and output sites and volumes;
- Storage and linkage interior structure;
- Storage capacities and physical characteristics;
- Flow pathways;
- Throughput rates;
- Response of storage and output to recharge;
- System response under various flow conditions.

The information to perform this karst drainage systems analysis can be established through five complementary approaches (Ford and Williams, 2007):

- Water balance estimation;
- Borefole analysis;
- Spring hydrograph analysis;
- Tracer studies;
- Modeling of karst aquifer.

While traditional techniques may be applied for water balance estimation, borehole analysis and modeling of karst aquifer, spring hydrograph analysis and water tracing require specific development to be applied to karst aquifers.

10.4.1 Spring Hydrograph Analysis

10.4.1.1 Karst Aquifer Hydrodynamic Functioning

Spring hydrograph analysis provides an insight into the characteristics of the aquifer from which it is originated, such as nature and operation of karst drainage systems and water storage (Bonacci, 1993). The classical methods aim at identifying an experimental law in (a part of) the spring hydrograph which allow to characterize separately the infiltration conditions and the phreatic zone, as well as to define the degree of karst development (Mangin, 1975a,b,c). It also allows prediction of spring flow which is essential to evaluate the groundwater resources and to define the best methods for exploiting and managing them (Mangin, 1994; Bakalowicz, 2005). The karst springs hydrographs responses to precipitation can be classified in three cases:

- Fast response spring: if the response time of the karst aquifer is fast with respect to the mean spacing between storms, the individual storm pulses will appear in the spring hydrograph and have the same shape as surface stream hydrograph;
- Intermediate response spring: the response time of the aquifer is comparable to the mean spacing between the storm, but the individual storm pulses are smeared out;
- Slow response: only the effects of seasonal wet and dry periods are observed in the hydrograph.

The response time of karst aquifers depends on various factors such as the contribution of allogenic recharge and internal runoff, the carrying capacity and internal structure of the conduit systems, and the area of the groundwater basin (White, 2002). The combined analysis of karst spring hydrographs and chemographs allows a more refined characterization of the karst drainage systems (Jakucs, 1959; Drake and Harmon, 1973; Bakalowicz, 1979, 1984; Sauter, 1992; Jeannin and Sauter, 1998; Dewandel et al., 2003).

Due to the characteristics of the kart terrains, two main types of flow are encountered:

- A slow flow or diffuse flow that occurs through karst fissures of small dimensions, in the laminar regime;
- A turbulent fast flow or conduit flow that occurs in karst fissures of large dimensions and through irregular conduit.

Depending upon the characteristics of karst fissures, karst spring hydrographs can be represented (Figure 10.1). Therefore, by analyzing the karst spring hydrographs, it is possible to identify the characteristics of the karst aquifer from which an outflow occurs (Bonacci, 1993). The hydrodynamic functioning of karst aquifers depends on:

- The state of development of the karst conduit network not only in the infiltration zone, but also in the phreatic zone;
- The partitioning of the infiltration in fast, slow and delayed conditions;
- The storage capacity of the phreatic zone.

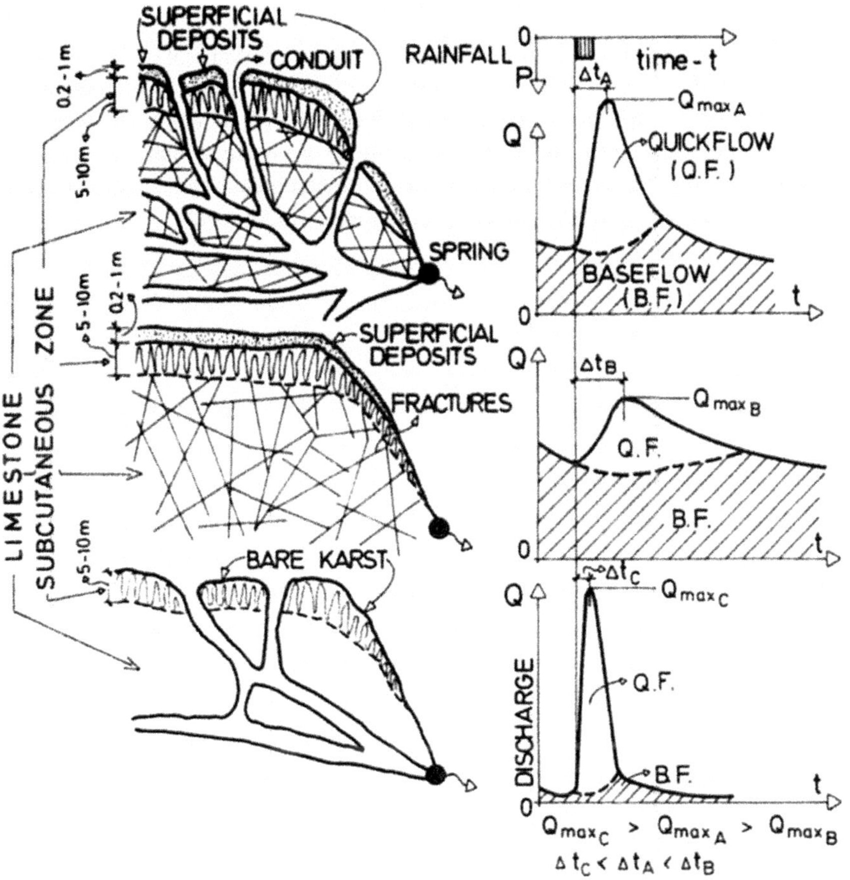

Fig. 10.1 Various forms of discharge hydrograph Q as a reaction to the same rainfall P for **a** combined, **b** diffuse, and **c** conduit type karst springs (From Bonacci, 1993. Reprinted from *Hydrol. Sci. J.* with permission of International Association of Hydrological Sciences. Copyright 1993)

10.4.1.2 Mangin's Equation

Various analytic equations can be used to evaluate the recession hydrographs (Padilla et al., 1994). Mangin's equation assumes that the recession is composed of both quickflow and baseflow. During a recession in a karst spring, Mangin (1975a,b,c) represents the discharge at time t (Q_t) as:

$$Q_t = \Phi_t + \Psi_t \tag{12}$$

where Ψ_t is an infiltration function translating the effects of surface recharge through unsaturated zone and saturated zone to the spring; and Φ_t can be described by Maillet's formula (1905):

$$\Phi_t = q_t^b = q_0^b e^{-\alpha t} \tag{13}$$

where q_t^b is the baseflow at time t; q_0^b is the baseflow extrapolated from t_i at the beginning of the recession; and α is the baseflow coefficient. Ebrahimi et al. (2007) applied the Maillet's equation and the Mangin's equation to recession curves of the Kalamsooz spring to study the hydrodynamic behavior of karst aquifers in Boroujerd, western Iran. The resulting graphical representation of both observed and simulated hydrographs are presented in Figure 10.2.

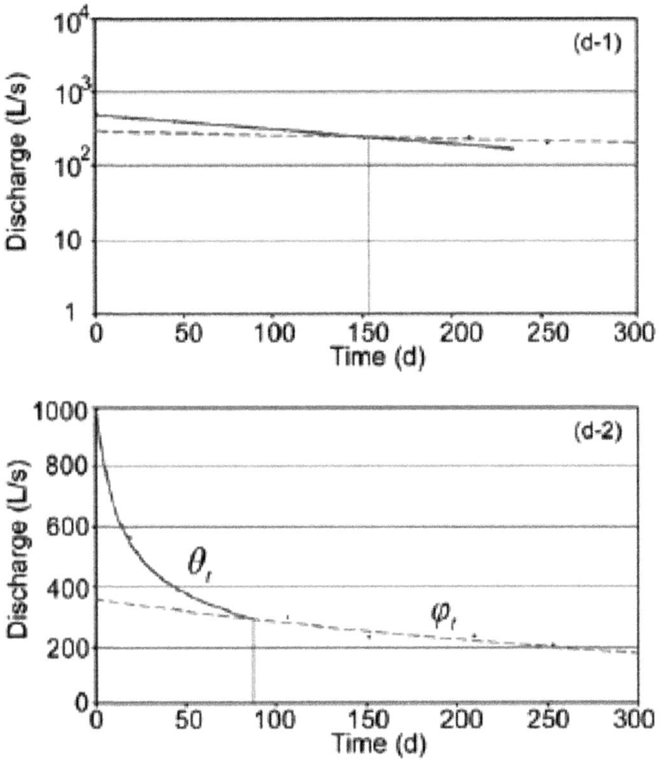

Fig. 10.2 Observed spring recession curve simulated by Maillet's equation (Eq. 12) and Mangin's equation (Eq. 13) for Kalamsooz springs (Reprinted from Ebrahimi et al., 2007 with permission of International Association of Hydrological Sciences. Copyright 2007)

10.4.2 Tracer Studies

Tracer studies in karst hydrology allow the delineation of catchment basin, the estimation of groundwater flow velocities, the determination of areas of recharge and the identification of pollution sources. Four classes of tracers are available for karst hydrology investigation (Ford and Williams, 2007):

- Artificial tracers: dyes and salts;
- Particulates: spores, bacteriophages and microspheres;
- Natural tracers: microorganisms, ions, environmental isotopes;
- Natural and artificial pulses of discharge, solutes and sediment.

10.4.3 Karst Aquifer Modeling

The modeling of karst aquifer systems aims at characterizing the system and to simulate its evolution. Two different and complementary approaches have emerged in the modeling of karst aquifer, a probabilistic approach and a parametric approach (Kiraly, 1998; Teutsch and Sauter, 1998; Kovacs, 2003). The probabilistic approach uses input event and output response combined with local observation of flow and transport processes. The parametric approach is based on theoretical concepts of aquifer structure and flow and transport mechanisms, from which is derived the simulated hydraulic behaviour of the aquifer. Four conceptual approaches are been used in karst groundwater flow modeling:

- Equivalent porous medium (EPM): EPM treats large areas as having uniform structural and hydraulic properties and therefore do not adequately represent the complex flow field in karst heterogeneous and anisotropic systems;
- Discrete fracture network: it describes flow within individual fractures or conduit without considering matrix characteristics;
- Double porosity continuum: dual-continuum models apply to karst aquifers represented by dual matrix-fissure, matrix-conduit or fissure-conduit groundwater systems. The porous matrix and fractured medium are treated as two-separated overlapping continua with their own hydraulic, geometric and flow characteristics. Such models are appropriate in systems when most of the flow paths are occurring in well-connected fractures and significant storage and exchange of water is achieved in and through matrix porosity; or when most of the most paths are provided by conduit that are well connected to fracture and significant storage and exchange of water is achieved in and through fracture porosity (Liedl and Sauter, 1998).
- Triple porosity: it integrates the most common and significant karst groundwater flow characteristics, being that storage is often dominant in rock matrix and fissure while flow is occurring through conduits. In triple porosity model, the flow is simulated as laminar in the porous matrix and fissures and as turbulent in the conduits. This approach is the ideal for karst hydrogeological modeling (Maloszewski et al., 2002).

10.5 Karst Water Resources Exploitation

10.5.1 Engineering

Karst aquifers have been exploited since time immemorial (Mijatovic, 1975; Burger and Dubertret, 1984). The exploitation of karst water resources are linked to the successful drilling of wells in karst terrains with high and sustainable yield as well as to the protection of vulnerable karst terrains to pollution. Exploitation methods vary due to karst terrains and karst aquifer characteristics. The main methods for karst water exploitation include (Pulido Bosch, 1999):

- Taking water directly from spring (Figure 10.3a);
- Excavating galleries in spring to collect water from aquifer (Figure 10.3b);
- Extracting water directly by manual methods when natural points of access are available in near surface karst saturated terrains (Figure 10.3c);
- Constructing a dam in the spring (Figure 10.3d);
- Constructing horizontal wells below a dam in the spring (Figure 10.3e);
- Pumping water from vertical wells (Figure 10.3f) or horizontal wells (Figure 10.3g) situated near, above or below the water source;
- Using galleries located below the altitude of the spring with sluices placed between the impervious and aquifer materials for water release control (Figure 10.3h);
- Combining the use of gallery and vertical wells drilled through impervious materials at the margins or into the aquifer materials themselves (Figure 10.3i);
- Collection devices based on the combined use of gallery and vertical wells drilled through impervious materials at the margins or into the aquifer materials themselves (Figure 10.3j);
- Opening of a large diameter well in the bordering impervious rocks with at a certain depth a gallery branches off, in the interior of which horizontal or vertical wells are drilled (Figure 10.3k);
- A gravity drain located at the same level as the gallery can complete the exploitation method based on the opening of a large diameter well in the bordering impervious rocks where a gallery branches off at a certain depth, in the interior of which horizontal or vertical wells are drilled (Figure 10.3l);
- Construction of a dam in the conduit of a submarine spring and drilling of a well inland (Figure 10.3m);
- Installation of a barrier at the gallery's outlet to the sea to prevent the entry of saltwater (Figure 10.3n);
- Construction of a well-gallery system above sea level to avoid salinization (Figure 10.3o);
- Construction of wells (Figure 10.3p).

The optimal underground tapping structure is the one placed directly into the main spring conduit or main underground flow zone. The hydrogeological mapping, geophysical investigations, boreholes and speleology are recommended tools for the optimal selection of the underground tapping structure.

10.5.2 Impacts

The impacts of exploitation of karst water resources can be direct or indirect. The direct impacts includes the depletion of piezometric levels occurring at the local or regional scale of the karst aquifer that can lead to overexploitation, the compartmentalization of the aquifer due to the lowering of piezometric level in irregular karst geomorphology, the deterioration in water quality, the aquifer salinization process, the alteration of river regimes, and the changes in wetlands (Pulido Bosch et al., 1989; Candela et al., 1991; Simmers et al., 1992; Pulido Bosch et al., 1995). The indirect impacts encompass the soil salinization, the progressive desertification, the changes in physical properties of the aquifers, the induced pollution from long distances and the sinkhole formation. The specific exploitation impacts related to karst aquifers—overexploitation, salinization and sinkhole—are further detailed below.

10.5.2.1 Overexploitation

The optimal exploitation capacities of karst aquifer resources may be performed through stochastic analysis. This approach was applied in case studies from the Serbian karst by Jemcov (2007). Water supply potentials were evaluated not only on the basis of groundwater budgets, but also included analyses of storage changes in karst water reservoirs under natural conditions and calculation of the potential expansion of currently tapped sources to define optimal exploitable regimes. The results obtained through these analyses are a significant contribution to avoid karst aquifer overexploitation (Jemcov, 2007).

The Carpatho-Balkan karst aquifer systems are the main sources for regional water supply in eastern Serbia. The main problem of these karst aquifer systems is a water deficiency during recessional periods, which results from fast discharge of karst groundwater through springs; a widespread problem for karst water supply aquifers (Jemcov, 2007).

Among the various karst aquifer management approaches, the concept of a water loan from storage in karst aquifer is based on the relatively fast replenishing of water during high-water period, commonly accomplished during one hydrological cycle. However, the major concern for tapping karst groundwater using this concept is overexploitation. Therefore, assessment of potential and available kart water resources for groundwater tapping is important.

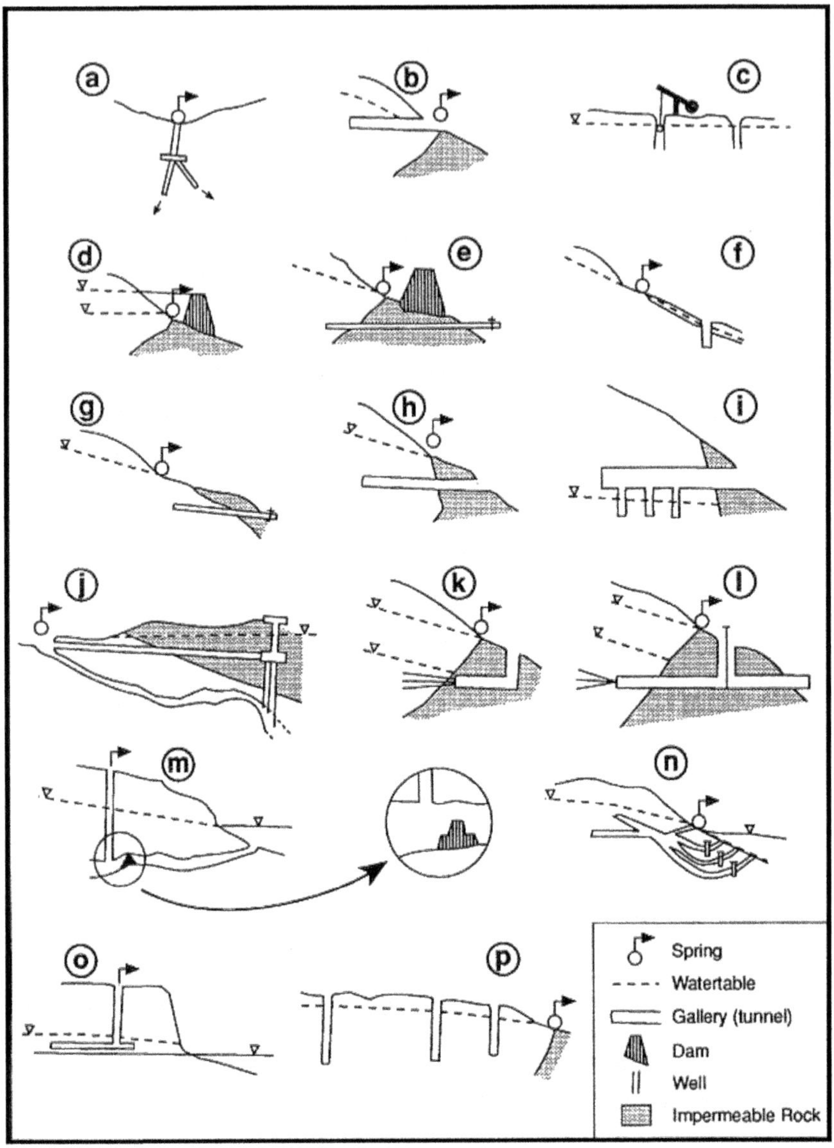

Fig. 10.3 Exploitation methods of karst aquifers (From Pulido Bosch, 1999. Reprinted from *Karst Hydrogeology and Human Activities*, David Drew and Heinz Hotzl, eds., with permission of A. A. Balkema. Copyright 1999)

The determination of the optimal exploitation capacity of the St. Petka source was performed using the estimated water resources stored in the karst aquifer under natural conditions as a baseline and compared with water resources variations in storage under artificial conditions (Jemcov, 2007). Simulated exploitation rates of the St. Petka spring allowed the determination of exploitation potential, limits and optimal values through various exploitation scenarios (Figure 10.4). As the optimal exploitation capacity does not necessary protect from karst water resources overexploitation, a conservative and sustainable management implies in this case a rational exploitation that encompasses ecological limitations and replenishment factor.

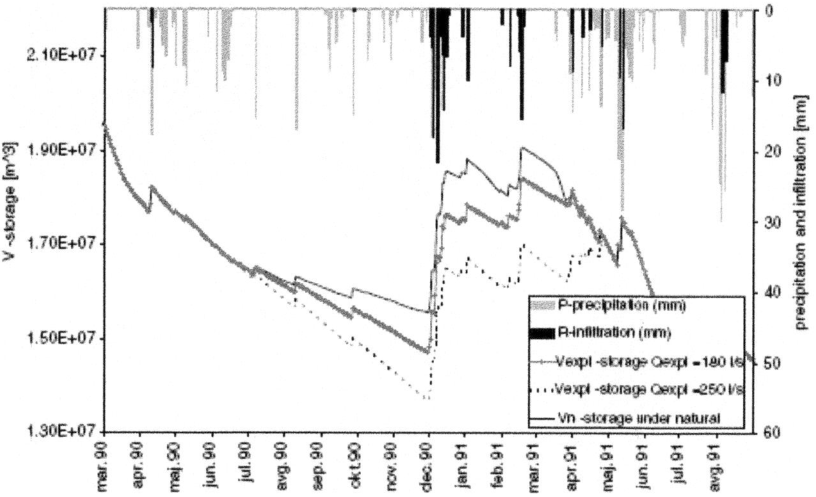

Fig. 10.4 Optimal exploitation capacity of karst aquifer systems: Simulated exploitation rates of St. Petka Spring (Reprinted from Jemcov, 2007 with permission of Springer. Copyright 2007)

10.5.2.2 Salinization

When exploiting a coastal karst aquifer, the decrease in the piezometric level lead to saltwater intrusion. The hydromechanics of littoral karst springs and the saltwater intrusion processes into the spring vary significantly due to coastal karst systems diversity. The research on saltwater intrusion in coastal karst system is extensive, however, studies are limited to the description of one particular coastal karst aquifer. The freshwater head above sea level changes in time and space due to hydrological and hydrogeological conditions as well as to the sea movement. Karst groundwater is restricted by constraints on the hydraulic head and transition zone thickness. A simple scheme representing the fresh and sea water behavior based on the Ghyben (1888/1889) and Herzberg (1901) in the case of karst aquifer is depicted in Figure 10.5 (Bonacci and Roje-Bonacci, 1997). No significant progress on the individual karst spring functioning has been achieved because of the highly heterogeneous conditions of karst aquifers and the many measurements needed to describe their overall functioning (Gjurasin, 1942, 1943; Breznik, 1976;

Dam and Sikkema, 1982; Sikkema and Dam, 1982; Johnston, 1983; Bonacci, 1987; Pavlin, 1990; Bonacci, 1995; Bonacci and Roje-Bonacci, 1997; Gilli, 1999; Blavoux et al., 2004; Pinault et al., 2004; Arfib et al., 2007, Fleury et al., 2007).

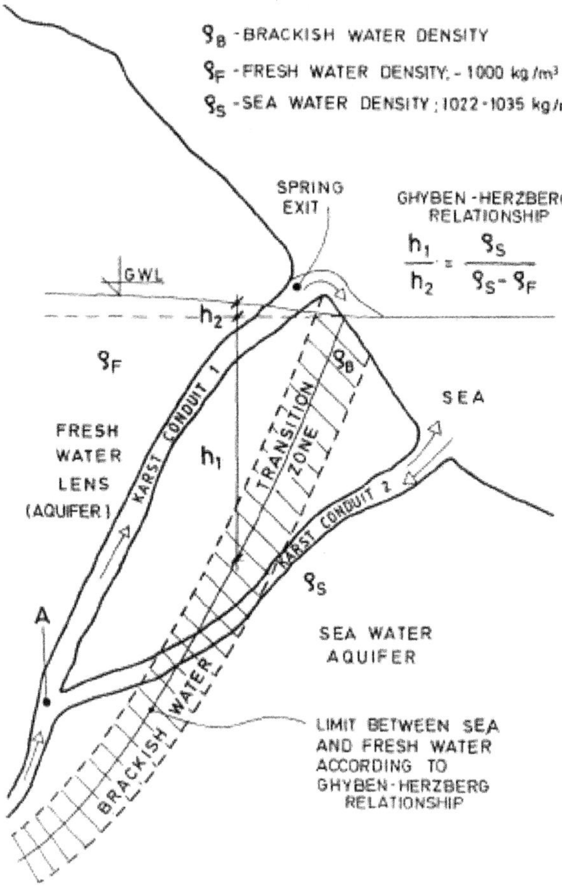

Fig. 10.5 Schematic presentation of the relationship between fresh and sea water in a homogeneous coastal karst aquifer (From Bonacci and Roje-Bonacci, 1997. Reprinted from *Hydrol. Sci. J.* with permission of International Association of Hydrological Sciences. Copyright 1997)

Figure 10.6 represents the daily water salinity measured in the Blaz Spring and piezometer B20 located on the Croatian sea coast (Bonacci and Roje-Bonacci, 1997). The salinity of brackish karst springs situated along the Adriatic coast varies from 10 to more than 18000 mg Cl L^{-1} with unfavorable distribution during the year. The intrusion of sea water into the Blaz Spring occurs each year during warm and dry summer or autumn seasons, following long periods without precipitation (Bonacci and Roje-Bonacci, 1997).

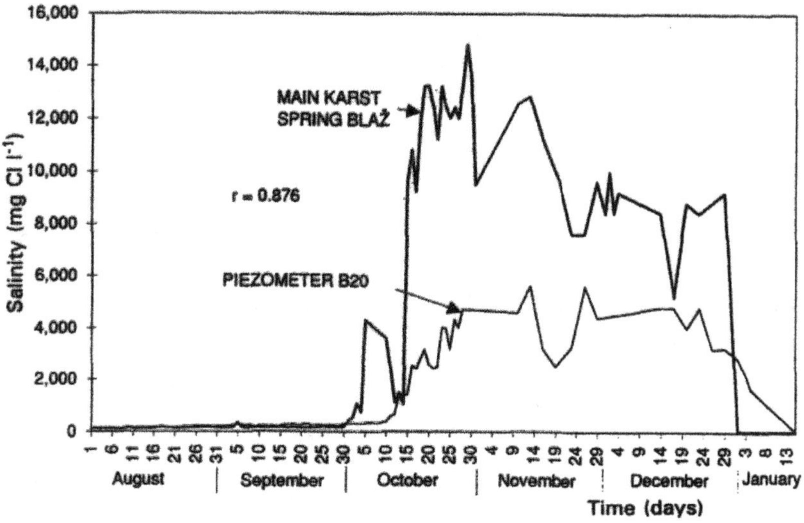

Fig. 10.6 Daily water salinity measured in the main karst Blaz Spring and piezometer B20 (From Bonacci and Roje-Bonacci, 1997. Reprinted from *Hydrol. Sci. J.* with permission of International Association of Hydrological Sciences. Copyright 1997)

10.5.2.3 Sinkhole

Karst collapse resulting from aquifer exploitation is widespread (LaMoreaux, 1991). When the piezometric level decreases, a risk of karst collapse occurs because of the subsoil resistance reduction resulting from the loss of the water that acted as a stabilizing element supporting part of the load. The exploitation may also causes the reactivation of conduits and karst hollows filled with decalcified sediments and the rapid removal of these materials which contributed to the pre-existing equilibrium, may induce the rock cover to collapse because of lack of support (Garay, 1986).

Sinkholes are closed depression in karst terrain. Sinkhole formation includes bedrock dissolution as well as bedrock or soil collapse resulting respectively from underlying cavities or piping. The main types of sinkholes observed are (White, 2007):

- Cover collapse sinkholes that form abruptly as a plug of soil falls into an underlying cavity;
- Reveling sinkholes that form gradually as loose, poorly cohesive soils drain away into the subsurface.

The sinkhole genesis and evolution in the Apulia region of Italy was investigated by Delle Rose et al. (2004). Three possible types of sinkhole mechanisms formation were identified:

- Collapse of a chamber in a natural cave or in man-made cavities;
- Slow and gradual enlargement of doline through dissolution;
- Settlement and internal erosion of filling deposites of buried dolines.

The karst system evolution and sinkhole development mechanisms are illustrated by Delle Rose et al. (2004) in Figure 10.7. The sinkhole formation consists

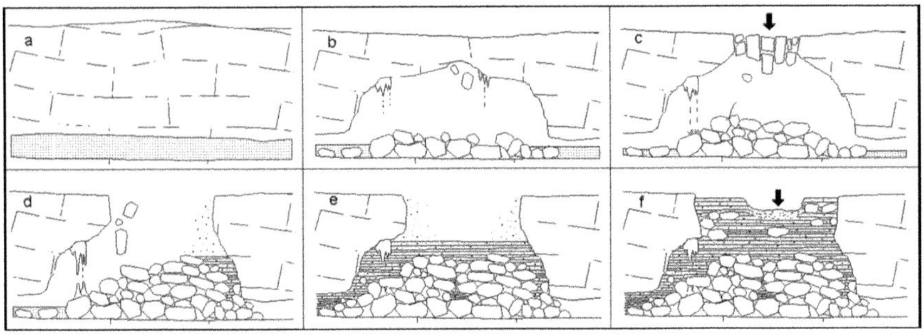

Fig. 10.7 Karst system evolution and sinkhole development (From Delle Rose, 2004. Reprinted from *Natural Hazards and Earth System Sciences* with permission of M. Parise. This work is licensed under a Creative Commons License, Copyright 2004, Author(s))

Fig. 10.8 Grotta della Poesia Grande, a collapse sinkhole along the Adriatic coast of Apulia, in southern Italy. The cave is part of a larger karst system, formed by two collapse sinkholes, caves and intervening flooded galleries. Rocks involved are pliocene calcarenites. The system is of great importance for archaeology since one of the cave hosts thousands of ancient inscriptions on its walls (Photo: courtesy of M. Parise)

of several phases. First, a karst cave begins to form in the phreatic zone (Figure 10.7a) and is developed through mass movements (e.g. rockfalls from the vault and walls), while at the same time, the land surface is lowered by surface erosion (Figure 10.7b), until the vault partially or totally collapse and a sinkhole is formed at the land surface (Figure 10.7c). Second, the sinkhole walls become unstable and may be subject to rockfalls, whilst the cave stream moves at greater depth as the karst base level is being lowered (Figure 10.7d) and the sinkhole is filled by terre rosse and continental deposits (Figure 10.7e). Finally, new minor sinkholes may be formed within the filling deposits, due to differential settlements, subsurface erosion, or piping (Figure 10.7f) (Delle Rose et al., 2004).

Most sinkhole in Apulia region probably result from a collapse of a chamber in a natural cave or man-made cavities. An example of the surface feature of the Apulian karst is the Grotta della Poesia Grande sinkhole (Figure 10.8).

10.5.3 Protection

The risk of groundwater contamination depends on the hazard (origin), the vulnerability of the system (pathway) and the potential consequences of a contamination event, i.e., its impact on the groundwater resource or source (target) (Daly et al., 2004). The Pan European Approach to vulnerability, hazard and risk mapping is based on an origin pathway-target model, which applies for both groundwater resource and source protection.

10.5.3.1 COST 620

COST 620, “Vulnerability and Risk Mapping for the Protection of Carbonate (Karst) Aquifers”, a working group chaired by Zwalhen (2003) in the frame of COST Organization (Co-operation in Science and Technology) proposes the most appropriate methods for studying and mapping the vulnerability in order to protect the resource according to the EU Water Framework Directive (WFD) published in 2000. COST 620 distinguished two concepts of groundwater vulnerability, following the definitions proposed by Vrba and Zaporozec (1994): intrinsic and specific vulnerability. Intrinsic vulnerability takes into account the hydrogeological characteristics of an area but is independent of the nature of the contaminants. Specific vulnerability additionally takes into account the properties of a particular contaminant (or group of contaminants) and its relationship to the hydrogeological system (Daly et al., 2002; Zwahlen, 2004). The hazard estimation concept as proposed by COST Action 620 considers three factors—type of hazard (toxicity, solubility and mobility), quantity release and likelihood of contamination—which control the degree of harmfulness of each hazard.

This new approach follows two methods developed and applied in Switzerland and France during the 1990s: the EPIK (Doerfliger et al. 1999) and RISKE (Petelet-Giraud et al., 2000) methods, which are based on a multicriterion analysis. The reference criteria are related to epikarst, protective cover, infiltration and the karst network in EPIK, and to the rock characteristics, infiltration, soil, the karst development and epikarst in RISKE.

10.5.3.2 Risk Assessment and Mapping

Risk assessment should represent not only the effects but also the consequences of groundwater contamination. Andreo et al. (2006) developed the first application of a Pan European Approach to vulnerability, hazard and risk mapping to the Sierra de Líbar, a karst hydrogeology system in Andalusia, Spain. The risk map to protect the karst groundwater of the Sierra de Líbar was created by overlaying the hazard and vulnerability maps (Andreo et al., 2006).

References

Andreo, B., Goldscheider, N., Vadillo, I., Vias, J. M., Neukum, C., Sinreich, M., Jimenez, P., Brechenmacher, J., Carrasco, F., Hotzl, H., Perles, M. J., and Zwahlen, F., 2006, Karst groundwater protection: First application of a Pan-European Approach to vulnerability, hazard and risk mapping in the Sierra de Líbar (Southern Spain), *Sci. Total Environ.* **357**:54–73.

Arfib, B., de Marsily, G., and Ganoulis, J., 2007, Locating the zone of saline intrusion in a coastal karst aquifer using springflow data, *Ground Water* **45**(1):28–35.

Badon-Ghyben, W., 1888/1889, Nota in verband met de voorgenomen putboring nabij Amsterdam (Notes on the probable results of well drilling near Amsterdam), Tijdschrift van het Koninklijk Instituut van Ingenieurs, The Hague, **9**:8–22.

Bakalowicz, M., 1975, Géochimie des eaux karstiques et karstification, Annales de Spéléologie **30**(4):581–589.

Bakalowicz, M., 1977, Etude du degré d'organisation des écoulements souterrains dans les aquifères carbonatés par une méthode hydrogéochimique nouvelle, *Comptes Rendus de l'Académie des Sciences*, Paris, D **284**:2463–2466.

Bakalowicz, M., 1979, Contribution de la geochimie des eaux à la connaissance de l'aquifère karstique et de la karstification, Univ. Paris-6, Thèse Doctorate Sciences, pp. 1–269.

Bakalowicz, M., 1984, Water chemistry of some karst environment in Norway, *Norsk Geografisk Tidsskrift* **38**:209–214.

Bakalowicz, M., 2005, Karst groundwater: a challenge for new resources, *Hydrogeol. J.* **13**(1):148–160.

Bakalowicz, M., and Mangin, A., 1980, L'aquifère karstique. Sa définition, ses caractéristiques et son identification, *Mémoires hors série de la Société géologique de France* **11**:71–79.

Blavoux, B., Gilli, E., and Rousset, C., 2004, Alimentation et origine de la salinité de la source sous-marine de Port-Miou (Marseille–Cassis. Principale émergence d'un réseau karstique hérité du Messinien, *Comptes rendus Geoscience* **336**(6):523–533.

Bogli, A., 1980, *Karst Hydrology and Physical Speleology*, Springer-Verlag, Berlin.

Bonacci, O., 1987, *Karst Hydrology*, With Special Reference to the Dinaric Karst Springer Series in Physical Environment, Vol. 2. Springer-Verlag, New York, 184p.

Bonacci, O., 1993, Karst spring hydrographs as indicators of karst aquifers, *Hydrol. Sci. J.* **38**(1):51–62.
Bonacci, O, 1995, Groundwater behaviour in karst: Example of the Ombla Spring (Croatia), *J. Hydrol.* **165**:113–134.
Bonacci, O., and Roje-Bonnaci, T., 1997, Sea water intrusion in coastal karst springs: Example of the Blaz Spring (Croatia), *Hydrol. Sci. J.* **42**:89–100.
Breznik, M., 1976, Mechanism and development of the brackish karstic spring Almyros Iraklion, *Annales Geologiques des Pays Helleniques* **28**:29–46.
Burdon, D. J., and Papakis, N., 1963, *Handbook of Karst Hydrogeology*, United Nations Special Fund, Institute for Geology and Subsurface Research, Athens, Greece.
Burger, A., and Dubertet, L., eds., 1984, *Hydrogeology of Karstic Terrains: Case Studies*, IAH, International Contribution of Hydrogeology 1, Heise, Hannover.
Candella, L., Gomez, M. B., Puga, L., Rebollo, L., and Villarroya, F., 1991, Aquifer overexploitation, in: *Proceedings of XXIII IAH Congress*, Canary Island, p. 580.
Cvijic, J., 1918, Hydrographie souterraine et évolution morphologique du karst, *Rev. Trav. Inst. Géogr. Alpine*, t. VI, 375–426.
Daly, D., Dassargues, A., Drew, D., Dunne, S., Golscheider, N., Neale, S., Popescu, I. C., and Zwahlen, P., 2002, Main concepts of the "European approach" to karst-groundwater-vulnerability assessment and mapping, *Hydrogeol. J.* **10**:340–345.
Daly D., Hotzl H., De Ketelaere D., 2004, Risk definition, in: *Vulnerability and Risk Mapping for the Protection of Carbonate (karst) Aquifers*, F. Zwahlen, ed., EUR 20912, Brussels: EC, Directorate-General XII Science Research and Development, pp. 106–107.
Dam, J. C. van, and Sikkema, P. C., 1982, Approximate solution of the problem of the shape of the interface in a semi-confined aquifer, *J. Hydrol.* **56**:221–237.
Delle Rose, M., Federico, A., and Parise, M., 2004, Sinkhole genesis and evolution in Apulia, and their interrelations with the antropogenic environment, *Nat. Hazards Earth Syst. Sci.* **4**:747–755.
Doerfliger, N., Jeannin, P. Y., and Zwahlen F., 1999, Water vulnerability assessment in karst environments: a new method of defining protection areas using a multi-attribute approach and GIS tools (EPIK method), *Environ. Geol.* **39**(2):165–176.
Drake, J. J., and Harmon, R. S., 1973, Hydrochemical environments of carbonate terrains, *Water Resour. Res.* **11**:958–962.
Dreybrodt, W., 1988, Processes in karst systems. physics, chemistry and geology, *Springer series in Physical Environment 4*, Springer, Berlin Heidelberg New York, pp. 1–288.
Dreybrodt, W., and Buhman, D., 1991, A mass transfer model for the dissolution and precipitation of calcite from solutions in turbulent motion, *Chem. Geol.* **90**:107–112.
Dreybrodt, W., Lauckner, J., Liu, Z., Swensson, U., and Buhman, D., 1996, The kinetics of the reaction $CO_2 + H_2O \rightarrow H^+ + HCO_3^-$ as one of the rate limiting steps of the dissolution of calcite in the system $H_2O - CO_2 - CaCO_3$, *Geochim. Cosmochim. Acta* **60**:3375–3381.
Dewandel, B., Lachassagne, P., Bakalowicz, M., Weng, P., and Al-Malki, A., 2003, Evaluation of aquifer thickness by analyzing recession hydrographs. Application to the Oman ophiolite hard-rock aquifer, *J. Hydrol.* **274**:248–269.
Ebrahimi, B., Pasandi, M., and Ahmadipour, M. R., 2007, Hydrodynamic behaviour of karstic aquifers in Boroujerd, western Iran, *Hydrol. Sci. J.* **52**(1):192–205.
Fleury, P., Bakalowicz, M., and Marsily, G. de, 2007, Submarine springs and coastal karst aquifers: A review, *J. Hydrol.* **339**:79–92.
Ford, D., and Williams, P., 2007, *Karst Hydrogeology and Geomorphology*, John Wiley & Son Ltd, England, pp. 1–562.
Garay, P., 1986, Informe geológico sobre la tierra de hindimineto de Pecheguer (Alicante), *Jorn. Karst Euskadi* **1**:323–333.

Gilli, E., 1999, Détection de sources sous-marines et précision de l'impluvium par mesure des variations de salinité. L'exemple de la source de Cabbé-Massolins (Roquebrune-Cap-Martin, France), *Comptes-rendus de l'Académie des Sciences de Paris*, Sciences de la Terre et des Planétes, **329**(2):109–116.

Gjurasin, K., 1942, Prilog hidrografiji primorskoja krsa (Contribution to the littoral karst hydrology), *Tehnicki vjesnik* **59**:107–112.

Gjurasin, K., 1943, Prilog hidrografiji primorskoga krsa (Contribution to the littoral karst hydrography), *Tehnicki vjesnik* **60**:1–17.

Herzberg, A., 1901, Die Wasserversorgung einiger Nordseebder (The water supply of parts of the North Sea coast in Germany), *Z. Gasbeleucht. Wasserversorg* **44**:815–819 and **45**:842–844.

Hoobs, S. L., and Smart, P. L., 1986, Characterization of carbonate aquifers: A conceptual base, in: *Proceedings of the 9th International Congress of Speleology*, Barcelona, Vol. 1, pp. 43–46.

Jakucs, L., 1959, Neue methoden der hohlenforsching in ungarm und ihre ergebnisse, *Hohle* **10**(4):88–98.

Jeannin, P. –Y., and Sauter, M., 1998, Analysis of karst hydrodynamic behaviour using global approaches: A review, *Bulletin d'Hydrogéologie, Neuchâtel* **16**:31–48

Jemcov, I., 2007, Water supply potential and optimal exploitation capacity of karst aquifer systems, *Environ. Geol.* **51**:767–773.

Jennings, J. N., 1985, *Karst Geomorphology*, Basil Blackwell, Oxford, pp. 1–293.

Johnston, R. H., 1983, The saltwater-freshwater interface in the tertiary limestone aquifer Southeast Altantic outer-continental shelf of the USA, *J. Hydrol.* **61**:239–249.

Kiraly, L., 1998, Modelling karst aquifers by the combines discrete channel and continuum approach, *Bulletin d'Hydrogéologie, Neuchâtel*, **16**:77–98.

Kiraly, L., 2003, Karstification and groundwater flow, *Speleogenesis Evol. Karst Aquifers* **1**:1–26.

Kovacs, A., 2003, Geometry and Hydraulic Parameters of Karst Aquifers: A Hydrodynamic Modeling Approach, Faculté des Sciences, Université de Neuchâtel, Thesis, pp. 1–131.

LaMoreaux, P. E., 1991, Environmental effects to overexploitation in a karst terrane, in: *Proceedings of XXIII IAH Congress*, Canary Island, pp. 103–113.

Liedl, R., and Sauter, M., 1998, Modeling of aquifer genesis and heat transport in karst systems, *Bulletin d'Hydrogéologie, Neuchâtel*, **16**:185–200.

Liu, Z., and Dreybrodt, W., 1997, Dissolution kinetics of calcium carbonate minerals in H_2O-CO_2 solutions in turbulent flow: The role of the diffusion boundary layer and the slow reaction $H_2O + CO_2 \rightarrow H^+ + HCO_3^-$, *Geochim. Cosmochim, Acta* **61**:2879–2889.

Maloszewski, P., Stichler, W., Zuber, A., and Rank, D., 2002, Identifying the flow systems in a karstic-fissured-porous aquifer, the Schneealpe, Austria, by modelling of environmental O-18 and H-3 isotopes, *J. Hydrol.* **256**(1–2):48–59.

Mangin, A., 1975a, Contribution à l'étude hydrodynamique des aquifères karstiques, Première Partie. Genéralités sur le karst et les lois d'écoulement, *Annales de Spéléologie* **29**:283–332.

Mangin, A., 1975b, Contribution à l'étude hydrodynamique des aquifères karstiques, Deuxième Partie. Concepts méthodologiques adoptés, *Annales de Spéléologie* **29**:495–601.

Mangin, A., 1975c, Contribution à l'étude hydrodynamique des aquifères karstiques, Troisième Partie. Constitution et fonctionnement des aquifères karstiques. *Annales de Spéléologie* **30**:21–124.

Mangin, A., 1994, Karst hydrogeology, in: *Groundwater Ecology*, J. Stanford, J. Gibert, and D. Danielopol, eds., Academic Press, San Diego, CA, U.S.A., pp. 43–67.

Marsaud, B., 1997, Structure et fonctionnement de la zone noyée des karsts à partir des résultats expérimentaux, PhD thesis, Université d'Orsay Paris XI, Documents du BRGM N°268.

Mijatovic, B. F., 1975, Exploitation rationelle des eaux karstiques, in: *Hydr. Terrains Karstiques*, AIH, pp. 123–145.

Padilla, A., Pulido-Bosch, A., and Mangin, A., 1994, Relative importance of base flow and quickflow from hydrographs of karst spring, *Ground Water* **32**(2):267–277.

Palmer, A. N., 2000, Hydrogeologic control of cave patterns, in: *Evolution of Karst Aquifers*, A. Klimchouk A, D. C. Ford, A. N. Palmer, and W. Dreybrodt, eds., Speleogenesis Nat Speleol Soc, Huntsville, pp. 77–90.

Palmer, A. N., 1991, Origin and morphology of limestone caves, *Geol. Soc. Am. Bull.* **103**:1–21.

Pavlin, B., 1990, Yugoslav littoral belt karst springs used for water supply, in: *Proc. 6th Int. IAEG Congress*, Balkema, Rotterdam, The Netherlands, pp. 1387–1394.

Petelet-Giraud, E., Doerflinger, N., and Crochet, P., 2000, RISKE: Méthode d'évaluation multicritère de la cartographie de la vulnérabilité des aquifères karstiques. Application aux systèmes des Fontanilles et Cent-Fonts (Hérault, Sud de la France). (RISKE: a multicriterion assessment method for mapping the vulnerability of karst aquifers. Application to Fontanilles and Cent-Fonts karst systems, Hérault, southern France), *Hydrogéologie* **4**:71–88.

Pinault, J. -L., Doerfliger, N., Ladouche, B., and Bakalowicz, M., 2004, Characterizing a coastal karst aquifer using an inverse modeling approach: The saline springs of Thau, southern France, *Water Resour. Res.* **40**, doi:10.1029/2003WR002553.

Plummer, L. N., Wigely, T. M. L., and Parkhurst, D. L., 1978, The kinetics of calcite dissolution in CO_2-water systems at 5° to 60°C and 0.0 to 1.0 atm CO_2, *Am. J. Sci.* **278**:179–216.

Pulido Bosch, A., 1999, Karst water exploitation, in: *Karst Hydrogeology and Human Activities, Impacts, Consequences and Implications*, D. Drew, and H. Hotzl, eds., A. A. Balkema, Rotterdam, International Contributions to Hydrology N°20.

Pulido Bosch, A., Castillo, A., and Padilla, A., eds., 1989, *La sobreexplotación de acuíferos, Temas Geológico-Mineros*, IGME, Madrid **10**:1–687.

Pulido Bosch, A., Morell, I., and Andreu, J. M., 1995, Hydrogeochemical effects on the groundwater mining of the Sierra de Crevillente (Alicante, Spain), *Environ. Geol.* **26**:232–239.

Sauters, M., 1992, Quantification and forecasting of regional groundwater flow and transport in a karst aquifer, *Tubinger Geowissenschaftlichen Abhandlungen, Reihe C* **13**:1–150.

Shuster, E. T., and White, W. B., 1971, Seasonal fluctuations in the chemistry of limestone springs: a possible means for characterizing carbonate aquifers, *J. Hydrol.* **14**(1): 93–128.

Sikkema, P. C., and Dam, J. C. van, 1982, Analytical formulae for the shape of the interface in a semi-confined aquifer, *J. Hydrol.* **56**:201–220.

Simmers, I., Villarroya, F., and Rebollo, L. F., eds., 1992, *Selected Papers on Aquifer Overexploitations*, IAH Selected Papers, Heise, Hannover **3**:1–392.

Smith, D. I., and Atkinson, T. C., 1976, The erosion of limestones, in: *The Science of Speleology*, T. D. Ford, and C. H. D. Cullingford, eds., Academic Press, London, pp. 151–177.

Svensson, U., and Dreybrodt, W., 1992, Dissolution kinetics of natural calcite minerals in CO_2-water systems approaching calcite equilibrium, *Chem. Geol.* **100**:129–145.

Teutsch, G., and Sauter, M., 1998, Distributed parameter modeling approaches in karst-hydrological investigations, *Bulletin d'Hydrogéologie, Neuchâtel* **16**:99–110.

Thrailkill, J. V., 1968, Chemical and hydrologic factor in the excavation of limestone caves, *Geol. Soc. Am. Bull.* **79**:19–46.

Vrba, J., and Zaporozec, A., 1994, *Guidebook on Mapping Groundwater Vulnerability*, Heinz Heise, Hannover, Germany, Int. Assoc. Hydrogeol., Int. Contrib. Hydrogeol. 16.

White, W. B., 1969, Conceptual models for limestone aquifers, *Ground Water* 7(3):15–21.

White, W. B., 1988, *Geomorphology and Hydrology of Karst Terrains*, Oxford University Press, New York.

White, W. B., 1999, Conceptual models for karstic aquifers, *Karst Water Inst. Spec. Pub.* **5**:10–16.

White, W. B., 2002, Karst hydrogeology: recent developments and open questions, *Eng.Geol.*, **65**:85-105.

White, W. B. 2007, Groundwater flow in karstic aquifers, in: *The Handbook of Groundwater Engineering*, J. W. Delleur, ed., CRC Press, Taylor & Francis Group, pp. 21.1–21.47.

Williams, P. W., ed., 1993, *Karst Terrains; Environmental Changes and Human Impacts*, Catena Supplement, 25, pp. 1–268.

Witherspoon, P. A., Wang, J. S. Y., Iwai, K., and Gale, J. E., 1980, Validity of cubic law for fluid flow in a deformable rock fracture, *Water Resour. Res*, **16**:1016–1024.
Zwahlen F, ed., 2004, *Vulnerability and Risk Mapping for the Protection of Carbonate (Karst) Aquifers*, EUR 20912. Brussels: European Commission, Directorate-General XII Science, Research and Development, pp. 1–297.

Chapter 11
Water Quality, Pollution and Conservation

Ülkü Yetiş*

Abstract This chapter presents the main water pollutants and their impacts on the environment as well as the water quality parameters (physical, chemical and biological) that are used in water resources assessment and management. The water quality of river, streams, lakes and groundwater is reviewed. Remediation and preventive measures are also examined. Water conservation practices and its effective use are analyzed for system users and system operators.

Keywords Water quality, contaminants, conservation measures, sag curve, eutrofication, nitrate, pesticides, chlorinated organic solvents, gasoline and oil constituents, heavy metals, pathogens

11.1 Introduction

Water is a renewable source which is naturally recycled. It exists on Earth as a solid, liquid or gas. Oceans, rivers, clouds and rain, all of which contain water, are in a frequent state of change (surface water evaporates, cloud water precipitates, rainfall infiltrates the ground etc.). However, the total amount of the Earth's water does not change. The circulation and conservation of Earth's water which is called the "hydrologic cycle" results in renewal of water resources and provides a continuous supply (Figure 11.1).

The demand for water has increased significantly across the world, with population growth, increasing wealth, intensification of agriculture and the advent of industrialization. Further increase in water demand (with respect to agricultural and domestic uses) is expected owing to warmer temperatures and drier conditions due to climate change in many regions of the World. The main uses of freshwater resources can be categorized into two as indicated in Table 11.1.

Ülkü Yetiş
Department of Environmental Engineering, Middle East Technical University, 06531 Ankara, Turkey
* To whom correspondence should be addressed. E-mail: uyetis@metu.edu.tr

C.J.G. Darnault (ed.), *Overexploitation and Contamination of Shared Groundwater Resources.*

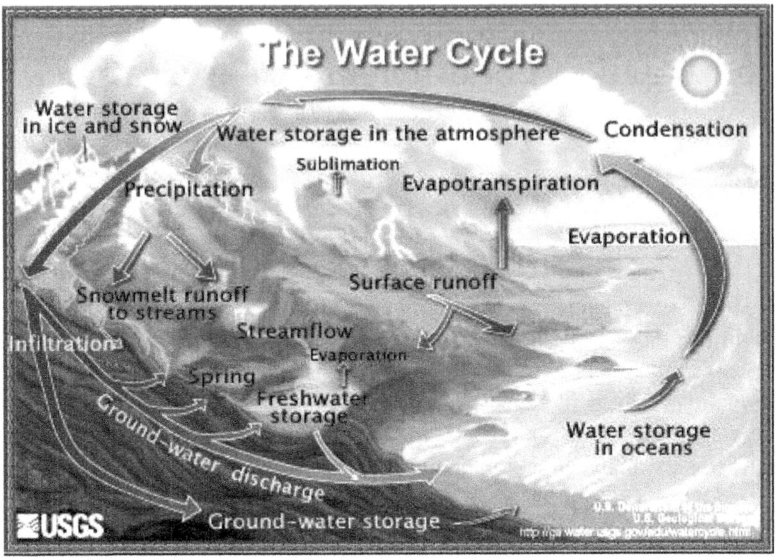

Fig. 11.1 Hydrologic cycle (USGS, 2006) (Source: http://www.usgs.gov)

Table 11.1 Main uses of freshwater resources (Kiely, 1997)

Abstraction	In Stream Uses
Domestic supply	Biological exploitation
Irrigation	Power generation
Industry-manufacturing	Transport/navigation
Industry-cooling	Recreation/amenity
Flushing off canals	Flood control
Diversion between catchments	Waste

Water resources are greatly influenced by all the uses indicated in Table 11.1. Domestic, industrial and agricultural use of water generates large volume of discharges which are given to water bodies, and therefore cause the deterioration of water quality. Similarly, stream uses also influence adversely the nature of aquatic resources and cause water pollution.

Water pollution is defined as "a large set of adverse effects upon water bodies (lakes, rivers, oceans, groundwater) caused by human activities" (Wikipedia, 2006). In essence, all water bodies contain many naturally occurring substances—mainly bicarbonates, sulphates, sodium, chlorides, calcium, magnesium and potassium which might originate from:

- Soil, geologic formations and terrain in the catchment area (river basin);
- Surrounding vegetation and wildlife;
- Precipitation and runoff from adjacent land;
- Biological, physical and chemical processes in the water;

In addition, natural phenomena such as volcanoes, storms, earthquakes etc. also cause major changes in water quality and the ecological status of water. However, these are not deemed to be pollution. The term "pollution" implies that the poor quality is due to human activities. Human activities cause more serious, longer term and larger scale problems. In general terms, water is considered to be polluted when it contains enough foreign materials to render it inappropriate for an intended use, such as for drinking, recreation, industrial use or fish farming.

11.2 Water Pollutants

Water pollution that varies in extent and variety of pollutant from one region to another has many causes and characteristics. Based on their health effects, wide range of pollutants discharged to surface waters can be grouped into broad classes. For example, the following list identifies eight types of pollutants:

- Pathogens;
- Oxygen-demanding substances;
- Nutrients;
- Suspended solids;
- Heavy metals;
- Toxic organics;
- Inorganic chemicals;
- Heat;
- Hazardous materials.

Each of these types of pollutants and their importance are indicated in Table 11.2. To estimate the hazard of different pollutants, not only their hazardous properties, but also other factors such as the volumes of their input into the environment, the ways and scale of their distribution, the patterns of their behavior in the water ecosystems, their ability to accumulate in living organisms, the stability of their composition and other properties should be taken into consideration.

11.2.1 Sources of Water Pollutants

Another classification of pollutants is according to the nature of their origin as either a point source or a nonpoint (dispersed) source pollutant. Point sources are generally man-made and mediated by man-made devices such as pipes and effluent outfalls. These sources are attributed to a defined location and include domestic sewage and industrial wastewaters which are collected and conveyed to a discharge point into the receiving environment. Domestic sewage consists of

Table 11.2 Major water pollutants and their impact on the environment

Type of Pollutant	Reason for Importance
Pathogens	Water-related diseases (e.g. gastro-intestinal, typhoid, shigellosis, hepatitis and cholera) are among the main health concerns in the world. Can directly affect humans by causing illness and possible death. Often contamination through contact with water or via food (e.g. via irrigated agriculture or via fish/shellfish).
Oxygen-demanding substances	The exhaustion of oxygen in water resources and the development of septic conditions, fish kills.
Nutrients	The growth of undesirable aquatic life and groundwater pollution. Although modest input of nutrients may have a positive effect on biodiversity as well as the total production of the ecosystem, an excess of nutrients will cause the ecosystem to accumulate excessive biomass. Possible results are massive algae blooms, reduced light penetration and fouling called as eutrification. Eutrification will stimulate the growth of algae, resulting in strong oxygen production during daytime. Respiration during the night, decomposition of organic matter, either from increased biological production or inputs from human activities and degradation of dead algae will lead to anaerobic conditions (fish kills). Eutrophication also stimulates the growth of nuisance and toxic algae (e.g. cyanobacteria red tides). Blooms of toxic blue–green algae severely restrict the use of the water. Many organisms can suffer degradation of their habitat as a result. In extreme situations fish may die.
Suspended solids	Suspended solids increase the turbidity of the water, reducing the available light for light depending organisms like seaweeds, sea grasses and corals. After sedimentation, suspended solids can cover benthic species and cause prevailing of anaerobic conditions
Heavy metals	Toxicity to aquatic life. Heavy metals such as mercury, cadmium etc. can concentrate in shellfish and fish tissues and damage the physiological processes and functions of reproduction, feeding and respiration, and resulting in unacceptable high concentrations for consumers. Heavy metals can interfere with microbiological processes in sewage treatment plants.
Toxic organics	Both organic and inorganic toxic even at extremely low concentrations, may be poisonous to fish and other aquatic organisms and microorganisms. Toxic substances with mutagenic properties can cause carcinogenic, mutagenic and teratogenic effects.
Inorganic chemicals	Crop damage, soil poisoning
Heat	Lower dissolved oxygen levels, changes in the activities and behavior of aquatic organisms
Hazardous materials	Oil spills, chemical spills etc. can cause fish kills.

wastes from homes, schools, office buildings and stores. The term municipal sewage includes domestic sewage along with any industrial wastewaters that are permitted to discharge into the sanitary sewers.

A dispersed source is a broad area from which pollutants reach a water body. There is no discrete or identifiable point of discharge, and pollution enters the environment by a multitude of pathways. Surface runoff from agricultural and urban areas is classified as a dispersed source pollutant. As the runoff moves, it picks up and carries away natural and human-made pollutants, finally depositing them into lakes, rivers, wetlands, coastal waters and even underground sources of drinking water. These pollutants include (USEPA, 2006a)

- Excess fertilizers, herbicides and insecticides from agricultural lands and residential areas;
- Oil, grease and toxic chemicals from urban runoff and energy production;
- Sediment from improperly managed construction sites, crop and forest lands and eroding streambanks;
- Salt from irrigation practices and acid drainage from abandoned mines;
- Bacteria and nutrients from livestock, pet wastes and faulty septic systems;
- Atmospheric deposition and hydromodification are also sources of non-point source pollution.

Point source pollutants are much easier to control than dispersed ones; those from point sources are collected and conveyed to a treatment facility where they can be removed from water. The point discharges from the treatment plant can easily be monitored by regulatory authorities. However, non-point sources are much more difficult to control. The only way of controlling these pollutants is to set appropriate restrictions on land use. In urban areas, the provision of reticulated sewerage systems and adequate street cleaning are important measures, while in farming and forestry areas, soil conservation practices and the controlled application of pesticides and fertilizers are necessary if pollution of waterways is to be avoided.

11.3 Water Quality Parameters

The impurities accumulated by water throughout the hydrologic cycle and as a result of human activities may be in suspended and dissolved form. Suspended materials consist of particles larger than molecular size. These particles are held in suspension by the buoyant and viscous forces. Dissolved particles consist of ions and molecules and detained in water by the molecular structure of water. Colloidal particles are small size suspended solids that usually exhibit many of the characteristics of dissolved substances. Size ranges of dissolved, colloidal and suspended particles in water are indicated in Table 11.3.

Table 11.3 Size ranges of particles in water

Type of Solid	Size Range
Suspended	>1 mm
Colloidal	1 mm and 0.001 mm
Dissolved	<0.001 mm

Determination of the quantity of these impurities, however, is not sufficient to define water quality. Depending on the selected use of water, many parameters have evolved to reflect the quality. These parameters are classified into three major classes: physical, chemical and biological parameters. Table 11.4 presents the typical parameters and their brief descriptions.

Table 11.4 Major water quality parameters

Parameter		Description
Physical		
	Total suspended solids, turbidity	Total suspended solids (TSS) concentration and turbidity both indicate the amount of solids suspended in the water, whether inorganic (e.g. soil particles) or organic (e.g. algae). The TSS test measures an actual weight of material per volume of water, while turbidity measures the amount of light scattered from a water sample (more suspended particles cause greater scattering). High concentrations of suspended solids affect light penetration and productivity, and cause lakes to fill in faster.
	Temperature	Temperature has a major influence on biological activity and growth. Elevated temperatures can cause a change in the species of fish and also result in serious depletion in dissolved oxygen concentrations.
	pH	Aquatic organisms differ as to the range of pH in which they grow. The pH of water determines the solubility and biological availability of chemical constituents.
	Secchi depth	Secchi depth is an indication of water clarity. Clarity which is affected by algae, soil particles and other materials suspended in the water can be used to determine a lake's trophic status.
Chemical		
	Chlorophyll	Chlorophyll is the green pigment in plants that allows them to photosynthesize. The increase in nutrients caused by pollution usually results in more algae and in turn eutrification, causing aesthetic problems.
	Organics	Organics may come from natural sources or may result from human activities. Dissolved organics are divided into two classes: biodegradable and non-biodegradable. Measurement of biodegradable organics is usually performed by the biochemical oxygen demand (BOD) test. Non-biodegradable organics may be quantified using chemical oxygen demand (COD) or total organic carbon (TOC) test.
	Dissolved oxygen	Oxygen is required by aquatic organisms for respiration and used by bacteria for the decomposition of organic matter. As a result, an oxygen-deficient environment can develop in lakes and rivers with excess pollution.
	Phosphorus	One of the main nutrients. Phosphorus can be measured as total phosphorus (TP) or as phosphate (PO_4) or orthophosphate (ortho-P). Phosphate and orthophosphate represent the fraction of TP that is available to organisms for growth.

(continued)

Table 11.4 (continued)

Parameter		Description
	Nitrogen	Another main nutrient. Nitrogen can be measured as total nitrogen (TN), total Kjeldahl nitrogen (TKN), nitrate-nitrogen (NO_3), nitrite-nitrogen (NO_2) or ammonia-nitrogen (NH_4). TKN represents the fraction of TN that is unavailable for growth. The remaining fractions, NO_3, NO_2 and NH_4, represent bioavailable forms of nitrogen.
Biological		
	Fecal coliform bacteria	Fecal coliform bacteria are relatively harmless microscopic organisms that live in large numbers in the intestines of man and warm-blooded animals. The presence of fecal coliform bacteria in aquatic environments indicates that the water has been contaminated with the fecal material of man or other animals. The presence of fecal contamination is an indicator of the presence of disease-carrying organisms, which live in the same environment as the fecal coliform bacteria.
	Benthic macroinvertebrates	Macroinvertebrates (e.g. insects in their larval or nymph form, crayfish, clams, snails and worms) that live in the bottom parts of waters are good indicators of watershed health. The basic principle behind the study of macroinvertebrates is that some are more sensitive to pollution than others. Therefore, if a stream site is inhabited by organisms that can tolerate pollution and the more pollution-sensitive organisms are missing a pollution problem is likely.

11.4 Water Quality in Rivers and Streams

Rivers and streams are shallow and narrow surface waters which flow rapidly downstream therefore carry algae and other solids and discourage the growth of rooted plants. The effect of pollution on a river depends both on the nature of the pollutant and the characteristics of the river. Volumetric flowrate of the river, the depth, the type of bottom, the types of aquatic life in the river and the surrounding vegetation and land-use patterns are the major stream characteristics affecting the extent of pollution. The nature and the amount of the pollutants reaching a river are also of primary importance.

The prevalent pollutants impacting rivers and streams are nutrients, pathogens, toxic substances, siltation and oxygen-demanding organics. Nutrient pollution which generally refers to elevated quantities of nitrogen and phosphorus mostly originates from domestic discharges, forestry operations and agricultural runoff, and can cause increased plant and algae growth resulting in a change in the populations of fish and other desirable aquatic species. Inadequately treated municipal wastewaters, agricultural and urban runoff and wildlife fecal material are the common sources of pathogens in rivers and streams. Silt is another key pollutant in rivers. The major contributors are the erosion of cultivated farmland, mining, urban runoff and construction sites. It can cause destruction of aquatic habitat and give damage to the food web that supports fish and other organisms in the river water. The discharge of oxygen-demanding organics into rivers and streams can cause a depletion of dissolved oxygen in downstream waters which can in turn cause severe problems for the aquatic life and result in dramatic changes in the ecological situation. Common sources of oxygen-demanding organics are inadequately treated municipal and industrial wastewaters, as well as urban and agricultural runoff.

Among the above-mentioned pollutants, oxygen-depleting organics and nutrients are so common and have a pronounced effect on almost all rivers and streams that they deserve special emphasis. This does not mean that these two group of pollutants are the most important pollutants in all rivers, but rather that no other pollutants have as much as overall effect on rivers.

Some pollutants like oxygen-demanding organics discharged into a stream can be removed by the natural degradation if it is below its self-purification capacity. In natural waters, to a limited extent, organic can be oxidized to simpler, more stable end-products by the biological life in the river and in this way the rivers can recover the adverse effects of pollutants naturally without being significantly or permanently affected.

Self-purification or natural assimilative capacity of a river depends on the strength and the amount of pollutants and also the volumetric flow rate of the river. Not as evident, but equally significant is of course the oxygenation rate or reaeration in the river. If high levels of organic matter are present in water, rate of oxygen utilization by microbes may exceed the oxygenation rate and use all available oxygen.

Another effective mechanism in the natural abetment of the pollutants is the dilution or flushing action of the flowing water. Dilution contributes to self-purification capacity simply by providing a decrease in the concentration of pollutants.

Meanwhile, it is important to note that not all pollutants can be assimilated in water by natural means. This is particularly true for non-biodegradable organics such as PCBs (polychlorinated biphenils), organotins etc. These compounds are mostly toxic to aquatic life, persistent in the nature and also bio-accumulative. For these substances, even the dilution in the flowing water can be ineffective if these substances are trapped in the river sediments.

11.4.1 Dissolved Oxygen (DO) Sag Curve

When biodegradable organic materials are discharged into a river, they exert oxygen demand and consume dissolved oxygen (DO) as they metabolize and consume organics through aerobic biodegradation. This causes a drop in the dissolved oxygen of water at the discharge point. The longitudinal profile of oxygen concentration is called an *oxygen sag curve* (Figure 11.2). The concentration of oxygen downstream a waste discharge is controlled mainly by the competing processes of oxygen depletion due to waste degradation and oxygen replenishment by aeration. The shape of the sag curve or the extent of sag depends on the level of pollution and the flow. Therefore, the shape can change seasonally, e.g. under high flow conditions, recovery of oxygen levels may be rapid. In cases of extremely heavy organic pollution, or very low streamflow, the oxygen in water may be completely depleted and septic conditions may prevail.

One of the major tools in water quality management in rivers is the ability to assess the self-purification capacity. This is done by developing a mathematical expression for the DO sag curve downstream from a waste discharge. The sources of oxygen and the factors causing oxygen depletion are quantified and used in writing a mass balance equation for oxygen. The major sources of oxygen are reaeration from atmosphere and photosynthesis of aquatic plants. Oxygen demand is exerted by oxygen-demanding substances discharged and already existing in the river water, non-point source pollution and oxygen consumption by the aquatic organisms and organisms living in sediments (benthic demand).

One equation used to describe DO sag curve is known as Streeter-Phelps equation. This equation is the simplest model for the DO profile in a river and is based on the assumption that the only two processes taking place are deoxygenation by the oxygen-demanding substances and oxygenation by the oxygen transfer at the surface of the water body. Rate of aeration is expressed by

$$d(DO)/dt = k_r r(DO_s - DO) = k_r D \qquad (1)$$

where DO_s = saturation DO concentration, DO = oxygen concentration in water, water, k_r = reaeration constant and D = DO deficit (=DO_s–DO). Reaeration constant k_r is very much dependent on the physical characteristics of the river.

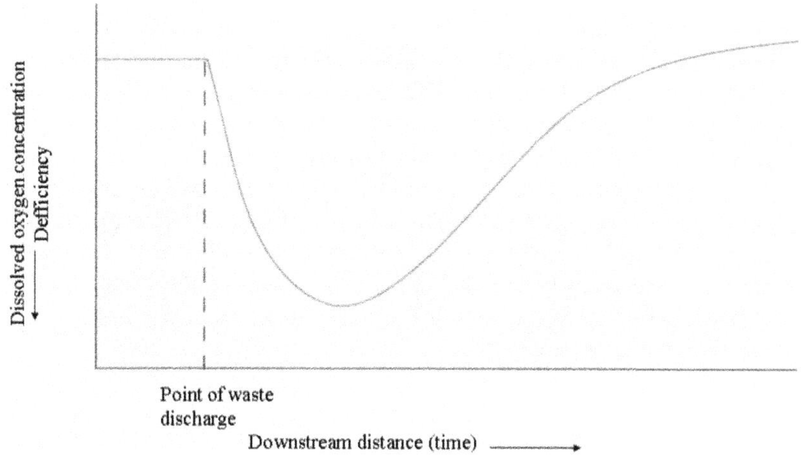

Fig. 11.2 Oxygen sag curve

The rate at which oxygen disappears from the stream as a result of microbial degradation of oxygen-demanding organic substances is exactly equal to the rate of increase in deficit and assumed to be proportional to the BOD remaining at that point, so that:

$$d(DO)dt = -dD/dt = k_d L_t \tag{2}$$

where k_d = deoxygenation rate; L_t = BOD remaining at some time after the waste enters the stream.

From the Eqs. (1) and (2) we can see that the oxygen deficit is a function of the competition between oxygen utilization and reaeration from the atmosphere.

$$dD/dt = k_d L - k_r D \tag{3}$$

where dD/dt = the rate of change of the oxygen deficit (D). By integrating Eq. (3), and using the boundary conditions (at t = 0, D = D_a and L = L_a; and at t = t, D = D_t and L = L_t), the DO sag curve equation is obtained:

$$D = (k_d L_a)/(k_k - k_d)(e^{-k_d t} - e^{-k_r t}) + D_a(e^{-k_r t}) \tag{4}$$

where D_t = oxygen deficit in river water after exertion of waste for time, t; L_a = initial ultimate BOD after river and wastewater have mixed; D_a = initial deficit after river and wastewater have mixed. Where $k_r = k_d$, Eq. (4) reduces to

$$D = (k_d t L_a + D_a)(e^{-k_d t}) \tag{5}$$

11.4.2 River Quality Management Strategy

The lowest point on the DO sag curve with respect to dissolved oxygen concentration is called the critical DO concentration and is of major interest because it indicates the worst condition. The use DO sag curve in water quality management is based on the determination of minimum DO that will protect the aquatic life in the stream (Davis and Masten, 2004). This value is generally set as DO standard to protect the stream. For a known river and waste discharge, the Streeter-Phelps equation can be solved to determine critical deficit. If this value is over the DO standard, it means the stream can assimilate the waste discharged without being seriously affected. In the opposite case, waste needs to be treated before discharge into the stream.

11.5 Water Quality in Lakes

Due to the physical characteristics, pollution problems in lakes and constructed reservoirs are different from the problems in rivers and streams. Lakes and reservoirs are stagnant water bodies without flushing action for incoming pollutants. Thus, in these water bodies, pollutants can remain for years and cause degradation of water quality. Lakes are also significantly influenced by seasonal temperature changes.

In lakes, water quality basically depends upon the nutrients' level rather than oxygen-consuming organic pollutants in rivers. Entrance of nutrients into lakes causes the excessive growth of algae and also rooted aquatic plants. When they die, they settle to the bottom where they are decomposed by bacteria and protozoa causing depletion of dissolved oxygen of the lake.

When excessive growth of algae occurs, an algal slimy mat which floats on the lake surface and interferes with the recreational use of the lake is formed. A lake suffering from algal blooms is also not a very good drinking water resource, because, water treatment costs are higher when water contains macroscopic algal particles.

Decaying plants and also silt carried to the lake by surface runoff and streams discharging accumulate as sediment at the bottom of the lake, decreasing the lake water depth gradually and causing a shift in the aquatic life.

Nutrient enrichment and gradual filling of a lake are natural processes and can be considered as inevitable aging of the lake called eutrophication. These processes along with the physical characteristics (shape, depth, geology, light, temperature and wind mixing) and biological productivity of a lake control the water quality. Lake productivity is primarily controlled by nutrients load to the lake. Lakes with low nutrients levels are called oligotrophic and characterized by clear water and low oxygen demand. Over the years, with the start of the accumulation of silty sediments at the bottom and also nutrients in lake water, the lake passes through a mesotrophic stage. At the next stage, nutrient levels are higher with algal blooms floating over the surface and the lake becomes eutrophic. In a eutrophic lake, water is turbid with a high oxygen demand and in turn with a low dissolved oxygen level at lower depths. At certain times of the year, the water at the bottom of the lake may even be devoid of dissolved oxygen. In some lakes the eutrophication process is so slow that it may take thousands of years without any major change in water quality. Other lakes may be eutrophic from the day they were formed or may become eutrophic rapidly. Although eutrophication is a natural process of aging, human activities and resulting pollution may markedly speed up the process. After eutrophication, with further aging, lakes become senescent and eventually marshes. Senescent lakes are with thick organic sediment layer and rooted water plants in great abundance.

11.5.1 Control of Eutrofication

Water quality management in a lake is primarily concerned with slowing eutrophication to its natural rate. The control of eutrophication is directly related to aquatic food chain (Figure 11.3). Although the process typically begins with sunlight-driven photosynthesis by algae and plants, balanced nutrition is also required to sustain life. Green plants capture energy from sunlight to convert inorganics (carbon dioxide, water and mineral compounds) into organic plant tissue. Lake photosynthesizers include algae and macrophytes. Together, they are the primary producers, because they create the organic material required by most other organisms for nutrients and energy. Oxygen, the waste product of photosynthesis, adds to the dissolved oxygen in lake. Besides light, algae and higher plants need oxygen, carbon dioxide and mineral nutrients to survive and grow. Carbon dioxide is virtually always available and comes from the weathering of carbonate rocks, diffusion from the atmosphere and from the respiration of organic matter by all of the organisms in the lake. The whole interaction of photosynthesis and respiration by plants, animals and microorganisms represents the food web. Food webs are usually very complex, and in any one lake ecosystem, hundreds of different species can be involved. These plants may die and decompose or be eaten by primary consumers—the second trophic level. This link in the food chain typically involves zooplankton grazing on algae, but also includes larval fish eating

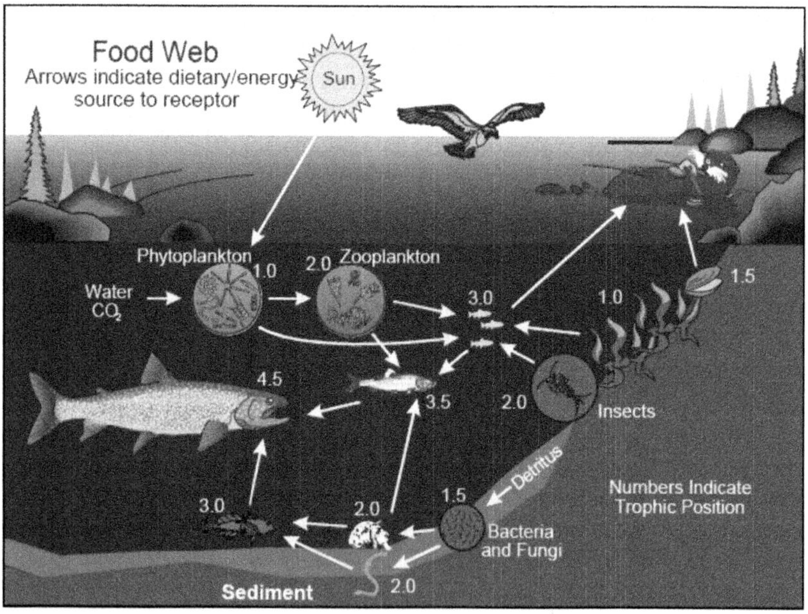

Fig. 11.3 Aquatic food chain (Source: http://www.epa.gov)

zooplankton and a variety of invertebrates that eat attached algae (periphyton) and higher plants. Other animals, such as small fish, secondary consumers (third trophic level) eat the primary consumers and thus are considered secondary consumers. Still larger consumers, such as large fish, ospreys and people are tertiary consumers (fourth trophic level). Thus, energy and nutrients originating from the photosynthetic production of biomass and energy cascade through the food web. There is some recycling of nutrients back up to the top of the cascade. Respiration, the oxidation of organic material, releases the energy that was originally captured from sunlight by photosynthesis. Both plants and animals respire to sustain their lives, and in doing so, consume oxygen. Microorganisms (bacteria and fungi) consume a large fraction of available oxygen in the decomposition of excreted and dead organic material.

Decomposers are sinks for plant and animal wastes, but they also recycle nutrients for photosynthesis. The amount of dead material in a lake far exceeds the living material. Detritus is the organic fraction of the dead material, and can be in the form of small fragments of plants and animals or as dissolved organic material. In recent years, scientists have recognized that zooplankton grazing on detritus and its associated bacterial community represent an additional important trophic pathway in lakes (WOW, 2004).

The amount of plant growth and normal balance of the food chain is controlled by the limitation of plant nutrients. Excessive discharges of nutrients into lakes promote blooms of blue green algae, and in turn cause eutrophication. The key to controlling lake eutrophication therefore lies in limiting nutrients. Of these, phosphorus is usually recognized as the limiting nutrient. The generally accepted limits for are 0.3 mg/L of ammonia plus nitrate nitrogen and 0.02 mg/L of orthophosphate phosphorus. Lake with annual mean total nitrogen and phosphorus concentrations greater than 0.8 mg/L and 0.1 mg/L, respectively, exhibit algal blooms and nuisance weed growths during most of the season (Hammer and Hammer, 1996). But even with very high oxygen nitrogen levels, if phosphorus concentrations are kept below 0.02 mg/L, excessive growth of algae usually do not occur.

Control of eutrophication in water bodies can be achieved by controlling the nutrient release, accelerating the outflow of nutrient material and sealing the lake bottom. Nutrient release control can be archived by diverting the nutrient-rich wastewaters or by limiting the phosphorus loading. Where water is used for human consumption, nitrogen can also be controlled using the wastewater treatment processes called nitrification–denitrification. Phosphate can be removed from sewage by biological and physico-chemical methods. If the majority of nutrients originate from non-point sources, then the control of point sources may be of limited value. Unfortunately, there are no ready solutions for the removal of nutrients from surface runoff. Waste minimization can be applied to the control of nutrients (phosphorus) from fertilizer fields by encouraging farmers to fertilize more often with smaller amounts and to take effective action to stop erosion that carries phosphorus bearing soil particles into lakes.

Accelerating the outflow of nutrients can be achieved by increasing the flow of water from the lake, removing the sediment by dredging and also by removing the vegetation. Sealing the lake bottom is provided to prevent the exchange of phosphorus between water and sediments. Membranes such as polyethene are placed on the bed of the lake and a layer of sand is spread on it.

11.6 Water Quality in Groundwaters

Groundwater makes up nearly 70% of all the world's freshwater; only 0.2% is found in lakes, streams or rivers and 30% is bound up in snow and ice on mountains and in the polar regions. As rivers and lakes tend to be supported by groundwater, it is not exaggerating to say that almost all the water we use for agriculture, industry and drinking water is either groundwater or has been groundwater at some point in the water cycle (UK Groundwater Forum, 2006).

Groundwater plays a number of very important roles in our environment and in our economies. As mentioned above, in the environment, it supports rivers, lakes and wetlands, especially through drier months when there is little direct input from rainfall. The flow of groundwater into rivers as seepage through the river bed, known as baseflow, can be essential to the health of wildlife and plants that live in the water.

Natural groundwater is generally of excellent quality, because of the natural removal of particulate matter and some other dissolved contaminants in the layers of soil through which the water slowly flows. Any bacterial contamination from surface sources or the soil is removed after groundwater has passed through some 30 m of saturated zone; in the unsaturated zone no more than 3 m may be necessary to purify water percolating through the zone (Forum Groundwater, 2006). The principal action is filtration but some other processes are also involved. Organic compounds, bacteria and viruses tend to be retained or absorbed on the matrix and may be degraded by microbial activity. Metals and other inorganic substances may also be absorbed, diluted by mixing, or may be modified or broken down into simpler products by chemical reactions. Although the purifying capacity of soil and rocks improve the water quality, aquifers do not have an infinite capacity for purifying contaminated water.

Nevertheless, even in areas far from human activity, groundwater quality may not be pure. Groundwater may have harmful concentrations of certain ions, e.g. iron, manganese, sulphate, hydrogen sulphide and, near coasts, sodium and chloride, due to natural processes. As groundwater flows through the solid matrix, it is in intimate contact with minerals in soil and rock deposits for long periods of time. This is why groundwater is usually harder than surface water. For the most part, natural contamination in groundwater poses no threat to public health. But, under some geological conditions, groundwater can contain dangerous concentrations of dissolved constituents like arsenic and fluoride.

Groundwater quality is of course not only a function of natural processes but also anthropogenic activities such as mining, agriculture, domestic and industrial wastes etc. Since demand for groundwater is expected to increase significantly in the near future due to the increase in urban populations, groundwater pollution, which often occurs in urban areas, represents a noteworthy threat to water supplies.

11.6.1 Groundwater Contamination

Contamination of groundwater is a relatively recent concern. People used to think that the underlying soil and rock remove pollutants from groundwater during the seepage. However, this supposition proved false when groundwater quality began to be monitored. It is clear that the natural purification capacity of soil and rock changes widely from one area to the other and some pollutants such as recalcitrant organics, reaching to groundwater do not decay or can not be flushed out naturally.

The potential sources of groundwater contamination are septic systems, disposal pits, deicing salts, landfills, underground storage tanks, storage lagoons, land application of sludge and wastewaters, fertilizers and other chemical used in agriculture, transport and transfer spills, improper wells, inactive mining sites, acid precipitation, saltwater intrusion (Chapter 9) and accidental disposal of hazardous substances (Figure 11.4). Table 11.5 presents the common sources of groundwater pollution along with their degree of importance. As indicated, the major problems with respect to groundwater contamination are hazardous waste landfill or impoundments and sanitary landfills. In many cases, the predominant source of groundwater pollution is the accidental spills of hazardous substances.

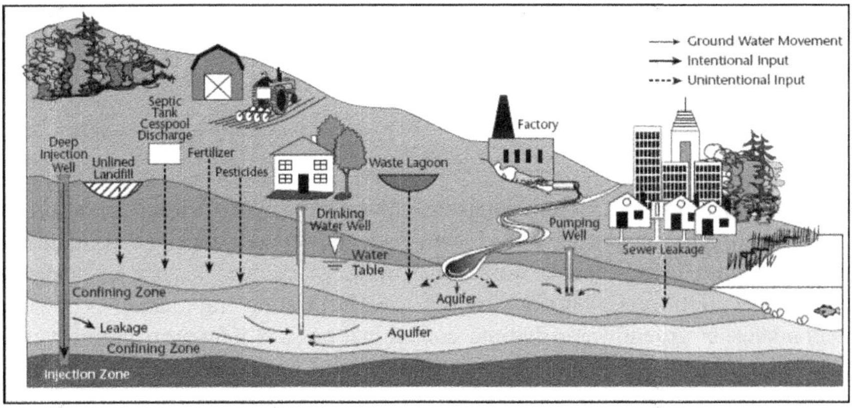

Fig. 11.4 Potential sources of groundwater contamination (Source: http://www.epa.gov)

The most common pollutants in groundwater are

- Nitrate;
- Pesticides;
- Chlorinated solvents;
- Gasoline and oil constituents;
- Heavy metals;
- Pathogens.

Table 11.5 Common sources of groundwater contamination (Nazaroff and Cohen, 2001)

Common Source	Rank[a]
Sources designed to discharge substances	
Septic tanks and cesspools	4
Injection wells of hazardous and non-hazardous waste	
Land application of wastewater and sludge	
Unplanned releases from treatment, storage and disposal facilities	
Hazardous waste landfill or impoundments	1
Sanitary landfills	2
Aboveground and underground storage tanks and pipes	5
Waste piles	
Releases that arise as a consequence of other activities	
Pesticide and fertilizer applications	3
Animal waste disposal	6
Mine drainage	7
Oil drilling	8
Saltwater intrusion	9
Irrigation return flow	10
Urban runoff	
Atmospheric deposition	
Sources designed to transport processes	
Pipelines	
Trucks, trains, ships	

[a] Ranking indicate estimated order of importance for improving groundwater.

11.6.1.1 Nitrate

Nitrate mainly originates from intensive agriculture. When virgin grassland is ploughed, organic matter is mixed with soil and in turn nitrogen is oxidized forming nitrate. Also causing nitrate in soil, is of course, the use of artificial nitrogenous fertilizers. Nitrate leaches by infiltrating rain and eventually accumulates in groundwater. Nitrate contamination is a long-term problem and remedial action is necessary. The cost of chemical treatment to remove it from groundwater is significant and disposal of water products from the process can also be difficult. An alternative course is to reduce the contamination at the source—the amount leaching from the soils. The scale of the problem can be reduced by better land management including.

- Reducing the use of artificial fertilizers;
- Reducing the extent of ploughing in the autumn;
- Sowing autumn crops early;
- Avoiding bare ground in the winter by sowing cover crops;
- Delaying the ploughing-in of crop residues;
- Carefully managing the disposal of farm wastes (UK Groundwater Forum, 2006).

11.6.1.2 Pesticides

Pesticides in groundwater originate from their use in agriculture, on roads and railways for weed control and use in industry for the control pests. Many factors determine whether a pesticide will reach groundwater, including its chemical properties, the soil type, the depth to groundwater and the pesticide management practices. By combining all these factors, the areas most vulnerable and the practices most conducive to pesticide contamination of groundwater can be determined (Table 11.6).

11.6.1.3 Chlorinated Organic Solvents

Chlorinated organic solvents such as trichloroethene (TCE) are another major group of pollutants that commonly occur in groundwater. These compounds which usually originate from dry cleaning and degreasing operations are usually denser than water and have relatively low aqueous solubilities, which led them to be referred to as dense non-aqueous phase liquids (DNAPL). If these substances are spilled into the subsurface, they may penetrate the water table and provide a long-term source of contamination. Other properties of these compounds in relation to their occurrence in groundwater are indicated in Table 11.7.

Table 11.6 Factors indicating the greatest likelihood of groundwater contamination by pesticides

Pesticide properties	
	High solubility
	Low adsorption
	Persistence
Soil characteristics	
	Sand and gravel
	Low organic matter content
Site conditions	
	Shallow depth to groundwater
	Wet climate or extensive irrigation
	Depressions or flat areas where water collects
Management practices	
	Poor timing with respect to climate
	Overapplication (rate too high or application too frequent)

Table 11.7 Properties of organic solvents

Properties	Implication
Low liquid viscosities	Easy movement into the subsurface
Low interfacial tensions	Easy movement of DNAPL into water-filled voids
High volatilities	Easy movement through the unsaturated zone as gases
Low absolute solubilities	Difficult removal from the groundwater zone
High solubilities relative to drinking water limits	Difficult clean up below MCL limits
Low partitioning to soils	Difficult retardation of their migration by the soil matrix
Low degradation rates	Persistence in the subsurface

11.6.1.4 Gasoline and Oil Constituents

Gasoline and oil constituents which are aromatic and aliphatic hydrocarbons, usually originate from spillage and leakage from petroleum refining facilities, storage (principally from underground storage tanks), handling and disposal of oils, gas stations, vehicle maintenance, accidents, runoff from agricultural yards, roads and urban and industrial estates. Some of the more common contaminants from gasoline and oil constituents are benzene, toluene, xylene, ethylbenzene (known as BTEX) and polycyclic aromatic hydrocarbons. These compounds are toxic and mostly known to be carcinogenic. A single spill has the potential to contaminate a very large volume of groundwater to levels in excess of the low concentration acceptable in drinking water. The wide distribution of gas stations almost in all parts of the World, and in some cases extensive networks of pipelines carrying petroleum products illustrates the potential risk to groundwater quality.

11.6.1.5 Heavy Metals

Heavy metals in groundwater are not only related to pollution or other human activities but also to natural processes. Naturally high levels of cadmium and arsenic can occur in areas where the bedrock contains high concentrations of metals; this applies to many types of shale and slate. Other heavy metals such as zinc, lead, chromium etc. may be present in groundwater due to contaminated runoff from urban areas. Metals may enter land and water via the atmosphere from smelters and other metal industrial plants and also from the use of leaded gas. The use of cadmium-contaminated phosphate fertilizer is another source of heavy metals in groundwater. Zinc being used as an anti-corrosion agent, may originate from galvanized steel products embedded in the ground. Other sources of zinc and arsenic are the spread of preservatives via waste spill at impregnating facilities or via wood products themselves, waste disposal sites and dumps of mining waste.

11.6.1.6 Pathogens

Pathogens in groundwater mainly originate from the fecal material of humans and other animals, and may be bacterial, viral or protozoan. Although filtered by natural processes, groundwater may contain coliform bacteria, E. coli, coliphage and human viruses. Viruses were found about ten times more often than fecal bacteria (Waterencyclopedia, 2006). Water containing fecal material may seep into the groundwater from the land surface or from underground sources of contamination. Major surface sources include

- Wastewater and biosolids from sewage treatment facilities that have been applied to land as a soil conditioner;
- Seepage from shallow artificial ponds (lagoons) used for processing sewage;
- Seepage from contaminated lakes and other surface-water bodies;
- Urban runoff;
- Feces from cattle and other livestock operations;
- Improperly constructed sanitary landfills where trash and garbage are disposed.

Fecal contamination can also reach groundwater from underground sources, such as improperly functioning septic tank systems, underground reservoirs for liquid household sewage (cesspools) or leaking underground sewer lines.

11.6.2 Remediation

Usually the groundwater pollution is localized, although in urban areas where there are many potential sources of pollution, contamination is widespread. If groundwater is contaminated by a relatively small-scale source of pollution such as a spill of hydrocarbons, the question is: how can the quality be restored or at least improved to an acceptable standard? The first concern is to determine the distribution of the contaminant in the aquifer, whether it is in the unsaturated zone or the saturated zone or both, and whether it is entirely in solution or there is a non-aqueous phase. Answers to these questions will suggest a possible remediation strategy (UK Groundwater Forum, 2006).

Despite the availability of the techniques listed in Table 11.8, restoring an aquifer to its original pristine state is difficult if not impossible. When a contaminated aquifer is pumped, the groundwater flows preferentially through the more permeable horizons and consequently these are cleaned relatively easily and quickly. However, contaminants that have penetrated into less permeable horizons and the smaller pore spaces may be virtually unaffected. When pumping stops, contaminants from these zones will diffuse into the permeable zones that have been cleared.

Table 11.8 Basic groundwater remediation methods (Source: UK Groundwater Forum, 2006)

Method	Description
Containment	Containment of the pollution to prevent it from migrating from its source. Several ways are available to contain groundwater contamination: physically, by using an underground barrier of clay, cement or steel; hydraulically, by pumping wells to keep contaminants from moving past the wells; or chemically, by using a reactive substance to either immobilize or detoxify the contaminant. When buried in an aquifer, zero-valent iron (iron metal filings) can be used to turn chlorinated solvents into harmless carbon dioxide and water.
Pump and Treat	Removal of the contaminated water by pumping and then treating it to an appropriate quality. The water can then be either re-injected into the aquifer or used directly for water supply.
Bioventing	Removal of the volatile fraction of a contaminant by a process called "bioventing". Soil gas is removed by vacuum pumping from a borehole. This circulates air which volatizes and assists in the degradation of organic contaminants. The method removes volatile contaminants from an aquifer that are present at less than the residual saturation level.
Bioremediation	Bioremediation dealing with organic contaminants is a treatment process that uses naturally occurring microorganisms to break down some forms of contamination into less toxic or non-toxic substances. The technique is intended to encourage bacteria to grow by adding nutrients to the contaminated zone. Organic matter is degraded into simple compounds such as water and carbon dioxide. By adding nutrients or oxygen, this process can be enhanced and used to effectively clean up a contaminated aquifer. Alternatively, specific bacteria can be introduced to metabolize a particular contaminant. The process can be encouraged by injecting air into the contaminated zone from a borehole. Some of the newest cleanup technologies use surfactants, oxidizing solutions, steam or hot water to remove contaminants from aquifers. This stimulates aerobic biodegradation. Under favourable conditions, bioremediation can occur as a result of natural processes in an aquifer. Indigenous microbes will degrade organic contaminants if a supply of nutrients is present for their metabolism. Because bioremediation relies mostly on nature, involves minimal construction or disturbance and is comparatively inexpensive, it is becoming an increasingly popular cleanup option.

11.6.3 Preventive Measures

After an aquifer has been contaminated, it is difficult to entirely define or isolate a contaminant plume. The difficulty of cleaning an aquifer emphasizes the need to prevent or minimize all forms of pollution. Prevention is the key to controlling groundwater pollution and includes finding the major sources of contamination, and learning to control them. It must be accepted that, prevention is cheaper than cure; and with existing technology, remedial actions are usually ineffective, it is often unrealistic to talk about a "cure" for groundwater contamination. Groundwater quality often cannot be restored within a reasonable time and at a reasonable cost. Natural decontamination that requires decades and perhaps centuries may occur if the pollutants are decay or flushed out of the aquifer. Depending on the type of pollutant, this natural cleanup may not occur. Nevertheless, wherever pollution occurs in an aquifer, it should be contained and treatment initiated, bearing in mind the use made of the groundwater, any health risk that exists and cost.

There are many different ways of preventing groundwater contamination which depend on where groundwater is located, where it is used and what dangers of contamination exist. Prevention measures basically include using land-use management plans applied on the local level by towns and cities. For example:

- Zoning ordinances that prevent residential or industrial development in areas where groundwater recharge occurs;
- Establishing strict regulations pertaining to the construction of septic systems, fuel and wastewater tanks, chemical storage tanks and solid waste landfills;
- Prudent application of pesticides and fertilizers in agricultural areas can also be effective in this regard.

A key step in developing zoning ordinance regulations for groundwater protection is the identification of practices, measures and structural facilities which can be used to protect groundwater. Measures and practices which can be installed at the time of initial project construction are particularly appropriate for incorporation into local zoning ordinances. Requirements for spill prevention and/or waste reduction plans may also be useful, especially for regulating medium large manufacturing establishments. Also important is the staff capability of preparing and implementing such plans. The particular types of measures or management practices necessary for groundwater protection vary among the different types of land uses. Density restrictions for unsewered development, closed holding tanks with off-site disposal of wastewater, spill prevention, waste reduction plans and water well isolation distances for underground storage tanks and above ground hazardous substance storage are some of the "best management practices" which may be incorporated in local zoning ordinances.

To prevent groundwater contamination from septic systems, a number of steps could be taken. Most states have standards for constructing new septic systems and repairing systems that fail to operate properly. Monitoring of groundwater and surface water detects areas where septic systems are contributing to water quality problems. New designs are being developed to better treat wastewater in situations where a standard septic tank and drainfield falls short of adequately treating wastewater. And finally, the heightened awareness of the risks from hazardous household chemicals and sewage, along with greater opportunities for safe disposal, should minimize such contaminants from entering groundwater through septic systems.

The key to the prevention of groundwater pollution from improper use of pesticides and fertilizers, is to avoid overapplication of these chemicals by following soil test guidelines as well as by practicing simple landscape maintenance techniques. This is achieved by developing a comprehensive nutrient management plan and using only the types and amounts of nutrients and pesticides necessary to produce the crop, applying nutrients at the proper times and with proper methods, implementing additional farming practices to reduce nutrient losses and following proper procedures for fertilizer storage and handling.

To help protect water wells against contamination, it is also important to use the natural protection that soil provides by maintaining adequate separation distances between wells and potential sources of contamination. The vulnerability potential of an aquifer to groundwater contamination is in large part a function of the susceptibility of its recharge area to infiltration. Areas that are replenished at a high rate are generally more vulnerable to pollution than those replenished at a slower rate. Unconfined aquifers that do not have a cover of dense material are susceptible to contamination. Bedrock areas with large fractures are also susceptible by providing pathways for the contaminants. Confined, deep aquifers tend to be better protected with a dense layer of clay material. Wells that connect two aquifers increase the chance of cross contamination between the aquifers. Soil overlying the water table provides the primary protection against groundwater pollution. Bacteria, sediment and other insoluble forms of contamination become trapped within the soil pores. Some chemicals are absorbed or react chemically with various soil constituents, thereby preventing or slowing the migration of these pollutants into the groundwater. In addition, plants and soil microorganisms use some potential pollutants, such as nitrogen, as nutrients for growth, thereby depleting the amount that reaches the groundwater. Just as any man-made filtering device can be overloaded, so can the natural filtering capacity of soil. Large amounts of potential pollutants concentrated in a small area can cause localized groundwater contamination, depending on the depth and type of soil above the water table (Purdue, 2006).

11.7 Water Quality Standards

Water quality standards define water quality goals of a water body by designating uses of the water and by setting criteria necessary to protect the beneficial uses, such as swimming, fishing, aquatic life habitat and agricultural and drinking water supplies. Water quality standards are fundamental tools that help protection of surface and groundwater resources and generally include the following components (USEPA, 2006b):

- *Designated uses* of the water body (e.g. recreation, water supply, aquatic life, agriculture);
- *Water quality criteria* to protect designated uses (numeric pollutant concentrations and narrative requirements);
- An *antidegradation policy* to maintain and protect existing uses and high quality waters;
- *General policies* addressing implementation issues (e.g. low flows, variances, mixing zones).

Designated uses are identified by taking into consideration the use and value of the water body for public water supply, for protection of fish, shellfish and wildlife, and for recreational, agricultural, industrial and navigational purposes. In designating uses for a water body, the suitability of a water body for the uses is examined based on the physical, chemical and biological characteristics of the water body, its geographical setting and scenic qualities and economic considerations. Each water body does not necessarily require a unique set of uses. Instead, the characteristics necessary to support a use can be identified so that water bodies having those characteristics can be grouped together as supporting particular uses.

Water quality criteria can include general narrative statements that describe good water quality and specific numerical concentrations that are known to protect designated uses such as swimming, drinking and the propagation and growth of aquatic life. Numerical criteria are for specific physical, chemical (toxics) and biological characteristics of the waters (e.g. minimum of 4.0 mg/L dissolved oxygen) and estimations of concentrations of pollutant and degree of aquatic life toxicity allowable in a waterbody without adversely impacting its designated uses. Narrative criteria include general protective statements known as the "free from". These type of criteria say that all waters shall be free from sludge, floating debris, oil and scum, color and odor producing materials and substances that are harmful to human, animal, plant or aquatic life. These numerical and narrative criteria describe water quality necessary to protect.

The antidegradation provisions describe the conditions under which water quality may be lowered in surface waters and identify the steps and questions that must be addressed when regulated activities are proposed that may affect water quality. Designated uses must be maintained and protected, and in no case may waters be degraded below the levels necessary to protect existing uses.

Water authorities may adopt policies and provisions regarding water quality standards implementation, such as mixing zone, variance and low-flow policies. In some cases, authorities, at their discretion, allow mixing zones for point source discharges considering the progressive dilution of the effluent plume. Or, they may wish to include a variance as part of a water quality standard. A variance may specify an interim water quality criterion which is applicable for the duration of the variance. Water authorities may also create general policies that address implementation issues such as low flows, variances and mixing zones.

11.8 How to Conserve Water and Use It Effectively[1]

Water use is usually defined and measured in terms of withdrawal or consumption that which is taken and that which is used up. Withdrawal refers to water extracted

[1] Sources: USEPA, 2006c; USGS, 2006.

from surface or groundwater sources, with consumption being that part of a withdrawal that is ultimately used and removed from the immediate water environment whether by evaporation, transpiration, incorporation into crops or a product or other consumption. Conversely, return flow is the portion of a withdrawal that is actually not consumed, but is instead returned to a surface or groundwater source from a point of use and becomes available for further use.

Water use can also be divided into offstream and instream uses. Offstream water use involves the withdrawal or diversion of water from a surface or groundwater source for domestic and residential uses, industrial uses, agricultural uses and energy development uses. Instream water uses are those which do not require a diversion or withdrawal from the surface or groundwater sources, such as: water quality and habitat improvement, recreation, navigation, fish propagation and hydroelectric power production. Globally, of the three major categories of freshwater use—for agriculture, industry and domestic—agriculture dominates. On a worldwide basis, agriculture accounts for about 69% of all annual water withdrawals; industry, about 23%; and domestic use, about 8%.

Water users can be divided into two basic groups: system users (such as residential users, industries and farmers) and system operators (such as municipalities, state and local governments and privately owned suppliers). These users can choose from among many different water use efficiency practices, which fall into two categories:

- Engineering practices: practices based on modifications in plumbing, fixtures or water supply operating procedures;
- Behavioral practices: practices based on changing water use habits.

In the following sections, water conservation practices for system users are presented first, followed by practices for system operators.

11.8.1 Practices for System Users

11.8.1.1 Residential Users

An engineering practice for individual residential water users is the installation of indoor plumbing fixtures that save water or the replacement of existing plumbing equipment with equipment that uses less water. Residential demands account for about three-fourths of the total urban water demand. Indoor use accounts for roughly 60% of all residential use, and of this, toilets use nearly 40%. Toilets, showers and faucets combined represent two-thirds of all indoor water use. In new construction and building rehabilitation or remodeling, there is a great potential to reduce water consumption by installing low-flush toilets. Showers that account for about 20% of total indoor water use, can be replaced by low-flow showerheads that provide the quality of service found in higher-volume models.

Faucet aerators, which break the flowing water into fine droplets and entrain air while maintaining wetting effectiveness, are inexpensive devices that can be installed in sinks to reduce water use. Aerators can be easily installed and can reduce the water use at a faucet by as much as 60% while still maintaining a strong flow.

Because flow rate is related to pressure, the maximum water flow from a fixture operating on a fixed setting can be reduced if the water pressure is reduced. Homeowners can reduce the water pressure in a home by installing pressure-reducing valves. The use of such valves might be one way to decrease water consumption in homes that are served by municipal water systems. For homes served by wells, reducing the system pressure can save both water and energy. A reduction in water pressure can save water in other ways: it can reduce the likelihood of leaking water pipes, leaking water heaters and dripping faucets. It can also help reduce dishwasher and washing machine noise and breakdowns in a plumbing system.

Domestic wastewater composed of wash water from kitchen sinks and tubs, clothes washers and laundry tubs is called gray water. Gray water can be used by homeowners for home gardening, lawn maintenance, landscaping and other innovative uses.

In landscaping, water conservation can be provided using plants that need little water, thereby saving not only water but labor and fertilizer as well. A similar method is grouping plants with similar water needs. Scheduling lawn irrigation for specific early morning or evening hours can reduce water wasted due to evaporation during daylight hours. Another water use efficiency practice that can be applied to residential landscape irrigation is the use of cycle irrigation methods to improve penetration and reduce runoff. Cycle irrigation provides the right amount of water at the right time and place, for optimal growth. Other practices include the use of low-precipitation-rate sprinklers that have better distribution uniformity, bubbler/soaker systems or drip irrigation systems.

Behavioral practices involve changing water use habits so that water is used more efficiently, thus reducing the overall water consumption in a home. These practices require a change in behavior, not modifications in the existing plumbing or fixtures in a home. Behavioral practices for residential water users can be applied both indoors in the kitchen, bathroom and laundry room and outdoors.

In the kitchen, for example, 10–20 gallons of water a day can be saved by running the dishwasher only when it is full. If dishes are washed by hand, water can be saved by filling the sink or a dishpan with water rather than running the water continuously.

Water can be saved in the bathroom by turning off the faucet while brushing teeth or shaving, or by taking short showers rather than long showers or baths etc. Toilets should be used only to carry away sanitary waste.

Households with lead-based solder in pipes that flush the first several gallons of water should collect this water for alternative nonpotable uses (e.g. plant watering).

Water can be saved in the laundry room by adjusting water levels in the washing machine to match the size of the load. If the washing machine does not have a variable load control, water can be saved by running the machine only when it is full. If washing is done by hand, the water should not be left running.

Outdoor water use can be reduced by watering the lawn early in the morning or late in the evening and on cooler days, when possible, to reduce evaporation. Allowing the grass to grow slightly taller will reduce water loss by providing more ground shade for the roots and by promoting water retention in the soil. Growing plants that are suited to the area ("indigenous" plants) can save more than 50% of the water normally used to care for outdoor plants.

11.8.1.2 Industries

Most water-based industries can realize many benefits (economic and environmental) from the practice of water conservation. By reducing their use of water, industries can protect the environment and gain a competitive edge by reducing their own cost of doing business. In many cases, very simple changes can lead to extensive reductions in water consumption. The key components of industrial water conservation are to set up a water conservation program and to run a survey of the plant operations to identify the points where water reuse and a reduction in water consumption are possible.

11.8.1.3 Farmers

There are technologies and management strategies available that conserve water while maintaining yield and production standards. These technologies and management strategies like improved irrigation scheduling and crop specific irrigation management often not only conserve water, but also save energy and decrease growers costs.

Accurate water measurement and soil moisture monitoring are key components of efficient on-farm water management practices. Irrigation flow meters can be used to help calculate the efficiency of irrigation systems, identify water loss from leaks in conveyance systems and to accurately apply only the necessary amount of water based on soil moisture levels and weather conditions. Soil moisture monitoring is used in conjunction with weather data and crop evapotransporation requirements to schedule irrigation. Growers can avoid runoff and maximize water use efficiency by applying the right amount of water at the right time. This is commonly called irrigation scheduling. Fields should be designed for efficient water use by grading land, creating furrow dikes to conserve rainwater and by retaining soil moisture through conservation tillage.

The key to maximizing irrigation efforts is uniformity. The producer has a lot of control over how much water to supply and when to apply it, but the irrigation system determines uniformity. Deciding which irrigation systems is best for your operation requires a knowledge of equipment, system design, plant species, growth stage, root structure, soil composition and land formation. Various types of irrigation systems are available and should be chosen carefully depending on the crop and the application (Table 11.9).

Flood irrigation is used when water is plentiful, runoff can be returned to the delivery system and is of high enough quality to be reused, and when targeted irrigation is unimportant. Flood is typically used when slopes are mild enough to prevent erosion, but steep enough to ensure water reaches the bottom end of the field without oversaturating the top end of the field. Although flood systems are usually less efficient than sprinkler, micro or drip systems, they may be operated with acceptable efficiency. However, evaporation losses may be extreme in arid climates, and growers lack flexibility in water application duration and frequency.

Drip irrigation is much more efficient than flood irrigation. This type of system can be the most water-efficient method of irrigation, if managed properly, since evaporation and runoff are minimized.

Sprinkler irrigation is used in a variety of agricultural applications for germination, irrigation, dust control, climate control and wastewater dispersal. The types of sprinkler systems commonly used include: micro-spray and/or micro-jet irrigation equipment and polyethylene tubes fitted with emission device outlets at various spacing.

Subirrigation requires fairly sophisticated, expensive equipment and management. Advantages are water and nutrient conservation, and labor-saving through lowered system maintenance and automation.

Table 11.9 Types of irrigation

Irrigation Technique	Characteristics
Flood irrigation	Water is pumped or brought to the fields and is allowed to flow along the ground among the crops. This method is simple and cheap. Traditional flood irrigation can mean a lot of wasted water!
Drip or trickle irrigation	Water is sent through plastic pipes (with holes in them) that are either laid along the rows of crops or even buried along their rootlines. Water is delivered at or near the root zone of plants, drop by drop. Evaporation is cut way down, and up to one-fourth of the water used is saved, as compared to flood irrigation.
Sprinkler irrigation	These systems have a long tube fixed at one end to the water source, such as a well. Water is piped to one or more central locations within the field and distributed by overhead high-pressure sprinklers or guns or by lower-pressure sprays.
Subirrigation (or seepage irrigation)	It is a method of artificially raising the water table to allow the soil to be moistened from below the plants' root zone. Water is delivered from below, absorbed upwards and the excess collected for recycling.

11.8.2 Practices for System Operators

11.8.2.1 Engineering Practices

- Metering: The measurement of water use with a meter provides essential data for charging fees based on actual customer use. Billing customers based on their actual water use has been found to contribute directly to water conservation. Meters also aid in detecting leaks throughout a water system. One way to detect leaks is to use listening equipment to survey the distribution system, identify leak sounds and pinpoint the exact locations of hidden underground leaks. As mentioned, metering can also be used to help detect leaks in a system;
- Leak detection: An effective way to conserve water is to detect and repair leaks in municipal water systems. Repairing leaks controls the loss of water that water agencies have paid to obtain, treat and pressurize. The early detection of leaks also reduces the chances that leaks will cause major property damage;
- Water main rehabilitation: A water utility can improve the management and rehabilitation of a water distribution network by using a distribution system database. Using the database can help to lower maintenance costs and can result in more efficient use of the water resource;
- Water reuse: Another practice that should be considered by water system operators who operate publicly owned treatment works is the reuse of treated wastewater. Some potential applications for water reuse include landscape irrigation, agricultural irrigation, aesthetic uses such as fountains, industrial uses and fire protection.

In addition to engineering practices, system operators can use several other practices to conserve water or improve water use efficiency.

11.8.2.2 Planning and Management Practices

- Pricing: Information and education promoting conservation do not appear to be effective by themselves in achieving a conservation goal without at the same time imposing significant price increases to provide a financial incentive to conserve water. Customers use less water when they have to pay more for it and use more when they know they can afford it. However, most people consider water to be a "free good" and are not willing to pay higher prices that reflect the true costs associated with the water delivered to their homes. Rate structures have the advantage of avoiding the costs of overt regulation, restrictions and policing while retaining a greater degree of individual freedom of choice for water customers. Water utility managers must establish and design water rates that meet revenue requirements and are fair and equitable to all customer classes in the water system;

- Retrofit programs: Retrofit programs are another tool system that operators can use to promote water use efficiency practices. Retrofitting involves the replacement of existing plumbing equipment with equipment that uses less water. The most successful water-saving fixtures are those which operate in the same manner as the fixtures they are replacing - for example, toilet tank inserts, shower flow restrictors and low-flow showerheads;
- Residential water audit programs: Residential water audit programs involve sending trained water auditors to participating family homes, free of charge, to encourage water conservation efforts. Auditors visit participating homes to identify water conservation opportunities, such as repairing leaks and low-flow plumbing, and to recommend changes in water use practices to reduce home water use;
- Public education: Public education programs can be used to inform the public about the basics of water use efficiency. Public education is an essential component of a successful water conservation program;
- Planning for resource protection: Monitoring and managing land use and waste disposal practices around water supply sources, can potentially reduce the need for new water supply development and keep water treatment costs to a minimum. Adverse effects on a water supply source can be lessened through land use controls such as land preservation, nonregulatory and regulatory watershed programs, environmental assessment requirements and zoning. The protection of a water source by a utility can range from simple sanitary surveys of a watershed to the development and implementation of complex land use controls. Water supply source protection should play an important role in the overall management of a municipal water utility. Contamination of a water source can result from point and non-point sources of pollution such as chemical spills, waste discharges or agricultural runoff from the improper use of fertilizers, insecticides and herbicides;
- Drought management planning: When less rain falls than usual, there is less water to maintain normal soil moisture, stream flows and reservoir levels and to recharge groundwater. Falling levels of surface waters create unattractive areas of exposed shoreline and reduce the capacity of surface waters to dilute and carry municipal and industrial wastewater. Water quality often decreases as water quantity decreases, adversely affecting fish and wildlife habitats. In addition, dry conditions make trees more prone to insect damage and disease, and increase the potential for grass and forest fires. A drought management plan should address a range of issues, from political and technical matters to public involvement. Managing a resource essential to people's welfare during disaster and dealing with the associated emotional, economic and physical consequences makes drought management a very challenging task.

References

Davis, M. L., and Masten, S. J., 2004, *Principles of Emvironmental Engineering and Science*, McGraw Hill, USA.

Hammer, M. J., and Hammer, M. J. Jr., 1996, *Water and Wastewater Technology*, Prentice Hall, USA.

Kiely, G., 1996, *Environmental Engineering*, McGraw Hill Limited, U.K.

Purdue (September 5, 2006), http://www.purdue.edu/dp/envirosoft/groundwater/src/quality3c.htm.

Nazaroff, W. W., and Cohen, L. A., 2001, *Environmental Engineering Science*, John Wiley, USA.

UK Groundwater Forum (August 20, 2006), http://www.groundwateruk.org.

USEPA (August 1, 2006a), http://www.epa.gov/owow/nps/qa.html.

USEPA (August 11, 2006b), http://www.epa.gov/waterscience/standards/about.

USGS (August 2, 2006), http://ga.water.usgs.gov.

Waterencyclopedia (August 2, 2006), http://www.waterencyclopedia.com/La-Mi/Microbes-in-Groundwater.html.

Wikipedia (August 1, 2006), http://en.wikipedia.org.

WOW, 2004, (September 1, 2006), http://WaterOntheWeb.org, Water on the Web—Monitoring Minnesota Lakes on the Internet and Training Water Science Technicians for the Future—A National On-line Curriculum using Advanced Technologies and Real-Time Data. University of Minnesota-Duluth, Duluth, MN 55812.

Chapter 12
Physical and Chemical Groundwater Remediation Technologies

Krishna R. Reddy*

Abstract Groundwater is the main source of drinking water as well as agricultural and industrial usage. Unfortunately, groundwater quality has been degraded due to improper waste disposal practices and accidental spillage of hazardous chemicals. Therefore, it is critical that the groundwater contamination be prevented and the contaminated groundwater at numerous sites worldwide be remediated in order to protect public health and the environment. This chapter provides an overview of relevant regulations, general remedial approach, and most commonly used physical and chemical groundwater remediation technologies. The remediation technologies include pump-and-treat, in-situ air sparging, in-situ flushing, and permeable reactive barriers. The process description, applicability, limitations and a case study for each of these technologies are also presented.

Keywords Groundwater, contamination, remediation, pump-and-treat, air sparging, flushing, permeable reactive barriers

12.1 Introduction

About 40% of the drinking water comes from groundwater, about 97% of the rural population drinks groundwater, and about 30–40% of the water used for agriculture comes from groundwater (Sharma and Reddy, 2004). Therefore, groundwater is a valuable resource and it must be protected from any pollution. The United States Environmental Protection Agency (USEPA) estimated that there are thousands of sites that have been contaminated in the United States and over 217,000 these sites require urgent remedial action. These sites include:

Krishna R. Reddy
Department of Civil and Materials Engineering, University of Illinois at Chicago, 842 West Taylor Street, Chicago, IL 60607, U.S.A.

* To whom correspondence should be addressed. E-mail: kreddy@uic.edu

C.J.G. Darnault (ed.), *Overexploitation and Contamination of Shared Groundwater Resources.*

- National Priorities List (Superfund) sites;
- Resource Conservation and Recovery Act (RCRA) Corrective Action sites;
- Underground Storage Tanks (USTs) sites;
- Department of Energy (DOE) sites;
- Department of Defense (DOD) sites;
- Various Civilian Federal Agencies sites;
- State and private parties (including brownfields) sites.

Contamination of groundwater has been a major concern at these sites. The contaminants encountered at these sites include organic compounds, heavy metals, and radionuclides. DOE sites contain mixed wastes, including radioactive wastes, while DOD sites contain explosives and unexploded ordnance. The cost to cleanup these sites is estimated to exceed US $270 billion. This chapter provides an overview of regulatory framework, general remedial approach, and different common physical and chemical remediation technologies for cleanup of polluted groundwater.

12.2 Relevant Regulations

The assessment and remediation of previously contaminated sites and the proper management of newly created hazardous wastes have been regulated through the passage of major environmental laws and regulations (Sharma and Reddy, 2004). In 1980, the United States Congress established the Superfund program, also known as the Comprehensive Environmental Response, Compensation and Liabilities Act (CERCLA), to provide the financial assistance needed for the remediation of abandoned hazardous waste sites that pose serious risk to the health and safety of the public as well as the welfare of the environment. The Superfund program is administered by the USEPA in cooperation with regional governmental agencies. In order to determine which sites are eligible to receive federal aid under the Superfund program, a ranking system has been established to allow for a quantitative rating of sites across the United States. Sites that score high enough on the USEPA's hazard ranking system are placed on the National Priorities List (NPL). The National Priorities List is a published list of hazardous sites that require extensive and long-term remediation, and that are deemed eligible to receive funding from the Superfund program. Superfund sites must comply with the stringent remediation codes, liability standards, and documentation required by the Superfund program. According to this program, purchasers of contaminated sites may be held responsible for damage caused by previous owners even if these sites were contaminated by legal activities at the time of occurrence. Additionally, Superfund regulations require that a contaminated site be remediated to very low contaminant levels such that risk to public health is minimized. Such an approach is often inflexible and does not take into account the intended use of the rehabilitated site.

In 1980, the United States Congress also promulgated the Resource Conservation and Recovery Act (RCRA) to control newly created hazardous waste from the "cradle-to-grave". These regulations provide criteria for defining hazardous waste, generator responsibilities, transporter's requirements, manifest systems, and treatment, storage and disposal facility (TSDF) requirements. The regulations also address problems that could result from underground tanks storing petroleum and other hazardous substances.

Many state governments are also assisting in cleanup of contaminated sites. Nearly half of the states in the United States offer some type of voluntary remediation program. The purpose of such programs is to encourage remediation of sites with possible contamination while preventing any increased liability for participating parties. When a remediation project is completed, many states will issue a statement releasing the participants from state liability for any contamination that may exist at the site. Often state agencies will offer assistance to project participants if they are subject to federal liability.

12.3 General Remedial Approach

A systematic approach for the assessment and remediation of contaminated sites is necessary in order to facilitate the remediation process and avoid undue delays. The most important aspects of the approach include site characterization, risk assessment, and selection of an effective remedial action (Sharma and Reddy, 2004). Innovative integration of various tasks can often lead to a faster, cost-effective remedial program.

12.3.1 Site Characterization

Site characterization is often the first step in a contaminated site remediation strategy. It consists of the collection and assessment of data representing contaminant type and distribution at a site under investigation. The results of a site characterization form the basis for decisions concerning the requirements of remedial action. Additionally, the results serve as a guide for design, implementation, and monitoring of the remedial system. Each site is unique; therefore, site characterization must be tailored to meet site-specific requirements. An inadequate site characterization may lead to the collection of unnecessary or misleading data, technical misjudgment affecting the cost and duration of possible remedial action, or extensive contamination problems resulting from inadequate or inappropriate remedial action. Site characterization is often an expensive and lengthy process; therefore, it is advantageous to follow an effective characterization strategy to optimize efficiency and cost.

An effective site characterization includes the collection of data pertaining to site geology, including site stratigraphy and important geologic formations; site hydrogeology, including major water-bearing formations and their hydraulic properties; and site contamination, including type, concentration, and distribution. Additionally, surface conditions both at and around the site must be taken into consideration. Because little information regarding a particular site is often known at the beginning of an investigation, it is often advantageous to follow a phased approach for the site characterization. A phased approach may also minimize financial impact by improving the planning of the investigation and ensuring the collection of relevant data. Phase I consists of the definition of investigation purpose and the performance of a preliminary site assessment. A preliminary assessment provides the geographical location, background information, regional hydrogeologic information, and potential sources of contamination pertaining to the site. The preliminary site assessment consists of two tasks, a literature review and a site visit.

Based on the results of the Phase I activities, the purpose and scope of the Phase II exploratory site investigation need to be developed. If contamination was detected at the site during the course of the preliminary investigation, the exploratory site investigation must be used to confirm such findings as well as obtain further data necessary for the design of a detailed site investigation program. A detailed work plan should be prepared for the site investigations describing the scope of related field and laboratory testing. The work plan should provide details about sampling and testing procedures, sampling locations and frequency, a quality assurance/quality control (QA/QC) plan, a health and safety (S&H) plan, a work schedule, and a cost assessment. Phase III includes a detailed site investigation in order to define the site geology and hydrogeology as well as the contamination profile. The data obtained from the detailed investigation must be adequate to properly assess the risk posed at the site as well as to allow for effective designs of possible remedial systems. As with the exploratory investigations, a detailed work plan including field and laboratory testing programs as well as QA/QC and S&H plans should be outlined. Depending on the size, accessibility, and proposed future purpose of the site, this investigation may last anywhere from a few weeks to a few years. Because of the time and the effort required, this phase of the investigation is very costly. If data collected after the first three phases is determined to be inadequate, Phase IV should be developed and implemented to gain additional information. Additional phases of site characterization must be performed until all pertinent data has been collected.

Depending on the logistics of the project, site characterization may require regulatory compliance and/or approval at different stages of the investigation. Thus, it is important to review the applicable regulations during the preliminary site assessment (Phase I). Meetings with regulatory officials may also be beneficial to insure that investigation procedures and results conform to regulatory standards. This proactive approach may prevent delays in obtaining the required regulatory permits and/or approvals. Innovative site characterization techniques are increasingly being used to collect relevant data in an efficient and cost-effective manner. Recent advances in cone penetrometer and sensor technology have enabled contaminated sites to be rapidly characterized using vehicle-mounted direct push probes. Probes are available for directly measuring contaminant concentrations in-situ, in addition to measuring standard stratigraphic data, to provide flexible, real-time analysis. The probes can also be reconfigured to expedite the collection of soil, groundwater, and soil gas samples for subsequent laboratory analysis. Noninvasive, geophysical techniques such as ground-penetrating radar, cross-well radar, electrical resistance tomography, vertical induction profiling, and high-resolution seismic reflection, produce computer-generated images of subsurface geological conditions and are qualitative at best. Other approaches such as chemical tracers are used to identify and quantify contaminated zones, based on their affinity for a particular contaminant and the measured change in tracer concentration between wells employing a combination of conservative and partitioning tracers.

12.3.2 Risk Assessment

Once site contamination has been confirmed through the course of a thorough site characterization, a risk assessment is performed. A risk assessment is a systematic evaluation used to determine the potential risk posed by the detected contamination to human health and the environment under present and possible future conditions. If the risk assessment reveals that an unacceptable risk exists due to the contamination, a remedial strategy is developed to assess the problem. If corrective action is deemed necessary, the risk assessment will assist in the development of remedial strategies and goals necessary to reduce the potential risks posed at the site.

The USEPA and the American Society for Testing and Materials (ASTM) have developed comprehensive risk assessment procedures. The USEPA procedure was originally developed by the United States Academy of Sciences in 1983. It was adopted with modifications by the USEPA for use in Superfund feasibility studies and RCRA corrective measure studies (USEPA, 1989). This procedure provides a general, comprehensive approach for performing risk assessments at contaminated sites. It consists of four steps:

1. Hazard identification.
2. Exposure assessment.
3. Toxicity assessment.
4. Risk characterization.

The ASTM Standard E 1739-95, known as the Guide for Risk-Based Corrective Action (RBCA), is a tiered assessment originally developed to help assess sites that contained leaking underground storage tanks containing petroleum (ASTM, 2002).

Although the Standard is geared toward such sites, many regulatory agencies use a slightly modified version for non-UST sites. This approach integrates risk and exposure assessment practices with site assessment activities and remedial measure selection. The RBCA process allows corrective action activities to be tailored for site-specific conditions and risks and assures that the chosen course of action will protect both human health and the environment.

12.3.3 Remedial Action

When the results of a risk assessment reveal that a site does not pose risks to human health or the environment, no remedial action is required. In some cases, however, monitoring of a site may be required to validate the results of the risk assessment. Corrective action is required when risks posed by the site are deemed unacceptable. When action is required, remedial strategy must be developed to insure that the intended remedial method complies with all technological, economic, and regulatory considerations. The costs and benefits of various remedial alternatives are often weighed by comparing the flexibility, compatibility, speed, and cost of each method. A remedial method must be flexible in its application to ensure that it is adaptable to site-specific soil and groundwater characteristics. The selected method must be able to address site contamination while offering compatibility with the geology and hydrogeology of the site.

Generally, remediation methods are divided into two categories: in-situ remediation methods and ex-situ remediation methods. In-situ methods treat contaminated groundwater in-place, eliminating the need to extract groundwater. In-situ methods are advantageous because they often provide economic treatment, little site disruption, and increased safety due to lessened risk of accidental contamination exposure to both on-site workers and the general public within the vicinity of the remedial project. Successful implementation of in-situ methods, however, requires a thorough understanding of subsurface conditions. Ex-situ methods are used to treat extracted groundwater. Surface treatment may be performed either on-site or off-site, depending on site-specific conditions. Ex-situ treatment methods are attractive because consideration does not need to be given to subsurface conditions. Ex-situ treatment also offers easier control and monitoring during remedial activity implementation.

12.4 Remedial Technologies

If groundwater contamination is confirmed and remedial action is deemed necessary following a thorough site characterization and risk assessment, one of many remedial technologies may be utilized for corrective action. The most common physical and chemical remediation technologies are pump and treat, in-situ air sparging, in-situ flushing, and permeable reactive barriers. The most common biotechnologies include monitored natural attenuation, bioremediation, and phytoremediation, but these methods are not within the scope of this chapter. Containment methods such as slurry walls and grout curtains are also used to control contaminant plumes within groundwater but are not discussed within this chapter (USEPA, 1995). Containment methods such as these are often used as interim measures prior to the final selection and implementation of a remedial method. Actual remedial methods are varied in their applications and their limitations; thus, it is essential to evaluate the benefits, drawbacks, and economic impact of each method as well as the site-specific soil, hydrogeologic, and contaminant conditions.

12.4.1 Pump and Treat

Until recently, the most conventional method for groundwater remediation has been the pump and treat method. With pump and treat as shown in Figure 12.1, free-phase contaminants and/or contaminated groundwater are pumped directly out of the surface. Treatment occurs above ground, and the cleaned groundwater is either discharged into sewer systems or re-injected into the subsurface (Cohen et al., 1997). Pump and treat systems have been operated at numerous sites for many years. Unfortunately, data collected from these sites reveals that although pump and treat may be successful during the initial stages of implementation, performance drastically decreases at later times. As a result, significant amounts of residual contamination can remain unaffected by continued treatment. Due to these limitations, the pump and treat method is now primarily used for free product recovery and control of contaminant plume migration.

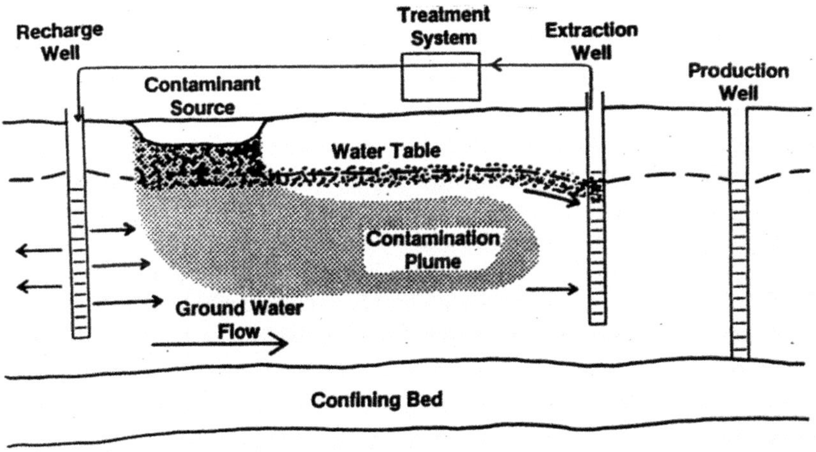

Fig. 12.1 Pump-and-treat system

Pump and treat requires simple equipment and it is effective for source zone removal where free-phase contamination is present. Some concerns with pump and treat include lingering residual contamination due to tailing and/or rebound, long time required to achieve remediation, biofouling of extraction wells and associated treatment stream that can severely affect system performance, high cost of treating large quantities of wastewater, and high operation and maintenance costs (USEPA, 1996). Tailing and rebound are attributed to presence of non-aqueous phase liquids (NAPLs), contaminant desorption, contaminant precipitation-dissolution, matrix diffusion, and groundwater velocity variation. The removal of NAPLs during pumping is attributed to dissolution of residual and pooled free-product, desorption and solubilization, and dissolution kinetics.

Numerous case studies are reported in the literature documenting the design and performance of pump-and-treat systems for groundwater remediation (FRTR, 1998a,b). For example, Fairchild Semiconductor Company in California manufactured chips, mother boards, and circuits for the emerging computer industry in the late 1960s. To maintain ultra clean conditions as a part of their manufacturing process, hundreds of gallons of solvent were used daily. Accidentally, hundreds of gallons of solvent have been spilled into the soil and underlying groundwater. The site soils consisted of alluvial deposits that are heterogeneous mixture of sand and gravel interbedded with silts and clays. The deposits are up to 1,500 ft thick. The upper aquifer zone occurs from the top of the saturated zone to the depth of approximately 165 ft below ground surface. Contaminants in the groundwater were TCE (trichloroethene), chloroform, 1, 1-dichloroethene, 1, 1, 1-trichloroethane, and vinyl chloride. The risk-based remedial objectives were:

- TCE: 5 μg/l for shallow aquifers;
- TCE: 0.8 μg/l for deep aquifers;
- Chloroform: 100 μg/l;
- 1,1- dichloroethene: 6 μg/l;
- 1,1,1-trichloroethane: 200 μg/l;
- Vinyl Chloride: 0.5 μg/l.

A network of extraction wells were designed to extract the groundwater. The groundwater was pumped to the surface and treated through an activated carbon process and re-injected into the ground to enhance hydraulic control and to flush the contamination zone. The extraction and treatment systems run continuously from January 1 through December 3, with the exception of brief shut downs for carbon change or routine maintenance. Monitoring the treatment included measuring groundwater elevations and collecting groundwater samples for analysis. Monitoring the pump system aims at maintaining a steady flow through extraction wells. The contaminant concentrations are steadily declining, but do not reached the remedial objectives. The remedial system is still in operation and the developments in Silicon Valley sparked the interest of Netscape Communications to lease 38.5 acres of the site.

12.4.2 In-situ Air Sparging

Air sparging, also known as biosparging, is an emerging remediation technology useful in the treatment of volatile organic contaminants. During the implementation of air sparging as shown in Figure 12.2, a gas, usually air, is injected into the saturated soil zone below the lowest known level of contamination. Due to the effect of buoyancy, the injected air will rise towards the surface. As the air comes into contact with the contamination, it will, through a variety of mechanisms, strip the contaminant away or assist in in-situ degradation. Eventually, the contaminant-laden air encounters the vadose zone, where it is often collected using a soil vapor extraction system and treated on-site (Reddy et al., 1995; Reddy and Adams, 2001).

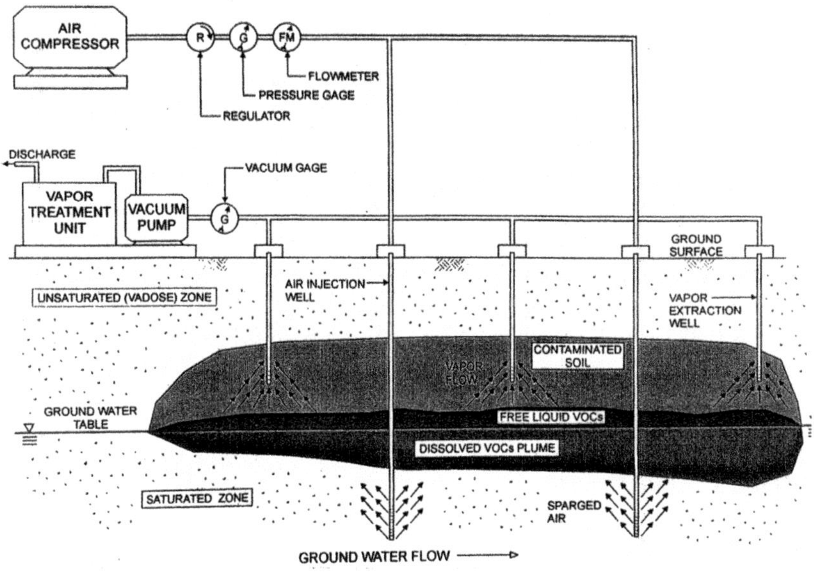

Fig. 12.2 In-situ air sparging system

This technology has been very popular because it causes minimal site disruption and reduces worker exposure to contaminants, it does not require removal, storage, or discharge consideration for groundwater, the equipment needed is simple and easy to install and operate, it requires short treatment time (1–3 years), and the overall cost is significantly lower than the conventional remediation methods such as pump and treat. However, there are several limitations of this technology. Contamination in low permeability and stratified soils poses a significant technical challenge to air sparging remediation efforts. Confined aquifers cannot be treated by this remediation technique. Air flow dynamics and contaminant removal or degradation processes are not well understood. If not properly designed, it could cause spreading of the contaminants into clean areas. It requires detailed data and pilot testing prior to its application.

Air sparging is based on the principles of air flow dynamics and contaminant transport, transfer and transformation processes (Reddy and Adams, 2001). Injected air moves through aquifer materials in the form of either bubbles or microchannels. In coarser soils such as fine gravels, air flow has been observed to be in the bubble form. In finer soils such as sands, the air flow has been observed to be in microchannel form. The density of bubbles or microchannels is found to be depended on the injected air flow rate. Soil heterogeneities are found to significantly affect the air flow patterns and the zone of influence. The transport mechanisms include advection, dispersion, and diffusion. The mass transfer mechanisms include volatilization, dissolution, and adsorption/desorption. Besides these, biodegradation is enhanced due to increased dissolved oxygen that can promote aerobic biodegradation.

Many sites have been successfully remediated using air sparging (Reddy and Adams, 2001). For example, Eaddy Brothers was a gasoline service station located in Hemingway, South Carolina. In September 1998, a release was reported from the station's underground storage tanks. Soil and groundwater at the site were found to be contaminated with MTBE, BTEX, and naphthalene. Concentrations are: MTBE 5,110,000 μg/L; benzene 226,000 μg/L; toluene 301,000 μg/L; ethylbenzene 280,000 μg/L, xylene 278,000 μg/L; and naphthalene 2,700 μg/L. Subsurface soils at the site consists of silty clays with inter-fingered thin clayey-sand lenses, and no confining units have been identified. The average hydraulic gradient is 0.005 with a calculated seepage velocity of 0.138 ft/year. The depth to groundwater is 2.5–17.9 ft below ground surface. The risk-based remedial objectives were:

- MTBE: 646 μg/L;
- Benzene: 191 μg/L;
- Toluene: 11,938 μg/L;
- Ethylbenzene: 9,426 μg/L;
- Xylene: 78,496 μg/L;
- Naphthalene: 418 μg/L.

Air sparging and soil vapor extraction units were installed. Ten vertical air sparging wells at a depth of 26 ft with 5 ft well screen were installed. Wells were connected to Kaeser SK-2 air sparge compressor operating at 70 psi. A total of 28 wells (on- and off-site) were used to monitor groundwater. Within a year, concentration dropped to 99% for MTBE, 99% for BTEX, and 96% for naphthalene. It took almost another year to drop the concentration of MTBE, benzene, and napthalene to the desired level. Air sparging was effective, fast, and easy to implement and monitor. The total cost for the cleanup of this site is US$ 197,515 which is relatively low compared to other means of remediation.

12.4.3 In-situ Flushing

Soil flushing involves pumping flushing solution into groundwater via injection wells as shown in Figure 12.3. The solution then flows down gradient through the region of contamination where it desorbs, solubilizes, and/or flushes the contaminants from the soil and/or groundwater. After the contaminants have been solubilized, the solution is pumped out via extraction wells located further down gradient. At the surface, the contaminated solution is treated using typical wastewater treatment methods, and then recycled by pumping it back to the injection wells (USEPA, 1991; Roote, 1997). Plain water or carefully developed solution (e.g., surfactant/cosolvent) are used as flushing solutions. However, one must select the type and concentration of flushing solution to optimize contaminant desorption and solubilzation.

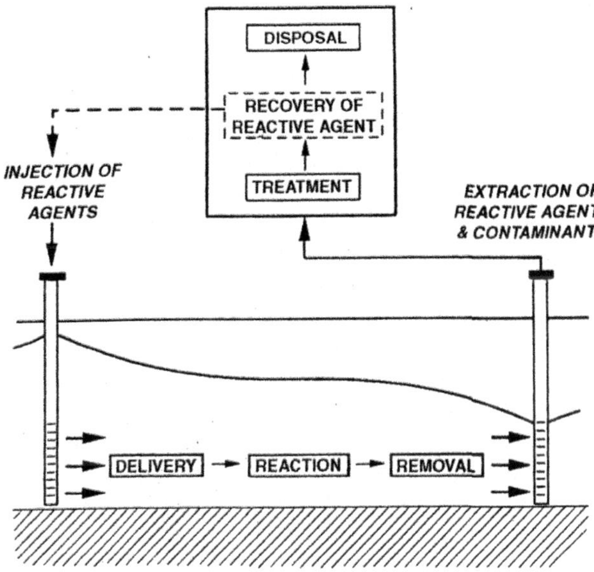

Fig. 12.3 In-situ flushing system

In-situ flushing causes less exposure of the contaminants to clean-up personnel and the environment. It is a simple and easy operation as compared to other technologies. It is applicable for a wide variety of contaminants, both organic and inorganic contaminants. It may be a slow process when heterogeneities such as soil layers or lenses of less permeable (less than 10^{-5} cm/s) or organic materials are located within the soil horizon. Since the contaminants are solubilized into the solution, they may be transported beyond the extraction well and unintentional spreading of the contamination may occur. Remediation times may be long and the effectiveness of the process largely depends on solution, contaminant, soil or groundwater interaction. Remediation depends strongly on the ability of the solution to desorb and solubilize the contaminant. The process may be costly with contamination located at large depths or with expensive solutions and long remediation times.

Marine Corps Base Camp LeJeune, Site 88, Building 25 was the location of a central dry cleaning facility. The site was contaminated with PCE and Varsol from storage and use during dry cleaning operations. PCE was present in groundwater at the site as DNAPL. Varsol—a petroleum distillate—was present as LNAPL. A demonstration of the surfactant-enhanced aquifer remediation system (SEAR) was performed under the U.S. Department of Defense Environmental Security Technology Certification Program (ESTCP). The target was to treat DNAPL in groundwater.

Shallow surficial aquifer existed at a depth of 16–20 ft. An order of magnitude difference existed in permeability between the shallower, more permeable zone (hydraulic conductivity of 10^{-4} cm/s) and the basal low permeability zone (hydraulic conductivity of 10^{-5} cm/s). The majority of DNAPL was present in a low permeability silty layer at base of the shallow aquifer, with about 105 gallons of DNAPL estimated to be present in the test zone. Contaminants found at the site include chlorinated solvents and total petroleum hydrocarbons (TPH); PCE was present as DNAPL, and Varsol was present as LNAPL. PCE concentrations in groundwater as high as 54 mg/L were monitored.

The test area was 20 ft wide by 30 ft long and 20 ft deep. Flushing solution consisted of surfactant, calcium chloride, and isopropyl alcohol. It was injected through three injection wells at a rate of 0.133 gallons per minute per well for 58 days. Six extraction wells removed subsurface liquids at a combined rate of 1 gpm. Above-ground treatment included gravity separation to remove separate phase DNAPLs. Evaporation to remove dissolved-phase contaminates, and ultra filtration (UF) to reconcentrate surfactant fluid prior to reinjection were implemented. Surfactant flush was followed by a 74 day water flush to remove injected chemicals and solubilized or mobilized contaminates. Partitioning interwell tracer test (PITT) was performed to demonstrate DNAPL removal and recovery of injected solution.

A total of 76 gallons of PCE was recovered during the demonstration with 32 gallons recovered as solubilized DNAPL and 44 gallons as free-phase DNAPL. DNAPL was effectively removed from the more permeable layer with DNAPL at a rate of 92–96%. DNAPL recovery from entire test zone (both layers) was 72%. Above-ground treatment system removed greater than 95% of extracted PCE, recovered 77% of surfactant and recovered 88% of isopropyl alcohol. The project reached an estimated 90% success level based on their initial goals. DNAPL was effectively removed from the more permeable layer with DNAPL remaining mostly in the lower permeable layer.

The results of the demonstration showed that aquifer heterogeneity has a strong influence on the performance of SEAR and that DNAPL source zone characterization is important because of the sensitivity of the technology to permeability contrasts. Total demonstration costs were US $3.1 million, including DNAPL source zone characterization, surfactant selection, well field installation, free-phase DNAPL removal equipment, pre-treatment PITT, technology application, surfactant regeneration, and indirect costs. Estimated total treatment cost for full-scale systems are US $12.8 million per acre.

12.4.4 Permeable Reactive Barriers

Permeable reactive barriers (PRBs) offer a passive approach for groundwater remediation. In general, a permeable wall containing an appropriate reactive material is placed across the path of a contaminant plume. As contaminated water passes through the wall, the contaminants are either removed or degraded (Figure 12.4). When designing a wall, not only must an appropriate reactive medium be chosen, but also wall dimensions must be designed to assure that the entire contaminant plume will be intercepted and enough residence time within the wall will be allowed for remediation to take place.

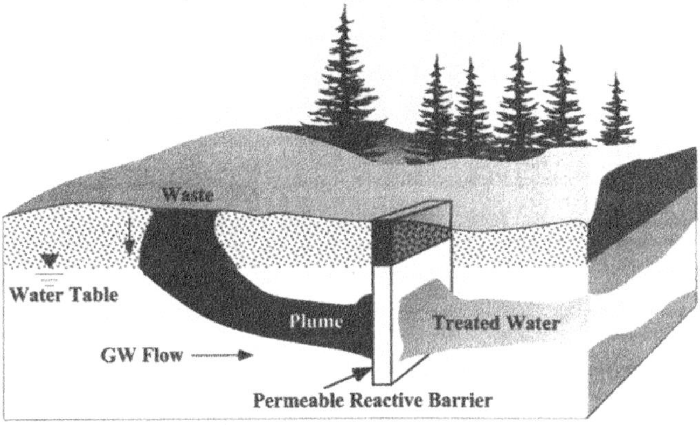

Fig. 12.4 Permeable reactive barrier

Typical reactant media contained in the barriers includes media designed for degrading volatile organics, chelators for immobilizing metals, or nutrients and oxygen to facilitate bioremediation (USEPA, 1998b). The media is often mixed with a porous material such as sand to enhance groundwater flow through the barrier. A permeable barrier may be installed as a continuous reactive barrier or as a funnel-and-gate system (Figures 12.5 and 12.6). A continuous reactive barrier consists of a reactive cell containing the permeable reactive medium (Figure 12.5). A funnel-and-gate system has an impermeable section, called the *funnel*, which directs the captured groundwater flow towards the permeable section, called the *gate*. The funnel walls may be aligned in a straight line with the gate, or other geometric arrangements of funnel-and-gate systems can be used depending on the site conditions (Figure 12.6). This funnel-and-gate configuration allows better control over reactive cell placement and plume capture. At sites where the groundwater flow is very heterogeneous, a funnel-and-gate system can allow the reactive cell to be placed in the more permeable portions of the aquifer. At sites where the contaminant distribution is very non-uniform, a funnel-and-gate system can better homogenize the concentrations of contaminants entering the reactive cell. A system with multiple gates can also be used to ensure sufficient residence times at sites with a relatively wide plume and high groundwater velocity (Figure 12.7). Figure 12.7a shows an example of a funnel-and-gate system with two gates emplaced with caissons, while Figure 12.7b shows an example of a funnel-and-gate system with two reactive media emplaced in series within the gate. PRBs are installed as permanent, semi-permanent, or replaceable units across the flow path of a contaminant plume.

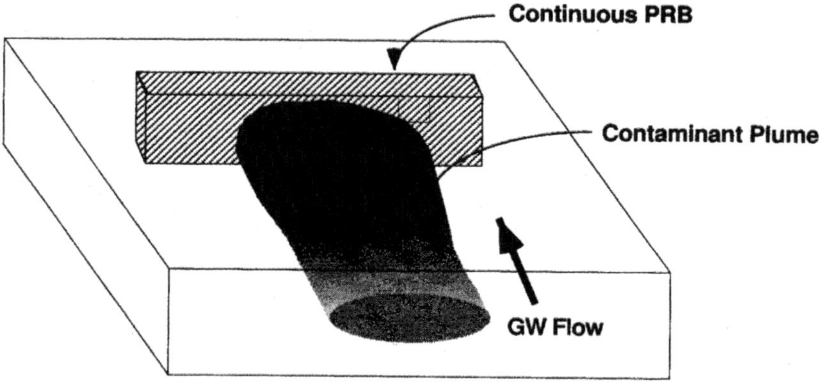

Fig. 12.5 Continuous permeable reactive barrier

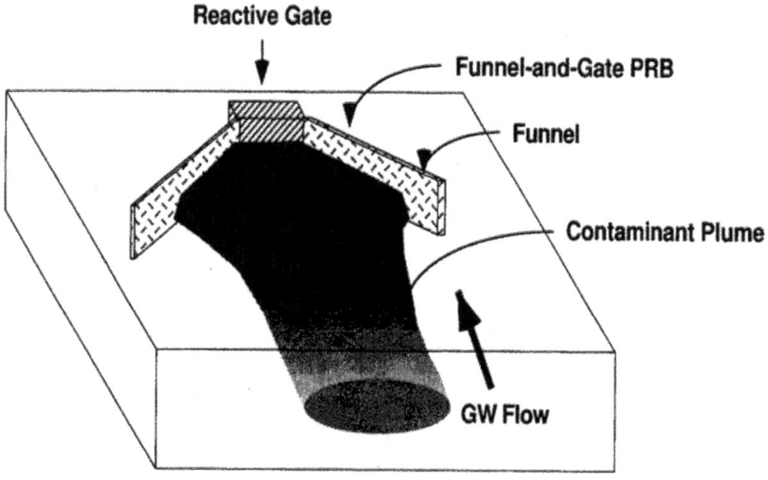

Fig. 12.6 Funnel-and-gate permeable reactive barrier

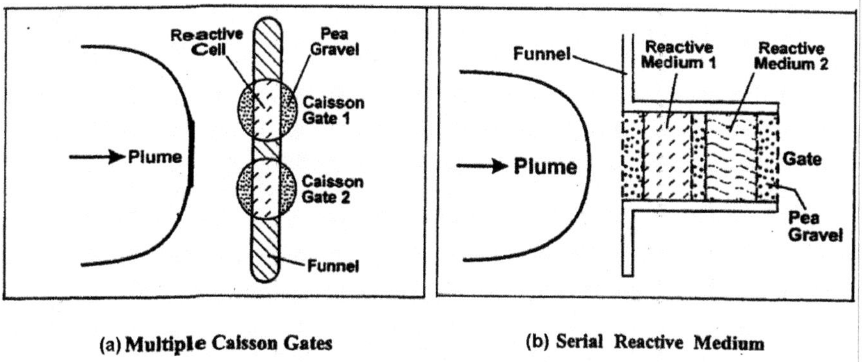

Fig. 12.7 Permeable reactive barrier configurations with multiple reactive cells

PRBs are often economically advantageous because:

- No pumping or above-ground treatment is required, the barrier acts passively after installation;
- No above-ground installed structures, so the affected property can possibly be put to productive use while it is being cleaned up;
- PRBs can be modified to treat several different types of contaminants;
- The reactive medium is often used up very slowly and has the potential to passively treat contaminated plumes for several years or decades;
- Very low operating costs other than site monitoring;
- No disposal costs or requirements for successfully treated wastes.

However, there are several limitations of PRBs which include:

- Lengthy treatment time relative to other active remediation methods (e.g., in-situ flushing, air sparging);
- Potential for loosing reactivity of the media, requiring replacement of the reactive medium;
- Potential for decrease in reactive media permeability due to biological clogging and/or chemical precipitation;
- Potential for plume bypassing the PRB as a result of seasonal changes in flow regime;
- Currently limited to shallow depths;
- Longevity of PRB performance is unknown.

General design approach includes site characterization, reactive media selection, treatability testing, PRB design using computer modeling, emplacement of PRB, and performance monitoring. Emplacement methods include conventional excavation, trenching machines, tremie tube/mandrel, deep soil mixing, high-pressure jetting, and vertical fracturing and reactant sand-fracturing. The monitoring parameters include contaminant concentration and distribution, by-products and reaction intermediates, groundwater velocity and pressure levels, permeability assessment of the reactive barrier, groundwater quality parameters (e.g., pH, redox potential, alkalinity), and dissolved gas concentrations (e.g., oxygen, hydrogen, and carbon dioxide).

Several studies have been reported where PRBs are used to treat contaminated groundwater (Sharma and Reddy, 2004). For example, the USCG facility Support Center included an electroplating shop which operated for more than 30 years, until 1984 (FRTR, 1997). In December 1988, a release was discovered during demolition of the former plating shop. Soil excavated beneath the floor of the shop was found to contain high levels of chromium. Subsequent investigation showed substantial groundwater contamination by chromium and chlorinated solvents. A full-scale PRB was constructed as part of an interim corrective measure. It was associated with a voluntary RCRA facility investigation where the electroplating shop was identified as a solid waste management unit under the facility's RCRA Part B permit. The barrier consisted of 450 tons of granular zero-valent iron placed into an underlying low conductivity layer at a depth of approximately 22 ft below ground surface. The required residence time in the treatment zone has been estimated as 21 h, based on a highest concentration scenario. The average velocity through the wall was reported as 0.2–0.4 ft/day. Analytical data from the first year of full-scale operation showed that the cleanup goal for Cr(VI) had been met but the goal for TCE had not. Cleanup goals for the site were based on primary drinking water standards: TCE (5μg/L) and Cr(VI) (0.1 μg/L). Cr(VI) concentrations were below cleanup goals in all down-gradient monitoring wells. However, TCE concentrations were above the cleanup goal in four of the six down-gradient wells. The reason for the elevated TCE concentrations in some of the down-gradient wells had not been identified. Estimated costs for the PRB were US $585,000, which corresponded to US $225 per 1,000 gallons of groundwater treated. By using a PRB rather than the typical pump and treat method, nearly US $4 million were saved in construction and long-term maintenance costs.

12.4.5 Bio-based Technologies and Treatment Trains

Monitored natural attenuation (MNA) is the use of natural attenuation processes within the context of a carefully controlled and monitored site cleanup approach to reduce contaminant concentrations, within a reasonable time frame, to levels that are protective of human health and the environment (ASTM, 1998; USEPA, 1998a). Unlike MNA which occurs naturally, bioremediation requires human intervention to create conditions that stimulate the growth of microorganisms to degrade/immobilize contaminants (Cauwenberghe and Roote, 1998). Phytoremediation uses plants to uptake or stabilize the contaminants, which is applicable to shallow aquifers remediation (Sharma and Reddy, 2004).

Using just one technology may not be adequate to remediate some contaminated sites with different contaminants and complex site conditions. Under such situations, different technologies are used sequentially or concurrently along with the primary treatment technology to achieve the remedial goals. Such use of multiple remediation technologies is often referred to as "treatment trains". Typical treatment trains used in contaminated sites include soil flushing followed by bioremediation, and pump and treat along with soil flushing or air sparging.

12.5 Conclusion

Groundwater is a valuable source of drinking water. It is also used extensively for agricultural and industrial applications. Remediation of contaminated groundwater is critical in order to protect human health and the environment. It is of the utmost importance to properly characterize the site, and such a characterization includes defining the site's geology, hydrology, and contamination, potential releases to the environment, and locations and demographics of nearby populations. Once the site has been characterized, a risk assessment of hazards at the site is performed and a suitable remedial action may be selected. In order to perform these different tasks in a fiscally responsible manner, it is important that the entire remedial planning, from initial site characterization efforts until the completion of site cleanup, follows a rational strategy. If contamination has been detected and risk posed by the contamination is unacceptable, an appropriate remedial technology must be selected and properly implemented. This requires a thorough understanding of not only the conditions within the subsurface, but also the advantages and drawbacks of the available remedial options. Such an understanding is necessary because improper implementation can often exacerbate site contamination. By possessing knowledge of the available technologies, remediation professionals will be better equipped to utilize proper judgment for the decisions regarding the remediation of contaminated sites. Several technologies exist for the remediation of contaminated groundwater. These technologies include pump and treat, air sparging, in-situ flushing, permeable reactive barriers, monitored natural attenuation, and bioremediation. Many of these technologies are used in combination or other innovative technologies are being developed. Remediation technology for a particular site is selected based on the site specific hydrogeologic and contaminant conditions, desired cleanup levels, remedial time, and cost.

References

ASTM, 1998, *Standard Guide for Remediation of Groundwater by Natural Attenuation at Petroleum Release Sites*, E-1943-98, West Conshohocken, Pennsylvania.

ASTM, 2002, *Annual Book of ASTM Standards*, West Conshohocken, Pennsylvania.

Cauwenberghe, L. V., and Roote, D. S., 1998, In Situ Bioremediation, Technology Overview Report TO-98-01, Groundwater Remediation Technologies Analysis Center, Pittsburgh, PA.

Cohen, R. M., Mercer, J. W., Greenwald, R. M., and Beljin, M. S., 1997, *Design Guidelines for Conventional Pump-and-Treat Systems*, Robert S. Kerr Environmental Research Laboratory, Ada, Oklahoma.

Federal Remediation Technologies Roundtable (FRTR), 1997, In Situ Permeable Reactive Barrier for Treatment of Contaminated Groundwater at the US Coast Guard Support Center, Elizabeth City, North Carolina.

Federal Remediation Technologies Roundtable (FRTR), 1998a, *Remediation Case Studies: Groundwater Pump and Treat (Chlorinated Solvents)*, Volume 9, EPA 542-R-98-013, Washington, D.C.

Federal Remediation Technologies Roundtable (FRTR), 1998b, *Remediation Case Studies: Groundwater Pump and Treat (Nonchlorinated Contaminants)*, Volume 10, EPA 542-R-98-014, Washington, D.C.

Reddy, K. R., and Adams, J. A., 2001, Cleanup of chemical spills using air sparging, Chapter 14, in: *Hand Book of Chemical Spill Technologies*, M. Fingas, ed., Mc-Graw Hill Company, pp. 14.1–14.29.

Reddy, K. R., Kosgi, S., and Zhou, J., 1995, A Review of in situ air sparging for the remediation of VOC-contaminated saturated soils and groundwater, *Hazardous Waste and Hazardous Materials* 12(2): 97–118.

Roote, D. S., 1997, *In-situ Flushing*, Technology Overview Report, Ground-Water Remediation Technologies Analysis Center.

Sharma, H. D., and Reddy, K. R., 2004, *Geoenvironmental Engineering: Site Remediation, Waste Containment, and Emerging Waste Management Technologies*, John Wiley & Sons, Hoboken, New Jersey.

USEPA, 1989, Risk Assessment Guidance for Superfund, Washington, D.C.

USEPA, 1991, Engineering Bulletin: In Situ Soil Flushing, EPA/540/2-91/021, Washington, D.C.

USEPA, 1995, Assessment of Barrier Containment Technologies, International Containment Technology Workshop, Baltimore, Maryland, August 29–31.

USEPA, 1996, *Pump-and-Treat Ground-Water Remediation: A Guide for Decision Makers and Practitioners*, EPA/625/R-95/005, Office of Research and Development, Washington, D.C.

USEPA, 1998a, Monitored Natural Attenuation for Ground Water, EPA/625/K-98/001.

USEPA, 1998b, Permeable Reactive Barrier Technologies for Contaminant Remediation, EPA/600/R-98/125, Office of Research and Development, Washington, D.C.

Chapter 13
Enhanced Aquifer Recharge

Krishna R. Reddy*

Abstract Enhanced recharge is defined as any engineered system designed to introduce and store water in aquifer. This chapter presents the benefits of enhanced aquifer recharge and describes the different methods to recharge aquifers. The applicability, advantages and limitations of each method are identified. Clogging of the recharge systems is a major concern and must be addressed during operation and maintenance of the systems. In addition to technical aspects, one must consider institutional and legal issues in selecting and developing recharge methods. A rational approach must be followed in the design, construction and operation of any recharge method. Numerous aquifer recharge projects have been successfully developed and implemented worldwide.

Keywords Aquifer, groundwater, depletion, recharge

13.1 Introduction

The basic concepts of groundwater hydraulics include vadose zone, unconfined aquifer and associated water table, confining units, and confined aquifer and associated potentiometric levels (e.g., Freeze and Cherry, 1979; Sharma and Reddy, 2004). If the discharge from any aquifer is less than or equal to the natural recharge, there is no concern with the depletion of groundwater in that aquifer. However, if the discharge from the aquifer exceeds the recharge, groundwater depletion occurs. The depletion of groundwater is also known as groundwater overexploitation and is mainly indicated by the decrease in water levels/potentiometric surface levels. The groundwater depletion can lead to short supply of water as well as environmental degradation.

Krishna R. Reddy
Department of Civil and Materials Engineering, University of Illinois at Chicago,
842 West Taylor Street, Chicago, IL 60607, U.S.A.

* To whom correspondence should be addressed. E-mail: kreddy@uic.edu

C.J.G. Darnault (ed.), *Overexploitation and Contamination of Shared Groundwater Resources*.

Enhanced recharge is defined as any engineered system designed to introduce and store water in aquifer. Enhanced recharge is also commonly known as artificial recharge or managed recharge. Enhanced recharge must be distinguished from incidental recharge. Incidental recharge is defined as recharge that reaches an aquifer from human activities not designed specifically for recharge. Some examples of incidental recharge sources include septic tank, leach fields, stormwater retention ponds, percolation from irrigation, and leaking water or wastewater pipes and tunnels. Incidental recharge may be source of groundwater contamination; therefore, it must be assessed carefully.

There are several benefits of enhanced recharge. The enhanced recharge can stabilize or raise groundwater levels, smooth out supply/demand fluctuation, reduce loss through evaporation and runoff, store water in aquifers for future use, and impede storm runoff and soil erosion. In addition, enhanced recharge should be an integral part of any water management strategy to improve water quality and smooth fluctuations, maintain environmental flows in streams/rivers, manage saline intrusion or land subsidence, and dispose/reuse of waste/stormwater.

Different sources may be available for the recharge water. These sources include excess surface water from streams, canals, lakes, reservoirs, stormwater runoff, treated effluents, reclaimed wastewater, imported water from other areas, groundwater from other aquifers, and potable water (often for storage and recovery). In the United States, monitoring data shows that the flow in many rivers and lakes exceeds the median flow; therefore, this overflow could be used as an excellent source of recharge water. Stormwater runoff water if properly collected could be another valuable source of recharge water. Treated effluents and reclaimed wastewater should be evaluated for their quality prior to deciding on their suitability as recharge water.

This chapter presents different enhanced recharge methods, various design and operation considerations, and institutional and regulatory issues. Next the chapter presents a rational methodology to design recharge systems. Finally, the chapter presents a few case studies demonstrating the successful completion of enhanced recharge of groundwater.

13.2 Enhanced Recharge Methods

The most common methods to enhance recharge groundwater can be grouped under (i) surface infiltration methods which involve percolation of recharge from or near the ground surface, (ii) vadose zone infiltration methods which involve percolation of recharge water at some depth below the ground surface, but within the vadose zone, and (iii) direct injection methods which involve direct injection of recharge water into the aquifer (ASCE, 2001; Bouwer, 2002; Topper et al., 2004; WDNR, 2006). In addition to these common methods, few other methods are also used. For example, artificial storage and recovery (ASR) method involves injection of recharge water directly into the aquifer using wells, but the recharge water is later recovered from the same wells. Riverbank filtration method is another method that gained much attention recently which uses well fields placed near surface water bodies with the intention of inducing surface water into the aquifer to provide some or all of the water produced by the well field. In developing countries, rainwater harvesting method, which involves connecting the outlet pipe from a guttered roof-top to divert rainwater to either existing wells or other recharge structures, is being used. More details on the common aquifer recharge methods are provided in this section.

13.2.1 Surface Infiltration Methods

In surface infiltration methods, recharge water is applied at the surface above an unconfined aquifer in man-made or natural depressions to infiltrate down to the underlying water table, ultimately causing the water table to rise. These methods are often preferred due to lower cost, greater simplicity, and lower operation and maintenance costs. However, these methods require that the soils from the infiltration location to the top of the aquifer possess high vertical permeability. The different surface infiltration methods, as depicted in Figure 13.1, include infiltration ponds or basins, infiltration ditches, stream channels, land applications, and in-channel systems (e.g., percolation ponds, leaky dams and recharge releases, sand storage dams, and subsurface dams).

Infiltration ponds or basins are essentially artificial depressions that receive water and allow the water to recharge an aquifer through the bottom of the structure. Existing excavations, such as gravel pits or leaky reservoirs, are often used for recharge basins. A series of recharge basins may be used to increase the recharge from a larger area.

Infiltration ditches are linear structures, such as canals and ditches, which are designed to leak water through their bottoms to recharge an aquifer. These can sometimes be built in areas where the topography or limits of available land preclude the use of infiltration ponds. Infiltration systems can be designed as lateral systems, dendritic systems, or contour systems. Lateral systems consist of small ditches which protrude at right angles from canals, flow control gates at head of the systems, furrow depths related to topography to maintain uniform velocity, and collection of runoff in a canal further down slope to routes water back in source stream. The dendritic systems divert flow from the main canal to a series of successive and smaller ditches, gates control flow to each ditch series, and bifurcation of ditches continues until all water has infiltrated. The contour systems consist of spreading water through ditches that follow land contours and return water through ditches to source stream at the end of the spreading area and at the lowest point.

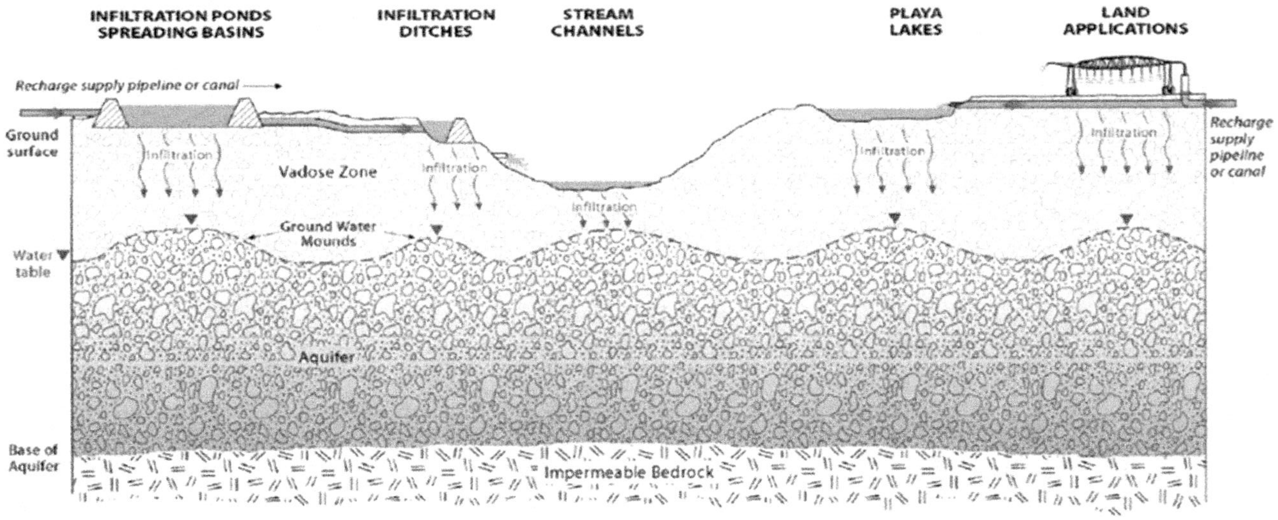

Fig. 13.1 Surface infiltration methods (Topper et al., 2004)

Stream channels can be used as infiltration ditches providing that the configuration of the water table allows the stream to infiltrate water into the ground. The stream channels may be modified with diversions, ditches, and check dams todecrease flow velocities and allow longer storage within the channel so that greater percolation can occur.

Land application includes a variety of methods where water is applied to the land surface. Recharge can occur if the application rate exceeds the evapotranspiration rate of the area. Common land application methods include: flooding, over irrigation, and some wastewater disposal systems that use irrigation methods. Flooding can essentially include diverting a stream and introducing recharge water into an impoundment area created by temporary dikes or embankments and allowing excess water to reenter into the source stream. Various irrigation techniques have been developed, such as overland flow, ditches and furrow, sub-irrigation, flooding or spray, which can readily be used to recharge the aquifers.

Various in-channel modifications can be made to enhance storage and increase infiltration of recharge water. As shown in Figure 13.2, these modifications include percolation ponds, leaky dams and recharge releases, sand storage dams, and subsurface dams.

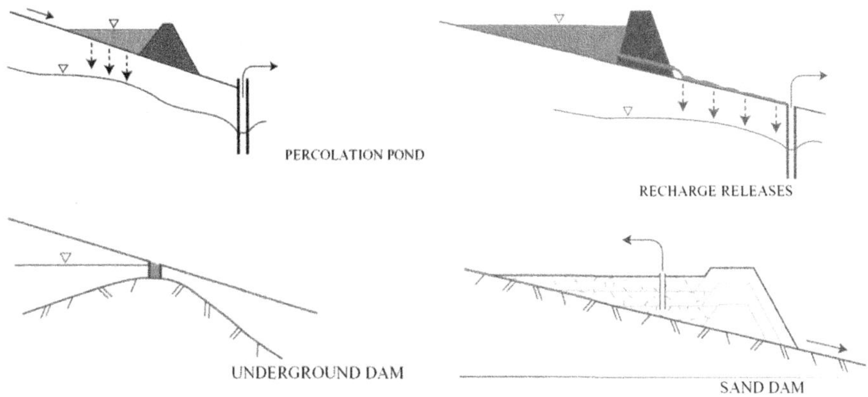

Fig. 13.2 In-channel modifications to enhance recharge (Gale, 2005)

Surface infiltration methods are applicable to unconfined aquifers with surface exposure and the aquifer materials may be alluvium, semi-consolidated, sediments at outcrop, and highly fractured bedrock. The advantages of these systems include initial low capital cost, simple maintenance, low operation and maintenance costs. Surface infiltration methods can use untreated surface water, and can co-exist with recreation use or wildlife habitat. Some disadvantages of these methods are that they require a near surface aquifer, and a permeable soil profile with high vertical permeability. High evaporation losses and groundwater vulnerability to surface contamination are disadvantages that make these method potentially incompatible with nearby land uses. Surface infiltration methods also require frequent maintenance to prevent clogging.

13.2.2 Vadose Zone Infiltration Methods

Vadose zone infiltration methods are used when near surface soils have low permeability or other land uses are not compatible with surface infiltration facilities. In these methods, recharge water is introduced at some deeper depth beneath the land surface (within the vadose zone) and then allowed to infiltrate into the unconfined aquifer. The most common methods are depicted in Figure 13.3 and include trenches and galleries, dry wells, infiltration shafts, and infiltration pits.

Infiltration trenches are excavated through shallow impermeable soils to facilitate recharge, use perforated pipes or permeable fill to infiltrate water. Infiltration galleries consist of multiple trenches. Trench systems can also be covered and used for other purposes such as parking lots, sports fields.

Dry wells are completed above the water table in the unsaturated zone. The wells can be completed with screens or slotted casing and are usually used to move water past a perching zone that is too deep for a trench system.

Infiltration shafts are large-diameter excavations drilled through a perching layer, but completed above the water table. The shafts can be completed with slotted casing or screens, or filled with permeable fill such as gravel and completed as an open hole. Infiltration pits are similar to infiltration shafts but are larger in diameter and may not be circular in shape.

Vadose zone infiltration methods are applicable to unconfined aquifers containing alluvium, semi-consolidated sediments at outcrop, or highly fractured bedrock. These methods are advantageous in that they can be used where surface layers of low permeability preclude surface infiltration, can co-exist with other surface urban uses such as parking lots and recreation facilities, and minimize evaporation losses. The disadvantages of these methods are higher initial capital costs, limited aerial extent, difficult to clean/maintain, and dependent upon near-surface geology.

13.2.3 Direct Injection Methods

Direct injection methods, as shown in Figure 13.4, involve injecting water directly into an aquifer using vertical wells. These wells are screened within the saturated portion of the aquifer and allow direct injection of recharge water. These wells may also be used for recovery purposes, if desired.

Vertical injection wells can be completed with slotted casing, well screens, or as open holes in competent formations. Injection wells minimize the vertical transit time of the water to the aquifer and can avoid unfavorable reactions between the water and soils or minerals in the unsaturated zone.

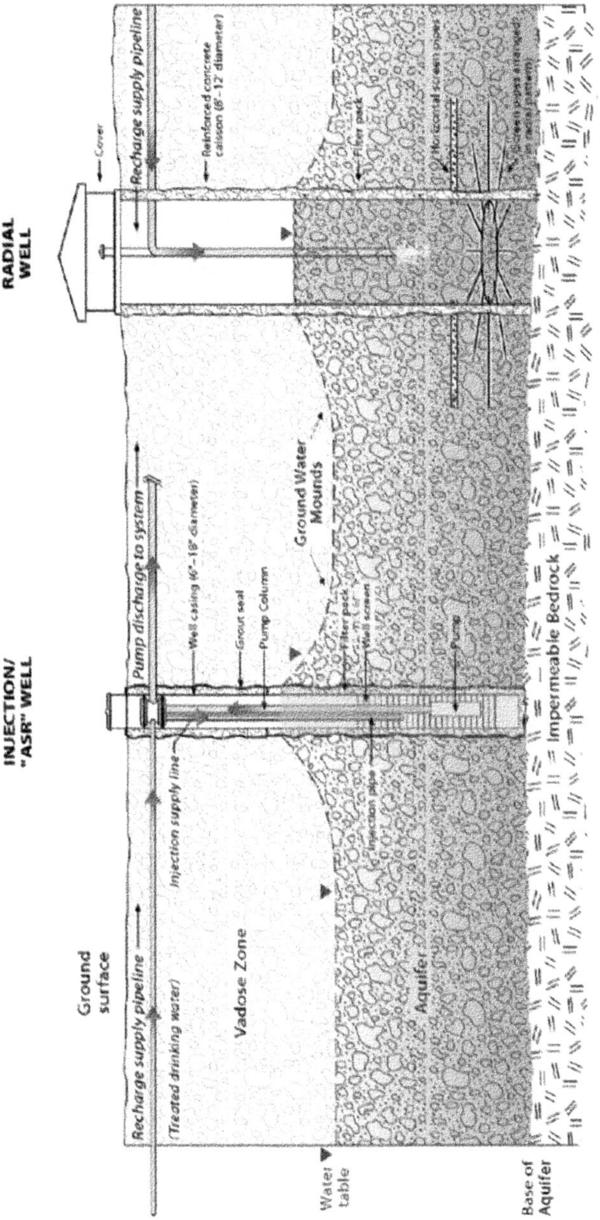

Fig. 13.3 Vadose zone infiltration methods (Topper et al., 2004)

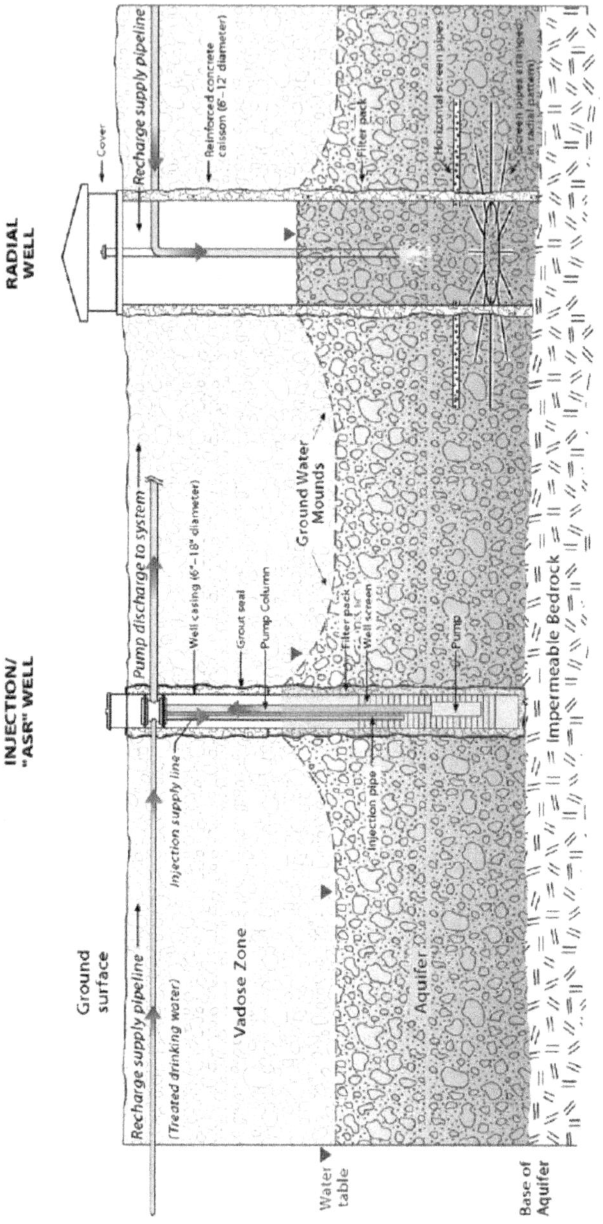

Fig. 13.4 Direct injection methods (Topper et al., 2004)

The direct methods are applicable to unconfined aquifers with limited surface exposure and confined aquifers consisting of deep alluvium and sedimentary bedrock aquifers. The advantages of these methods are that they can be used where vertical permeability is limited, occupy small surface areas, can fit in with most land use patterns, and can utilize existing water supply infrastructure. The disadvantages of these methods are high capital costs when existing infrastructure is not available, high energy requirements, high operation and maintenance costs. It also requires frequent pumping to remove clogging, pretreatment to drinking water standards, and tight control over source water quality. Additionally, contamination from recharge would be difficult to remediate.

13.3 Design, Operation, and Other Considerations

In planning and designing any recharge systems, one must consider the following: (1) hydrogeological considerations, (2) source water considerations, (3) operation/maintenance considerations, particularly clogging of recharge systems, (4) institutional and management issues, and (5) legal and regulatory issues.

13.3.1 Hydrogeological Considerations

As depicted in Figures 13.5 and 13.6, the effectiveness of recharge systems is extremely dependent on the site's hydrogeology. Recharge system must be located with in or up-gradient of the aquifer. The type of recharge systems depends on the type of the aquifer, confined or unconfined. The depth to the top of the aquifer, the permeability of geologic formations overlying the aquifer, and the heterogeneous geologic conditions control the distribution of recharge water and the amount of infiltration into the aquifer. It must be recognized that raising the water level in an aquifer can have undesired negative impacts on surrounding structures and land, and these negative impacts must be properly evaluated.

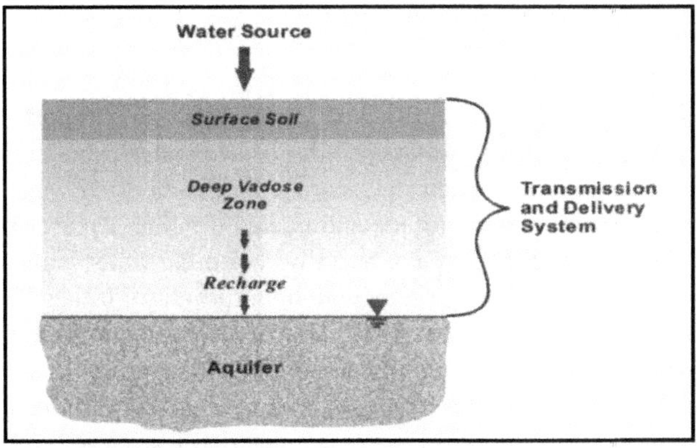

Fig. 13.5 Hydrogeologic considerations (Stevens, 2003)

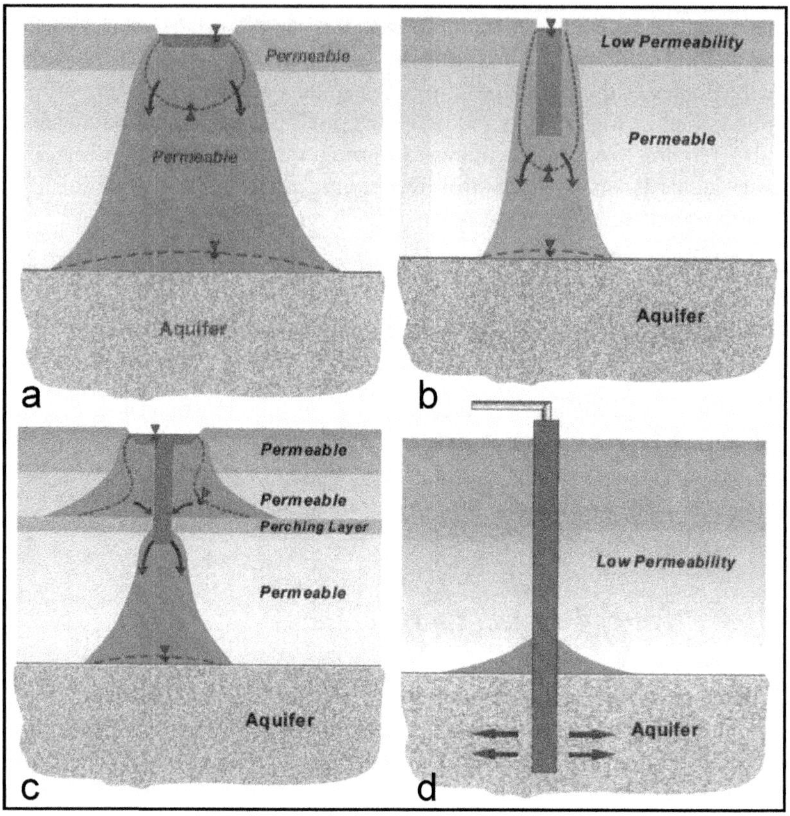

Fig. 13.6 Impacts of hydrogeologic conditions on recharge distribution (Stevens, 2003). **a** Permeable surface soil and vadose zone; **b** Low permeable surface soil and permeable vadose zone; **c** Perched low permeable layer in vadose zone; **d** Low permeable surface layer and vadose zone.

13.3.2 Source Water Considerations

Both availability (quantity) and quality of source water are important considerations. The sources of recharge water include surface water from streams, canals, lakes, reservoirs, reclaimed wastewater, storm runoff, imported water from other areas, groundwater from other aquifers, and treated drinking water. The quality of source water depends on the source. Generally, when the water is most abundant, the source is of poor quality (e.g., rivers with high turbidity). It then requires some form of treatment prior to recharge (e.g., simple sedimentation for river water). Wastewaters source requires a high cost treatment. Contamination of high quality groundwater, increase in solutes through mixing and dissolution, and adverse geochemical reactions due to source water characteristics are also possible.

13.3.3 Operation and Maintenance Considerations

Clogging of recharge systems is a critical issue. Silt or fine particles collected at the bottom of the recharge basin may cause clogging. Biological growth including bacteria and algae may also induce biological clogging. Other processes that may be responsible of clogging include air entrainment, swelling clay, and chemical precipitation (Figure 13.7).

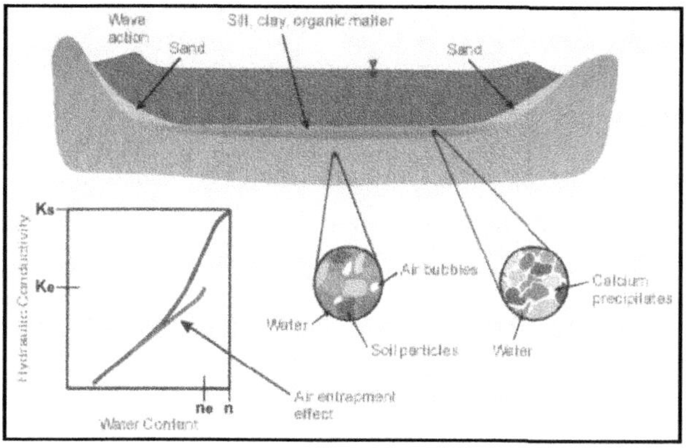

Fig. 13.7 Clogging of recharge systems (Stevens, 2003)

The extent of clogging depends on recharge water, soil, native groundwater, and design of recharge system. Cleaning varies from daily to periods of several years. Clogging layer must be removed by scraping off the material or raking or tilling the layer to enhance the permeability. The wave action to wash the side walls of the basin as well as dryout and cleaning are also used to remove clogging.

13.3.4 Institutional and Management Issues

Many institutional and management issues arise when dealing with recharge systems. These include water rights, land ownership, who pays and who benefits, who manages, private vs. public, etc.

13.3.5 Legal and Regulatory Issues

Legal and regulatory issues depend on the state and the country in which the recharge system is to be implemented. Many countries through their environmental agencies regulate stormwater infiltration systems, wastewater infiltration systems, and injection systems.

13.4 Recharge System Design Methodology

A phased, systematic approach is needed for the design of recharge systems. Typical phases include:

- Phase I—Preliminary Activities;
  - Data collection
  - Develop a hydrogeological conceptual model
  - Understand the hydrology (including meteorology)
  - Quantify the components of the water balance
  - Estimate aquifer storage capacity
  - Assess the quality of groundwater and source water
  - Environmental assessment
  - Public involvement
- Phase II—Field Investigation and Test Program;
  - Infiltration tests
  - Subsurface investigations
  - Water quality testing
  - Environmental assessment
- Phase III—Design;
  - Preliminary and final recharge system designs
  - Hydraulic analysis/Groundwater modeling
  - Pilot tests
  - Economic analysis
  - Environmental assessments
  - Public involvement
  - Engineering reports
- Phase IV—Construction and start-up;
- Phase V—Operation, maintenance, project review, and project modifications;
- Phase VI—Closure.
  - Sampling for residual contamination and eliminating pathways for groundwater contamination

13.4.1 Selected Case Studies

13.4.1.1 Case Study #1

The Dorz-Sayban plain is located 115 km to the southeast of Larestan, Iran. The overexploitation of groundwater caused significant drawdown of water table (1.5 m/year) and deterioration of groundwater quality. 3,500 hectares of land are irrigated using groundwater in this plain. To decrease the rate of drawdown of the water table, five floodwater-spreading systems for recharge of groundwater were designed and constructed in the region (Figure 13.8). Inflow and outflow from the recharge system were measured for nine flood events. 83.5% of total inflow of 886,000 m^3 recharged the aquifer, and 70% of suspended load has settled in the system.

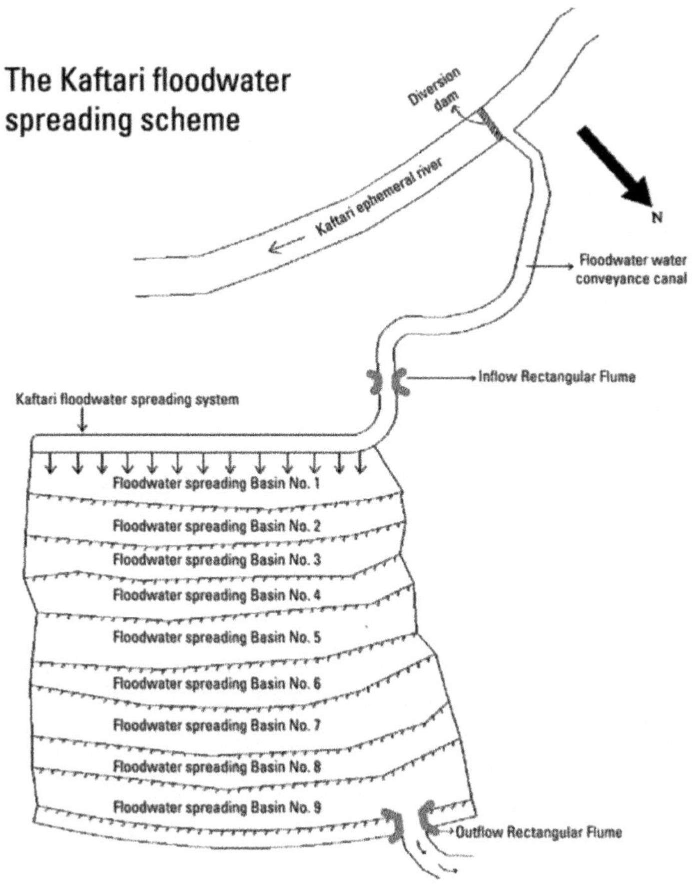

Fig. 13.8 Aquifer recharge water system used in Iran (Gale, 2005)

13.4.1.2 Case Study #2

Leaky Acres is a groundwater recharge facility operated near the City of Frenso, California. The groundwater recharge system was built in 1970 and consists of 26 ponds covering 200 acres. The surface soils are sandy, but two low-permeability aquitards are present at depth of 30–60 ft that are the limiting factor controlling the recharge rate. The water table is normally about 105 ft below ground surface. An average of 18 mgd of surface water from the local river is recharged to the groundwater at the site. The recharge system is taken out of service for 45 days each year to remove sediment and plugged materials from the beds.

13.5 Conclusion

Enhanced aquifer recharge should be an integral part of water management strategy. Surface and vadose zone infiltration methods are preferred for recharging shallow aquifers because of ease of operation and maintenance, and low cost. Direct injection methods are used to recharge deep confined aquifers, and recharge water in such case should meet drinking water standards. A systematic approach is recommended for the design of recharge systems. Clogging of recharge systems should be addressed through maintenance programs. Legal and regulatory issues should be properly addressed. Numerous successful enhanced aquifer recharge projects are reported worldwide.

References

American Society of Civil Engineers, (ASCE), 2001, Standard Guidelines for Artificial Recharge of Ground Water.

Bouwer, H., 2002, Artificial recharge of groundwater: Hydrogeology and engineering, *Hydrogeology Journal* **10**:121–142.

Freeze, R. A., and Cherry, J. A., 1979, *Groundwater*, Prentice Hall, Upper Saddle River, New Jersey.

Gale, I., 2005, *Strategies for Managed Aquifer Recharge (MAR) in Semi-arid Areas*, UNESCO, Paris.

Sharma, H. D., and Reddy, K. R., 2004, *Geoenvironmental Engineering; Site Remediation, Waste Containment, and Emerging Waste Management Technologies*, John Wiley & Sons, Inc., Hoboken, New Jersey.

Stevens, D., 2003, Presentation to the Groundwater Resource Association of California, San Jose, California.

Topper, R., Barkmann, P., Bird, D., and Sares, M., 2004, Artificial Recharge of Ground Water in Colorado—A Statewide Assessment, Colorado Geological Survey Department of Natural Resources.

Wisconsin Department of Natural Resources, (WDNR), 2006, Groundwater Recharge and Management, in: *State-of-the-Art of Water Supply Practices,* Chapter VI, SEWRPC Technical Report No.43.

Part V
Approaches to Shared Groundwater Resources Management: Conflict Prevention and Resolution

Chapter 14
Shared Groundwater Resources Management

Christophe J. G. Darnault*

Abstract Transboundary water sharing between states results in conflicts originated from water use, water quantity and quality issues. To avoid conflicts between countries that share transboundary groundwater resources, cooperation and negotiation, prevention, equitable water allocation and utilization, as well as exchange of information and data between states are essential. This chapter presents the groundwater resources management principles based on sustainable exploitation strategies that established a dynamic balance between the aquifer natural flows and storage and the flow imposed by exploitation. The transboundary groundwater resources models representative of the most common situations worldwide are presented and the management options for shared groundwater resources including separate, coordinate and joint management, delegation of responsibility, as well as sequential and flexible management are reviewed.

Keywords Shared groundwater resources management, overexploitation, pollution, transboundary groundwater models

14.1 Introduction

The exploitation and utilization of groundwater resources integrate a supply and demand management approach. The management of groundwater resources is based on extraction rates that do not to exceed recharge rates and exploitation strategies that established a dynamic balance between the aquifer natural flows and storage and the flow imposed by exploitation. The problems encountered in

---

Christophe J. G. Darnault
University of Illinois at Chicago, Department of Civil and Materials Engineering, 842 W. Taylor St., ERF# 2071, Chicago, IL 60607, U.S.A.

* To whom correspondence should be addressed. E-mail: darnault@uic.edu

C.J.G. Darnault (ed.), *Overexploitation and Contamination of Shared Groundwater Resources.*

the exploitation of groundwater resources are quantitative—overexploitation, and/or qualitative—contamination or degradation. The management of renewable resources such as groundwater relies on sustainability (UNESCO, 1992). When these groundwater resources are shared, the complexity of their management is enhanced and various management options are necessary.

14.2 Groundwater Resources Management

14.2.1 Physical Conditions: Flow and Storage Management

The groundwater and surface water interactions and the relations between groundwater basins are important considerations for the analysis of water inflow and outflow from an aquifer. The primary functions of an aquifer are water conveyance and storage, therefore, aquifer exploitation and management principles focus on flow and storage. An aquifer provides a renewable resource managed as a flow which is regulated by the variations in storage (Margat, 2004) (Figure 14.1.) The effect of the aquifer storage and the storage variation on the water production rate is important for aquifer exploitation strategy.

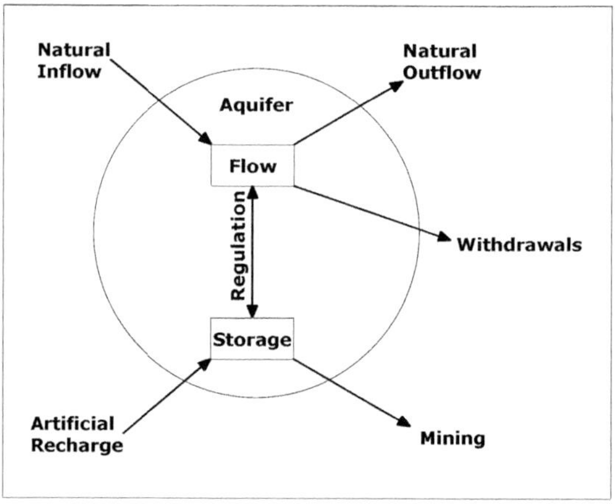

Fig. 14.1 Role of flow and storage in groundwater resources management (Modified from Margat, 2004)

14.2.2 Socio-Economic Conditions: Management Actors and Objectives

14.2.2.1 Management Actors

Actors with direct or indirect impacts do not have resource management as main objective and the effects of their acts accumulate and are felt within the system as a whole. Management authorities may define relevant and collective management objectives, but do not have any means of direct action on groundwater resources, besides indirect regulatory and economic instruments (Margat, 2004). The physical unity of an aquifer system and the multiplicity of actors with different and diverging interests result in an aquifer management based on pressure or incentives.

Economic actors may have the practical and financial means as well as the legal right to exploit the aquifer and find it to be a profitable source of water supply. Economic actors are direct users or producers and distributors of water as a commercial economic product. In addition to groundwater operators, many actors can directly or indirectly influence the aquifers exploitation strategy and its properties. The analysis of the system of economic actors involved: individual objectives, modes of action and actual or potential conflicts (Margat, 2004).

14.2.2.2 Management Objectives

The objectives and decision-making criteria of individual actors and management authorities are different. Individual actors are interested in micro-economic production and consumption objectives. Management authorities take decisions concerning the aquifer as a whole and with general interest objectives. Margat (2004) defined the economic management objectives as:

- Conflicts prevention and arbitration between individual exploiters of the same aquifer system;
- Conservation of the capacities and accessibility of groundwater resources;
- Conservation of the potential of groundwater resources in terms of quantity and quality for the present and future exploiters, or the prevention of excess exploitation damaging renewal possibilities and water properties, i.e. over-exploitation prevention;
- Allocation of groundwater resources, with priority being given to present and future demands, including reservations;
- Intensifying the use of resources if it is considered to be underexploited, by using groundwater instead of other water resources to satisfy certain demands.

14.2.3 Management Constraints and Criteria

Groundwater resources management is subject to constraints that depend on internal and external exploitation effects under different criteria (Margat, 2004). The internal effects and the inevitable repercussions of exploitation on the exploitability of the resource lead to internal constraints deriving from cost/benefit analyses based on the exploiter criteria. The external effects produce external constraints originating from arbitration between the interests and objectives of the users of the water produced and those of other actors.

14.2.3.1 Internal Constraints

Internal constraints determine the exploitation limits based on criteria linked to resource characteristics and water demand. Exploiters and management authority criteria are not defined at the same levels and may be different (Margat, 2004). The most important considerations for the exploiter are local productivity, exploitation methods, water quality produced, security of production, and direct production costs minimization. For the management authority, the most important considerations are: resource conservation in terms of both quantity and quality, equitable distribution of the conditions of access and productivity, and overall exploitation cost optimization.

14.2.3.2 External Constraints

The external constraints are based on reduction or prevention of external effects which are detrimental to other economic actors, and to the environment. The management of groundwater resources can impose constraints for resource protection and conservation, resulting in external constraints not only on the surface water development and use, but also on the land and subsurface occupation, and therefore on pollution prevention activities (Margat, 2004).

14.2.4 Management Decision Methods

The management decisions regarding an aquifer depend on whether the management is direct or indirect. Decisions may be strategic or tactical, based on predictive management techniques that rely on groundwater resources hydrodynamics, exploitation objectives and optimization (Margat, 2004).

14.2.4.1 Modeling and Predictive Management

Modeling is a tool for investigation and analysis, for checking consistency between observed data, hypotheses and calculated results, as well as for optimization of data collection (Pfannkuch, 1975). Models simulate behavior, forecasting the consequences of actions planned on the aquifer system, including exploitation scenarios. Modeling results indicate the feasibility and are used to compare data from different variants of the exploitation.

Forecasts of changing drawdown rates of the piezometric levels, decreasing discharge of natural outlets, and impact on salinity are important for groundwater exploitation (Margat, 2004). Modeling of production costs in terms of the depth variable and comparison of future states of the system within the established constraints are valuable for the sustainable development and management of groundwater resources.

14.2.4.2 Forecasting Unit Production and External Costs

Unit production costs and their evolution are essential for comparative feasibility studies of projected exploitation plans through predictive cost/benefit analyses (Margat, 2004). Depending on the types of impacts, external costs are not all calculable in monetary units but their magnitude can be indicated as stated by Margat (2004):

- Decreases in productivity of previous drilled wells can be evaluated through compensation costs, alternative solutions, or production losses costs;
- Degraded water properties can be evaluated in terms of costs resulting from remedial measures or from alternative sources of supply;
- Land subsidence or drainage harmful to vegetation can be evaluated through costs of remedial measures or reductions in land value.

14.2.4.3 Optimization Methods

Modeling enables different exploitation options and scenarios to be compared in terms of costs and results (Margat, 2004). Groundwater resources management is based on multiple objectives and criteria. A hierarchy of preferences has to be formalized for the criteria that will enable the establishment of an optimal solution or best compromise. The interconnection between a model representing the hydrogeology of an aquifer and a model of the decision-making system including constraints, criteria and preferences represents also a management model of an aquifer (El Magnouni and Treichel, 1994).

14.2.4.4 Management Control

The management authority should have regularly updated data on the condition of the aquifer and on the amount of water withdrawn to verify that changes in the exploited system correspond to simulated forecasts and plans agreed upon (Margat, 2004). Therefore, monitoring of variables of condition (e.g. piezometric levels) and decision variables (e.g. withdrawals) is necessary.

14.2.5 Exploitation Strategies

The groundwater resources exploitation occurred in the industrialized nations during 1950–1975 and in the developing countries during 1970–1990. This fast and uncontrolled emergence of groundwater resources exploitation generated socio-economic development, but today, it is encountering sustainability issues (Villholth, 2006). Each aquifer requires a unique exploitation strategy corresponding to its specific development objectives resulting from particular hydrogeological conditions and socio-economic constraints. When establishing an aquifer extraction regime, any pumping regime, even based on the sustainable yield principle, will reduce the natural outflow and affects the water table (Cohen, 1994). Sustainable yield principle application deals with several groundwater resources issues: temporal and spatial discrepancies, aquifer uncertainty, effect on water quality and groundwater-land use interactions (Haddad et al., 2001). Because of the variety and complexity of aquifer systems, management conditions are based on physical characteristics—the flow storage ratio and aquifer/surface stream relationships (Margat, 2004). The ratio of average flow and average storage, expressed in terms of duration of overall replenishment, varies from less than one year for thin and shallow alluvial aquifers, to the order of 10^4 to 10^5 years for deep aquifers containing fossil water. The aquifer/river relationship ranges from a very strong, continuous and permanent connection—thin alluvial aquifers with unclogged connection with river, to complete independence—deep confined aquifers and coastal aquifers. The groundwater abstraction rates encountered in some part of the world are not sustainable in the long term (Figure 14.2) and overexploitation of groundwater resources lead to various degrees of aquifer degradation and ecological impacts (Morris et al., 2003; Foster and Chilton, 2003). Three types of exploitation strategy based on the dynamic balance established between natural flows and storage of an aquifer and flows imposed by exploitation are possible and characterized by Margat (2004) as:

- A dynamic balance strategy: this strategy imposes an average withdrawal rate that is less than or equal to the average inflow, without prejudice to possible seasonal or annual variations. The aquifer storage decreases during an initial phase of imbalance and then stabilizes to assume a regulatory role.

- A temporary non-equilibrium strategy: this strategy has an average withdrawal rate higher than the average inflow, irrespective of whether withdrawals are increasing or stabilized. A time lag will then occur prior to equilibrium restoration. The water produced results mostly from withdrawals from storage. Groundwater storage can be partially recovered and stabilized through withdrawals reduction, artificial recharge and induced flow.
- A depletion or permanent non-equilibrium strategy: this strategy is the mining of groundwater at a withdrawal rate above the average inflow rate. Withdrawals from storage provide most of the water produced. In the long-term, the exploitation is limited by the drawdown becoming excessive. The equilibrium is not regained as the recharge of the groundwater storage is too slow and the storage capacity maybe irreversibly degraded through aquifer compaction due to pressure decrease.

Overexploitation is an excessive exploitation, a management error or failing which has to be prevented or stopped. Overexploitation has harmful consequences both in terms of water quantity and water quality issues. Overexploitation can exceed the limits of groundwater resources renewal to the detriment of future yields and dependent ecosystems (Margat, 2004).

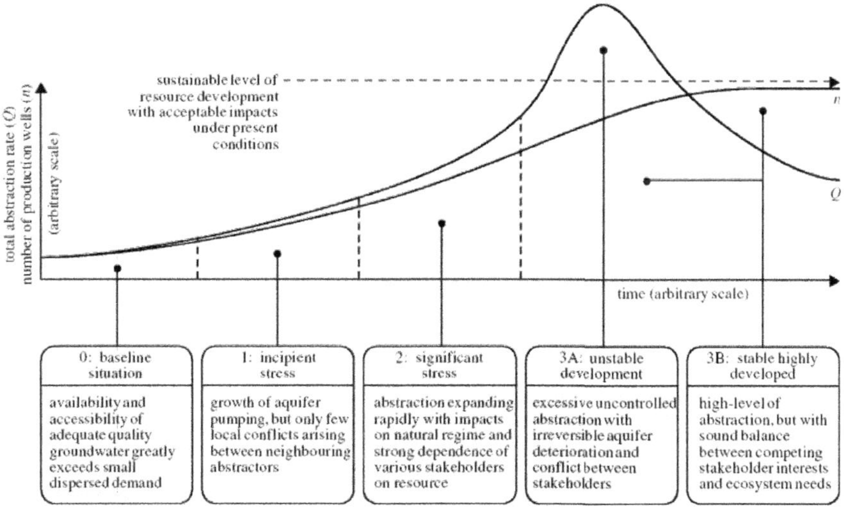

Fig. 14.2 Stages of groundwater resources development and their corresponding management needs (Reprinted from Foster and Chilton, 2003 with permission of the Royal Society. Copyright 2003)

14.3 Transboundary Groundwater Resources Models

Transboundary groundwater resources models have been developed for management and international law (Barberis, 1986; Eckstein and Eckstein, 2005). Eckstein and Eckstein (2005) have developed six conceptual models (Models A–F) representative of most common aquifers.

14.3.1 Model A

The model A (Figure 14.3) developed by Eckstein and Eckstein (2005) simulates an unconfined aquifer hydraulically connected to a river, both of which flow along an international border. The river represents the international border between the two states. If a state overpumps the aquifer section underlying its territory, the resulting cone of depression may extend across the river. The pumping state will draw water from the aquifer section underlying the nonpumping state and transport any pollution present in the aquifer section underlying the nonpumping state by groundwater flow to the aquifer section underlying the pumping state. If the river is gaining, water will flow from the aquifer into the river and any degradation present in one or both sides of the aquifer sections will impact the river. Any negative characteristic present in a losing river may impact both sections of the aquifer. A river can be loosing and gaining at different points along its course with the same aquifer and the river-aquifer interaction is accentuated with climate and seasons changes.

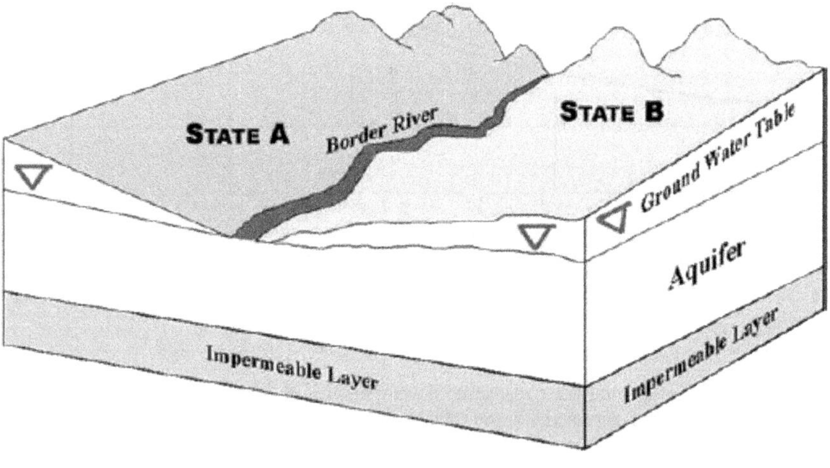

Fig. 14.3 Model A (Reprinted From Eckstein and Eckstein, 2005 with permission of the National Ground Water Association. Copyright 2005)

14.3.2 Model B

Model B (Figure 14.4) illustrates an unconfined aquifer intersected by an international border and linked hydraulically with a river intersected by the same international border (Eckstein and Eckstein, 2005). Water in the river and the aquifer flows down-slope from State A to State B. The transboundary problems will result either from pollution in State A being transported into State B through either the river or the aquifer; or from State A overpumping and therefore reducing the flow into State B. If the river is gaining upstream in State A and loosing downstream in State B, any negative characteristic found in one of the aquifer sections in State A could flow into the river and then into the aquifer on both sides of the river in State B.

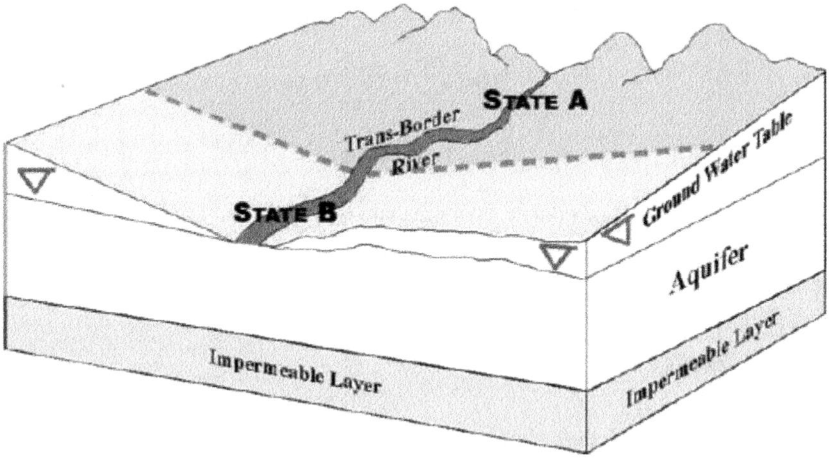

Fig. 14.4 Model B (Reprinted from Eckstein and Eckstein, 2005 with permission of the National Ground Water Association. Copyright 2005)

14.3.3 Model C

Model C (Figure 14.5) depicts an unconfined aquifer flowing across an international border that is hydraulically linked to a river that flows entirely in the territory of one state (Eckstein and Eckstein, 2005). Model C shows a gaining river–aquifer relationship in which groundwater recharged in State A flows into State B and then into the gaining river. This relationship can change and trans-boundary problems could be originated from State B to State A. Overpumping in State A could form a cone of depression that would locally reverse groundwater flowing from State A to State B within a distance limited by the cone of depression, therefore causing any negative characteristic located in the groundwater underlying State B to be transported to State A.

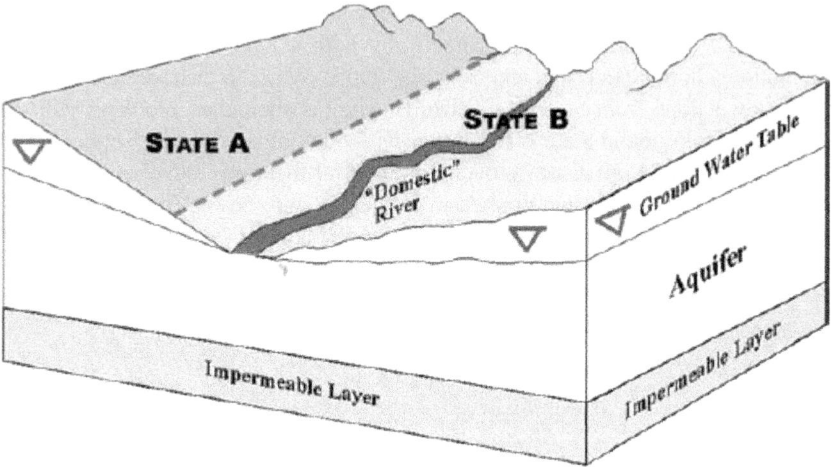

Fig. 14.5 Model C (Reprinted from Eckstein and Eckstein, 2005 with permission of the National Ground Water Association. Copyright 2005)

14.3.4 Model D

Model D (Figure 14.6) describes an unconfined aquifer that is not only completely located within the territory of one state but also hydraulically linked to a river flowing across an international border (Eckstein and Eckstein, 2005). The trans-boundary problems are resulting from the river volume and the water quality flowing from State A to State B. State A has the responsibility to ensure both the water quantity and quality of the river.

14.3.5 Model E

Model E (Figure 14.7) illustrates a confined aquifer unconnected hydraulically with surface water that traverses an international boundary or that is located entirely in another state (Eckstein and Eckstein, 2005). Overpumping in one or both states could impact the aquifer section along the border between the two states. Any negative characteristic found in the aquifer underneath one of the

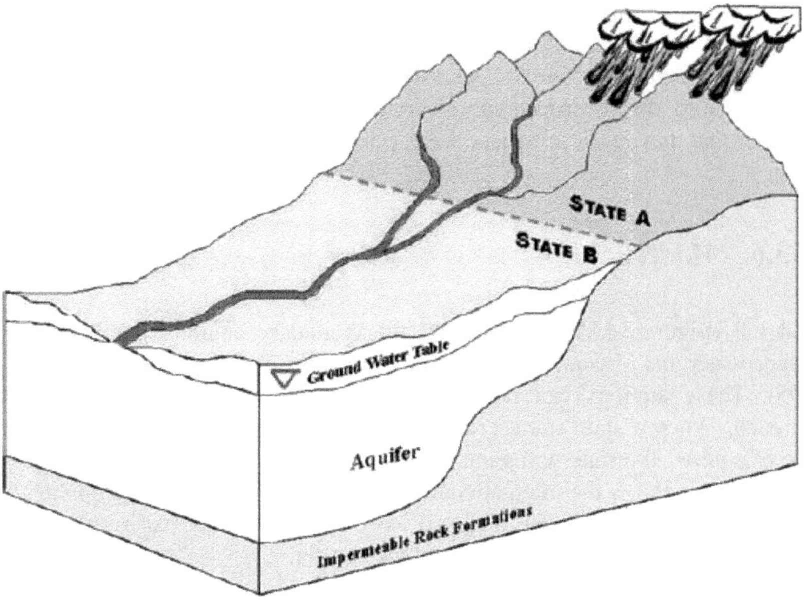

Fig. 14.6 Model D (Reprinted from Eckstein and Eckstein, 2005 with permission of the National Ground Water Association. Copyright 2005)

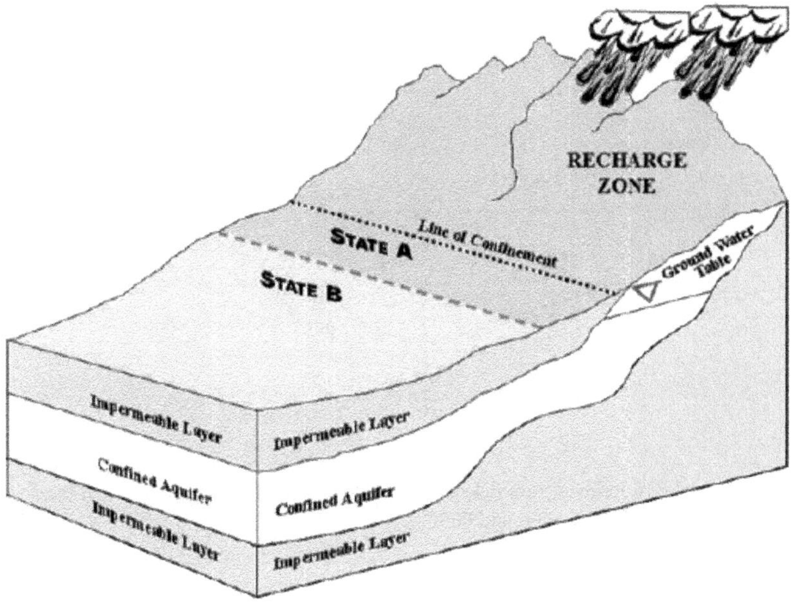

Fig. 14.7 Model E (Reprinted from Eckstein and Eckstein, 2005 with permission of the National Ground Water Association. Copyright 2005)

states could be transported to the other as a result of natural flow (i.e. from State A to State B) or a cone of depression locally reversing natural flow (within a distance limited by the cone of depression). Transboundary problems may arise as State A could divert surface runoff from recharging the aquifer or undertake activities that may result in surface water pollution in the recharge zone.

14.3.6 Model F

Model F (Figure 14.8) represents all transboundary aquifers unrelated to any surface water and disconnected from the hydrologic cycle (Eckstein and Eckstein, 2005). These aquifers can not be recharged and exploited in a sustainable approach. When a state starts producing groundwater from a water well in this type of aquifer, the state will generate an ever-expanding cone of depression that may encroach across the international border. If the states do not completely stop pumping, the aquifer will be depleted. Also, pollution may just result from the stagnant and nonrecharging character of the aquifer.

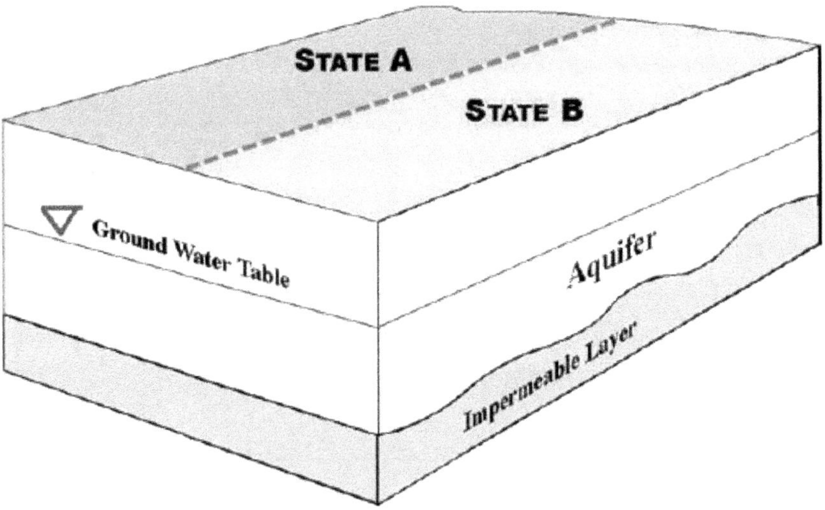

Fig. 14.8 Model F (Reprinted from Eckstein and Eckstein, 2005 with permission of the National Ground Water Association. Copyright 2005)

14.4 Shared and Transboundary Groundwater Resources Management

Shared and transboundary water resources management involve water rights and allocations. Although legislations and norms regulate the use of groundwater in society, these are absent in transboundary systems. Agreements on the water quantity that can be exploited need to be established regardless of water allocations, as it is the prime and critical concern regarding the welfare of today and future generations. In a shared aquifer system, the key issues related to its management described by Haddad et al. (2001) are how to determine the total extraction, how the measures needed to protect the aquifer be decided upon and enforced; how decision-making, and institutional structure will be implemented. In societies where water use has to be governed, institutional structure exists to govern and enforce the property rights, to adjudicate in case of disagreements and to ensure that water is used according to the sustainable yield constraint. No such institutions exist at the international level however, the UN International Law Commission (ILC) has adopted in 2006 a set of draft articles on the law of transboundary aquifers (Chapter 3). The management of shared aquifer can be a separate management, a coordinate management, a joint management or a delegation of responsibility; as well as an integration of the various management types (Haddad et al., 2001).

14.4.1 Separate Management

Separate management is the first option for shared aquifer management (Haddad et al., 2001). Under separate management, each party set its own development objectives and exploitation policies within its aquifer section by itself (e.g. well drilling, extraction rates, water quality standards, monitoring and data collection, modeling, water allocation, water rights and use). If the total amount extracted from the aquifer by each party is less than the aquifer recharges rates and the hydrogeological interdependence between the parts of the aquifers controlled by each party is limited, this management option may be the most efficient. However, if these conditions are different, this option does not prevent the unsustainable use of the aquifer, therefore, aquifer depletion and pollution will rendered the aquifer non-exploitable. Even if one party is concerned, it may be unable to affect the trends if other parties control critical areas. As separate management does not interfere with the party sovereignty, requires no special action and has minimal transaction costs, it is the default option for shared aquifer management.

14.4.2 Coordinated Management

The second option for shared aquifer management is for each party to manage the aquifer in its territory with some coordinated activities (Haddad et al., 2001). Coordination of any of the element of aquifer management that is beneficial to each party (e.g. extraction rates, data collection and modeling), while each party retains its right to manage separately all other activities, such as policies development is valuable. In a coordinated management, each party retains full authority of its part of the aquifer. When the needs and conditions for better management and economics are encountered, the parties may then coordinate their management activities (Feitelson and Haddad, 2001). However, coordinated management is not sufficient if a substantial interdependence exists between the parties in terms of aquifer or when water demand is higher than recharge rates.

14.4.3 Joint Management

A third option is the joint management of a shared aquifer (Haddad et al., 2001). In a joint management, the parties establish a unique institutional structure which manages the aquifer for a sustainable and conservative exploitation of the aquifer.

14.4.4 Delegation of Responsibility

A fourth possibility is the delegation of responsibility for the management of the aquifer to an external body (Haddad et al., 2001). The delegation of responsibility may be achieved through a regional or international body, or a privately owned corporation where the privatization of some aspects of aquifer management (e.g. monitoring, extraction and selling of water), leaving the parties in a role of regulators.

14.4.5 Sequential Flexible Management

Shared aquifer management regimes should be sequential and flexible to provide decision-makers with options to face various scenarios in a continuous changing environment (Feitelson and Haddad, 2001). Each management rationale is a response to a specific problem. The cooperation level may increase from a co-ordinated management framework to a joint management regime. The sequential approach is an evolving management process with many decision points that is improved over time.

The cooperative management of an aquifer between two parties relies on the understanding that unless cooperation is achieved through some coordinated activities, both parties will lose in the long term, as aquifer deterioration may occur. The problems encountered in shared aquifer systems that requires some level of cooperation in the management regimes are contamination and/or depletion of the aquifer, crises, inefficient water use, inefficient water supply and the need for comprehensive–integrative management (Feitelson and Haddad, 2001).

14.4.5.1 Aquifer Protection

To prevent contamination and/or depletion of an aquifer, or to mitigate it, the parties sharing and managing the aquifer have to cooperate to protect it. The rationale for this management approach is the aquifer protection (Feitelson and Haddad, 2001).

Maintaining the water quality and storage capacity of the aquifer is a common interest of all parties sharing the aquifer. This interest should lead to a joint management of the aquifer through several stages (Feitelson and Haddad, 2001). The first stage consists in a cooperative effort concerning qualitative and quantitative monitoring of aquifer as well as data management. The second stage of this cooperation effort is to address the main threats to the aquifer. More comprehensive and long-term issues regarding aquifer protection can be addressed later and would include for example the development of standards for water quality, wastewater treatment and reuse, and the coordination of research and development for solutions to aquifer threats.

14.4.5.2 Crisis Management

Aquifer management becomes controversial during crisis as policies from all parties need to be coordinated to avoid aquifer damages and exploitation interruption. This is the crisis management approach (Feitelson and Haddad, 2001). Crises management integrates crisis recognition, agreements on strategies to face the crisis and implementation of the crisis management scheme. The types of possible crises are defined by Feitelson and Haddad (2001) as:

- Sudden crises such as toxic materials spilling, discovery of hazardous materials in drinking water coming from wells;
- Cumulative crises resulting from cumulative effects of certain trends or natural events (e.g. droughts);
- Overexploitation by one side above the quantities agreed.

14.4.5.3 Efficient Water Use

The need for efficient water use increases with the water demand. Efficient water use is achieved through water transfer from low-value use to high-value use and the elimination of wasteful use. Physical and legal structures need to be established in order to achieve water transfer at feasible cost, thereby increasing water use efficiency and avoiding additional pumping (Lee and Jouravlev, 1998). These water transfers are usually attained through water markets which imply the properly pricing of water leading to the use of water in the most efficient manner and the elimination of wasteful use. (Winpenny, 1994; Dinar et al., 1997; Feitelson and Haddad, 2001).

Water markets will define the water rights and allocations based on the priority of domestic use and water quality differences (Berck and Lipow, 1994). The price of allocated water will be set by the free exchange of equity through use or property right to use of water, either for a limited time period (leasing) or in perpetuity (sale) (Feitelson and Haddad, 2001).

14.4.5.4 Efficient Water Supply

As public water supply is often inefficient, may have limited access to economies of scale, and aquifer overexploitation may develop, its privatization may improve its efficiency (Feitelson and Haddad, 2001). Private sector participation in shared aquifer management may impact the system efficiency as well as facilitate problems resolution that originated from lack of confidence between parties involved in long-standing conflicts (Lee and Jouravlev, 1997).

14.4.5.5 Comprehensive–Integrative Management

An aquifer should be managed in a comprehensive and integrative approach involving all parties. The goal of an integrative structure is to cover all aspects of aquifer management comprehensively by incorporating protection, crisis management, water use and water supply to achieve equitable and sustainable aquifer development and management (Feitelson and Haddad, 2001).

14.4.5.6 Comparative Benefits and Costs

Feitelson and Haddad (2001) performed a comparative analysis of the benefits and costs of each shared groundwater resources management options. In their analysis the groundwater resources are differentiated through their level of water interdependence between the parties across the boundaries and the aquifer use ratio. The aquifer use ratio is defined as total freshwater used to average annual recharge.

When the aquifer use ratio and level of interdependence increase, the damages resulting from no cooperation in aquifer management also increase (Figure 14.9). As the aquifer use ratio augments, water use will be transferred from less to more productive uses under the constraint of sustainable yield. Transaction costs do not rise as the use ratio increases.

When the aquifer use ratio is low, the marginal value product of water is low, the risk of aquifer pollution is low, and the costs of non-cooperation are low. As the aquifer use ratio increases, the costs of non-cooperation augment, the shadow value of water increases, and the coordinating management between the parties takes place to be able to realize the marginal value product of water. Once the use ratio is above one, the aquifer is overexploited and potential costs of non-cooperation increase exponentially as aquifer threats escalate and more complex management structures may be required (Feitelson and Haddad, 2001).

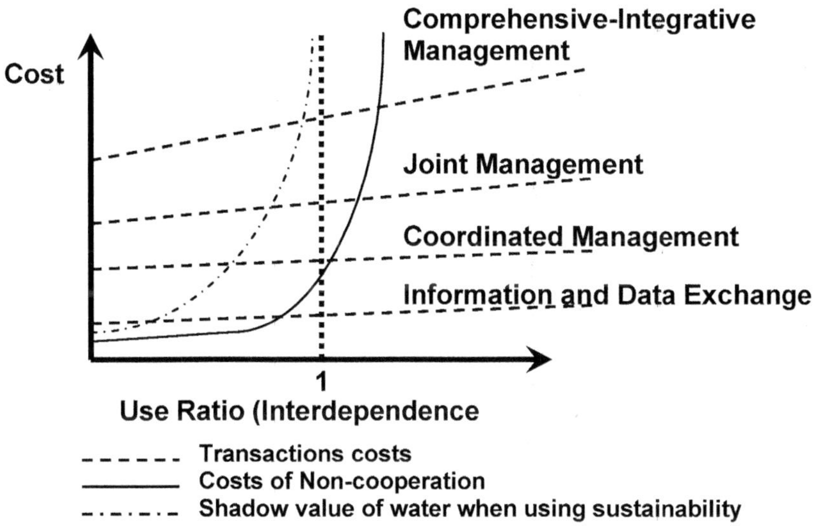

Fig. 14.9 Comparative benefits and costs of transboundary water resources management options (Modified from Feitelson and Haddad, 2001)

Many options are available for more intensive and comprehensive management of the aquifer, however the transaction costs will be higher. Therefore, the management of shared and transboundary groundwater resources will be developed and implemented gradually over time as the need for more intensive and comprehensive management rises.

References

Barberis, J. A., 1986, International ground water resources law, *Food and Agricultural Organization Legislative Study* **40**:67.

Berck, P., and Lipow, J., 1994, Real and ideal water rights—the propects for water-rights reform in Israel Gaza and the West Bank, *Resources and Energy Economics* **16**:287–301.

Cohen, P., 1994, Some comments about managing groundwater, in: *Joint Management of Shared Aquifers: The First Workshop*, E. Feitelson and M. Haddad, eds., The Harry S. Truman Research Institute for the Advancement of Peace and the Palestine Consultancy Group, Jerusalem.

Dinar, A., Rosengrant, M. W., and Meinzen-Dick, R., 1997, *Water Allocation Mechanisms—Principles and Examples*, Policy Research Working Paper 1779, The World Bank, Washington D.C.

Eckstein, Y., and Eckstein, G. E., 2005, Transboundary aquifers: Conceptual models for development of international law, Ground Water **43**(5):679–690.

El Magnouni, S., and Treichel, W., 1994, A multicriterion approach to groundwater management, *Water Resources Research* **30**:1881–1896.

Feitelson, E., and Haddad, M., 2001, A sequential flexible approach to the management of shared aquifers, in: *Management of Shared Groundwater Resources: The Israeli-Palestinian Case with an International Perspective*, E. Feitelson and M. Haddad, eds., Kluwer Academic Publishers, Boston, U.S.A., pp. 455–473.

Foster, S. S. D., and Chilton, P. J., 2003, Groundwater: The processes and global significance of aquifer degradation, *Philosophical Transactions of the Royal Society of London B* **358**:1957–1972.

Haddad, M., Feitelson, E., and Arlosoroff, S., 2001, The management of shared aquifers, in: *Management of Shared Groundwater Resources: The Israeli-Palestinian Case with an International Perspective*, E. Feitelson and M. Haddad, eds., Kluwer Academic Publishers, Boston, U.S.A., pp. 3–23.

Lee, T. R., and Jouravlev, A., 1997, Private Participation in the Provision of Water Services: Alternatives Means for Private Participation in the Provision of Water Services, United Nations Economic Commission for Latin American and the Carrabean, Serie Medio Ambiente Y Desarrllo 2, Santiago, Chile.

Lee, T. R., and Jouravlev, A., 1998, Prices, Properties and Markets in Water Allocations, United Nations Economic Commission for Latin American and the Carrabean, Serie Medio Ambiente Y Desarrllo 6, Santiago, Chile.

Margat, J., 2004, Groundwater management, in: *Groundwater Studies, An International Guide for Hydrogeological Investigations*, V. S. Kovalevsky, G. P. Kruseman, and K. R. Rushton, IHP-VI Series on Groundwater No.3, UNESCO, Paris, pp. 319–338.

Morris, B., Lawrence, A. R., Chilton, P. J., Adams, B., Caylow, R. C., and Klinck, B. A., 2003, *Groundwater and Its Susceptibility to Degradation: A Global Assessment of the Problems and Options for Management*, UNEP Early Warning & Assessment Report Series RS. 03-3, Nairobi, Kenya.

Pfannkuch, H. O., 1975, *Study of Criteria and Models Establishing Optimum Level of Hydrogeologic Information for Groundwater Basin Management*, Water. Res. Research Center, Univ. Minnesota, Minneapolis, Bull. 81, pp. 165.

United Nations Educational Scientific and Cultural Organization, (UNESCO), 1992, Ground Water; Managing the Invisible Resource, Environmental Development Briefs No.2, London, U.K.

Villholth, K. G., 2006, Groundwater assessment and management: Implications and opportunities of globalization, *Hydrogeology Journal* **14**:330–339.

Winpenny, J., 1994, *Managing Water as an Economic Resources*, Routledge, London.

Chapter 15
Sustainable Development and Integrated Management of Water Resources

Christophe J. G. Darnault*

Abstract This chapter presents the principles of sustainable development and integrated management of water resources in the context of water sharing between society and nature as well as in transboundary systems. The concept of sustainable development of water resources is presented with a focus on the model elements and directionality, the perspective of ecological economics, as well as the development of a systematic framework for the evaluation of water resources through the development of criteria and indicators. Integrated management and the four 1992 Dublin principles that define integrated water resources management are reviewed. Since, integrated water resources management provides methods and tools for an effective and equitable water allocation, the example of the water poverty index is also described. The application of sustainable development and integrated management to the transboundary water resources is also discussed.

Keywords Water scarcity, water stress, water sharing, equitable water allocation, sustainable development, integrated water resources management, indicators and criteria for water management, ecological economics, transboundary water resources, water poverty index

15.1 Introduction

The world is enduring serious water crises that may result in water conflicts. The causes are mostly originating from unsustainable development of water resources after water scarcity. In addition, the application of integrated management to water resources is essential to avoid conflicts. This chapter presents the principles of sustainable development and integrated management of water resources in the

Christophe J. G. Darnault
University of Illinois at Chicago, Department of Civil and Materials Engineering, 842 W. Taylor St., ERF# 2071, Chicago, IL 60607, U.S.A.
* To whom correspondence should be addressed. E-mail: darnault@uic.edu

C.J.G. Darnault (ed.), *Overexploitation and Contamination of Shared Groundwater Resources*.

context of water sharing between society and nature as well as in transboundary systems.

15.2 Sustainable Development and Water Resources

15.2.1 General Concept

The concept of sustainability was first introduced by Thomas Robert Malthus (1766–1834) in his book titled "An Essay on the Principle of Population" published in 1798. Malthus stated that "This natural inequality of the two powers of population, and of production in the earth, and that great law of our nature which must constantly keep their effects equal, form the great difficulty that to me appears insurmountable in the way to the perfectibility of society." As it was stated by the World Commission on Environment and Development (WCED) in 1987, "in its broadest sense, the strategy for sustainable development aims to promote harmony among human beings and between humanity and nature". According to the World Commission on Environment and Development (WCED, 1987) "Humanity has the ability to make development sustainable – to ensure that it meets the needs of the present without compromising the ability of future generations to meet their own needs" and therefore the pursuit of sustainable development needs:

- "A political system that secures effective citizen participation in decision making;
- An economic system that is able to generate surpluses and technical knowledge on a self-reliant and sustained basis;
- A social system that provides solutions for the tensions arising from disharmonious development;
- A production system that respects the obligation to preserve the ecological base for development;
- An administrative system that is flexible and has the capacity for self-correction".

15.2.2 Model Elements and Directionality

The implementation of sustainable development as an ideological paradigm implies the integration of three elements characterized by Young (1992) and Flint (2003) as (Figure 15.1):

- Economic vitality: the development that protects and/or enhances natural resources through improvements in management practices/policies, technology, efficiency, and life-style changes;
- Ecologic integrity: the understanding of natural system processes in landscapes and watersheds in order to guide design of sound economic development strategies that preserve these natural systems;
- Social Equity: the guarantee of equal access to jobs (income), education, natural resources, and services for all people—total societal welfare.

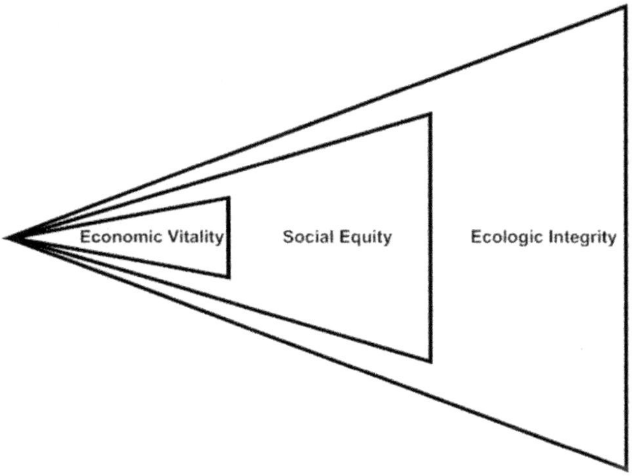

Fig. 15.1 Sustainability model: elements and directionality (Modified from Kranz et al., 2004)

Sustainable activities encompass analysis of the consequences of present actions on future public and environmental health through examination of the connections between environmental, economic, and social dimensions. In understanding the three overlapping triangles, it is also important to observe the directionality to each triangle's dependence on the others (Figure 15.1). All life depends on natural resources. However, economy and society are no less important to humanity than ecology (Wackernagel and Rees 1996). Rather, a directionality of dependence exists. Sustainable development considers directionality, where economic and cultural activities are integrated into natural processes in a cyclic approach not to degrade the environment upon which depends economic prosperity and social stability.

15.2.3 Ecological Economics

The perspective of sustainability through ecological economics is based on systems idea (Capra, 2002; Costanza, 2001). Lant (2004) developed a system of sustainable development in which natural, human, intellectual, and manufactured capital are transformed continuously, one into another, through market economy (Figure 15.2). Solar energy is the driving force of the system which evolves by interactions among its interdependent elements to release heat as waste (Capra 1996).

The market economy is analyzed by neo-classical economics. The ecological economics approach to sustainability demonstrates that economic growth may occur in positive, neutral, or negative ways with respect to sustainability (Lant, 2004). Sustainable economic growth occurs when a society increase capital efficiency with which various forms of capital are utilized to produce goods and services. Neutral economic growth occurs when social or natural processes are incorporated within the market economy without necessarily improving the effectiveness of the ecological-economic system. Unsustainable economic growth occurs when an increase in the output of goods and services comes at the expense of reductions in the value of natural capital, intellectual capital, and/or manufactured capital that exceed the value of the additional goods and services produced, and therefore undermining the entire systems' ability to recreate itself over time (Lant, 2004).

Water has essential roles in the ecological-economic process (Figure 15.2). Water is a raw material, a factor of production of a number of marketable commodities, some of which are themselves factors of production of other final goods. Because of its contribution to human health, treated potable water for domestic use is enormously valuable in producing human capital. Water in oceans, rivers, lakes, wetlands, soil, and other components of the hydrologic cycle is a, if not the, critical factor of production of ecosystem services (Lant, 2004). The

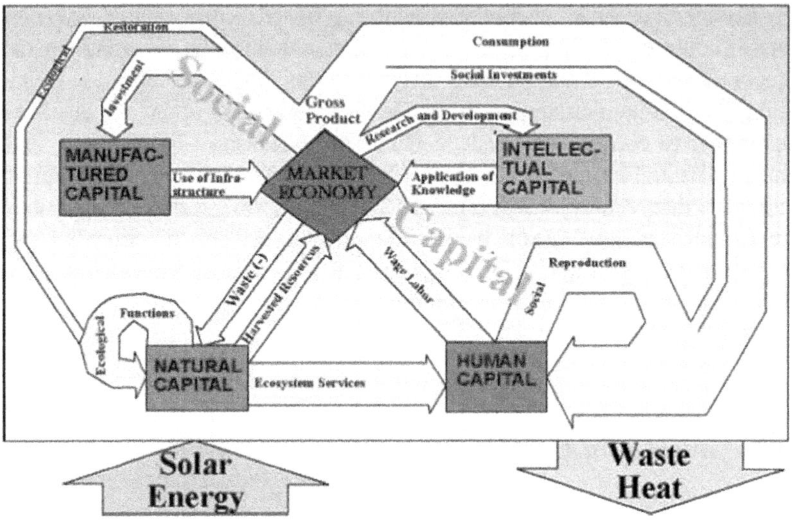

Fig. 15.2 A dynamic systems conceptualization of sustainable development (Reprinted from Lant, 2004 with permission of Universities Council on Water Resources. Copyright 2004)

marginal value of water must be evaluated among its uses to assess not only the highest and best uses of water, but also the contribution of water in sustainable development (Lant, 2004). Ecological economics provide an approach for evaluating reforms required for a more sustainable management of water resources.

15.2.4 Criteria, Indicators, and Evaluation

Kranz et al. (2004) developed system concepts linked to information concepts for the capital maintenance concept of sustainability to be used in the development of criteria and indicators for sustainable water resources management (Figure 15.3). Water resources systems needs are identified with a set of criteria related to categories of capital (Kranz et al., 2004).

Indicators are selected for each criterion using systems model to identify and represent the critical components and processes for each category of capital (Figure 15.4). In developing indicators to measure sustainability, Kranz et al. (2004) intend to determine the capacity of different capitals to maintain resources over time, especially the extent to which the capacity of social, economic, and natural capital is being maintained or enhanced. The indicators of stressors, investments, and capital are essential to assess the opportunities transmitted to future generations (Figure 15.4).

Flint (2004) developed a conceptual model to evaluate water resources through the development of criteria and indicators represented by a theoretical model (Figure 15.5). Interactions between environmental, social, and economic capital and processes that affect them are perceived in this model for better decision-making

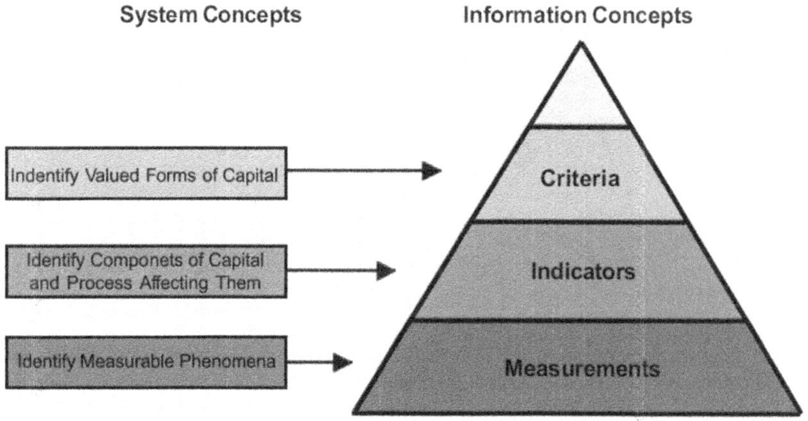

Fig. 15.3 Linkage between system concepts and information concepts in sustainable water resources (Reprinted from Kranz et al., 2004 with permission of Universities Council on Water Resources, Copyright 2004)

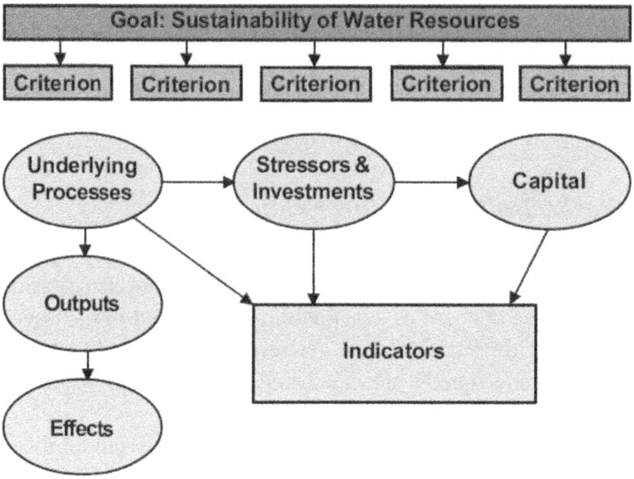

Fig. 15.4 Generic systems model linking goals, criteria, and indicators (Reprinted from Kranz et al., 2004 with permission of Universities Council on Water Resources, Copyright 2004)

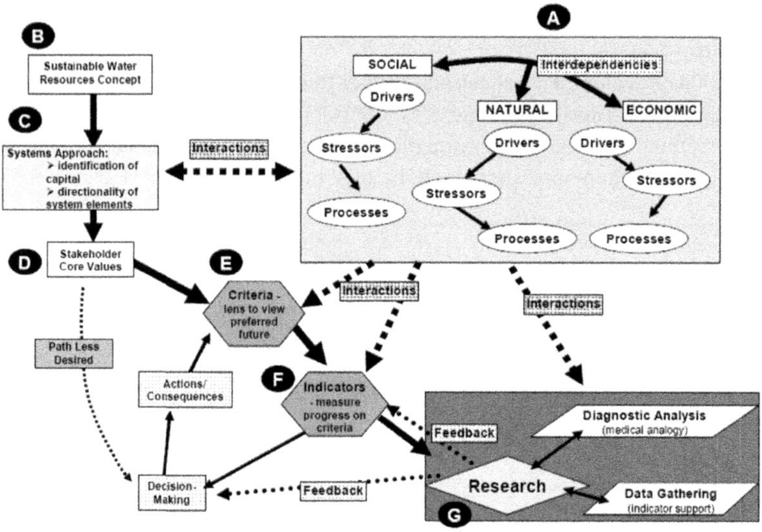

Fig. 15.5 Systematic framework for evaluating water resources through the development of criteria and indicators (Reprinted from Flint, 2004 with permission of Universities Council on Water Resources. Copyright 2004)

regarding sustainable water resources development (Figure 15.5A) (Flint, 2004). Figure 15.5B shows the development of a conceptual view of sustainable water resources representing the water use interdependencies in a multi-dimensional way among natural, social, and economic systems (Flint, 2004). The economic vitality is achieved in the context of the enhancement and preservation of ecological integrity, social well-being, and security for all. Figure 15.5C presents the approach to categorize the key forms of natural, social, and economic capital that need to be sustained to identify stakeholder core values with regard to water. The development of achievable goals for sustainable water resources reflecting different stakeholder core values should be formulated to address fundamental principles that underlay the conservation, protection, remediation, and longevity of water resources (Figure 15.5D). Criteria establishing the conditions to protect and maintain all the perceived beneficial uses of water assets need to be defined (Figure 15.5E). These criteria provide a lens through which the preferred future status of water (i.e., characteristics that best define water resources sustainability) may be evaluated (Flint et al., 2002). Indicators to measure sustainability progress are necessary to determine to which degree sustainability goals have been reached as well as to better understand past trends so that decision-makers may influence future directions of development (Figure 15.5F). From this conceptual model evolves also the need for research (Figure 15.5G) (Flint, 2004).

15.3 Integrated Management of Water Resources

Integrated water resources management (IWRM) aims at satisfying the freshwater needs of all countries for their sustainable development (UNCED, 1992b). IWRM is based on the principle that water is an integral part of the ecosystem, a natural resource and a social and economic good, whose quantity and quality determine the nature of its utilization. Therefore, water resources have to be protected by

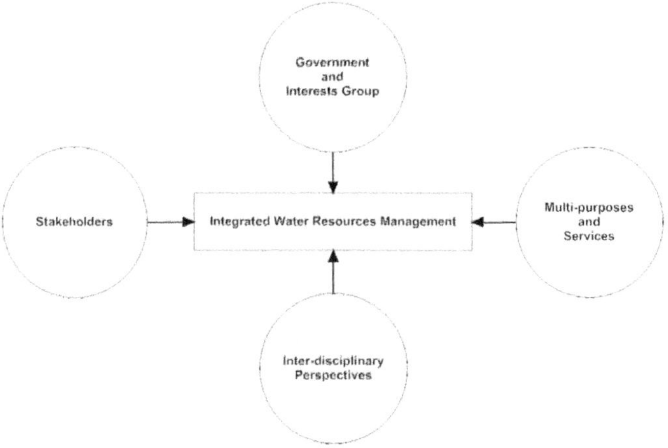

Fig. 15.6 Integrated water resources management: components and structure (Modified from Water Encyclopedia, 2007)

considering the functioning of aquatic ecosystems and the perenniality of resources to satisfy and reconcile needs for water in human activities. In developing and using water resources, priority has to be given to satisfy basic needs and safeguard ecosystems.

Integrated water resources management is the practice of making decisions and taking actions while considering human and nature concerns in a multi-faceted milieu of how water should be managed (Figure 15.6). IWRM uses structural measures and nonstructural measures to control natural and human-made water resources systems for beneficial uses.

15.3.1 Water Demand and Water Scarcity

Demand for water resources of sufficient quantity and quality for human society continue to intensity as population increases and socio-economic development accelerates (Flint and Houser, 2001). However, the capacity of the Earth to meet this demand is in decline because of overexploitation and pollution (Flint, 2004). Scarcity implies diminishing resources and/or a pressure on the supply of available resources from an increasing demand. Water scarcity involved inventories of available water resources combined with a socio-economic analysis of the driving forces for increasing demands and their evolution over time. Water scarcity projections are based on threshold levels of 1,000–2,000 m^3 per person per year to designate water stress (Falkenmark, 1997). Figure 15.7 shows how human and aquatic ecosystem proportions would change between 1950 and 2050 under the scenarios where human needs were met at either 1,000 or 2,000 m^3 $person^{-1}$ $year^{-1}$

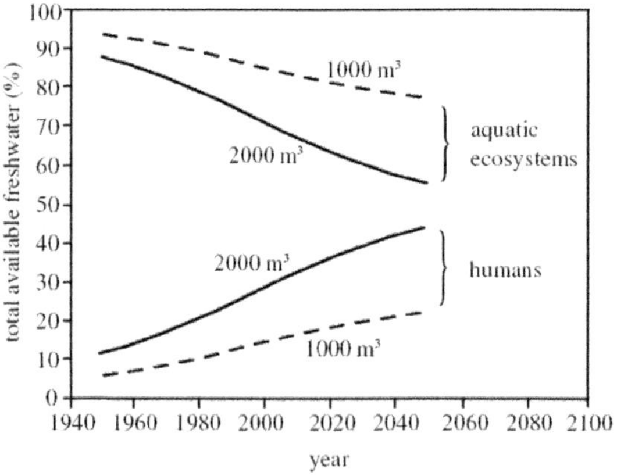

Fig. 15.7 Change in human water requirements and aquatic ecosystem water availability with time assuming two levels of human use: 2,000 m^3 $person^{-1}$ $year^{-1}$ (continuous lines) and 1,000 m^3 $person^{-1}$ $year^{-1}$ (dashed line) (Reprinted from Wallace et al., 2003 with permission of The Royal Society. Copyright 2003)

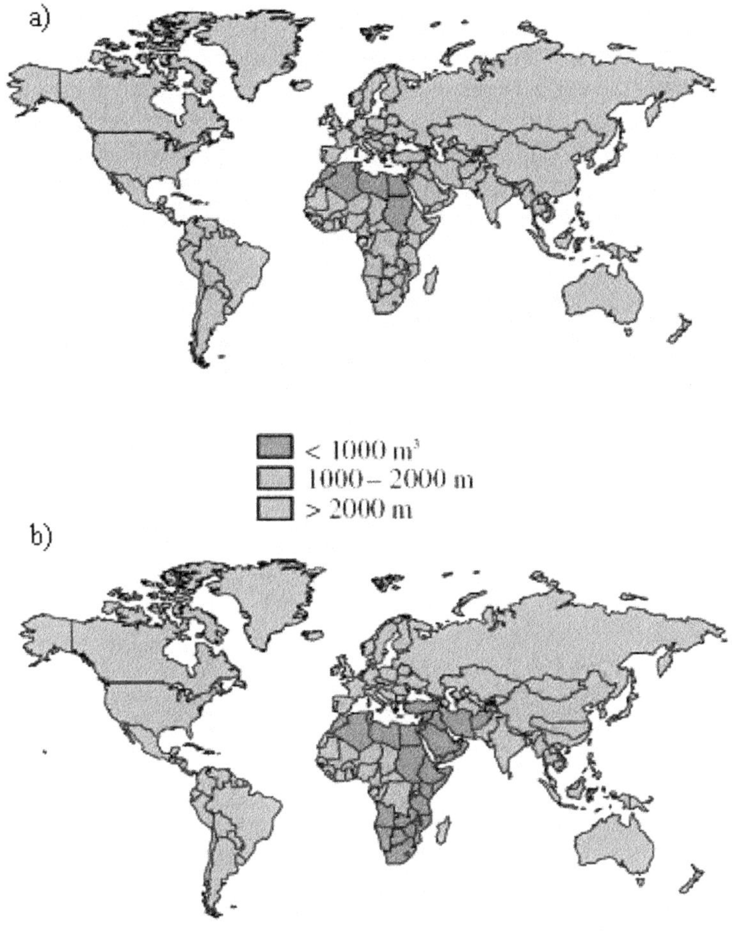

Fig. 15.8 Per capita annual renewable freshwater **a** now and **b** 2050 (Reprinted from Wallace et al., 2003 with permission of The Royal Society. Copyright 2003)

(Wallace et al., 2003). Early projection of water scarcity shows that some degree of scarcity currently exists throughout Southern Africa and the Middle East (Figure 15.8a) and that by 2050, 67% of the world's population (6.5 billion people) may experience some water scarcity (Figure 15.8b) (Wallace et al., 2003). Because humans use water from rivers, aquifers—the blue water system—the ecosystems primarily affected by human water use are aquatic and riparian zone ecosystems. Figure 15.9a,b gives a global view of where the pressures on aquatic ecosystems are and may be greatest now and in 2050. Figure 15.9a shows that, at present, around one-third of the land area of the world would have less than 50% of the available water resources to support aquatic ecosystems and most of this area would be less than 25%, a figure below which severe environmental impacts may occur. Figure 15.9b shows that by 2050, 40% of the land area will have less than 25% of the available water resources to support aquatic ecosystems. The number of areas where there is a danger of a water barrier and severe environmental degradation increases by 2050 to include Western Africa, Eastern Africa, and Central Asia (Wallace et al., 2003). Future global water scarcity will affect most of the world's population and population growth will dominates future water scarcity. Implications for aquatic ecosystems will be widespread in many parts of the world.

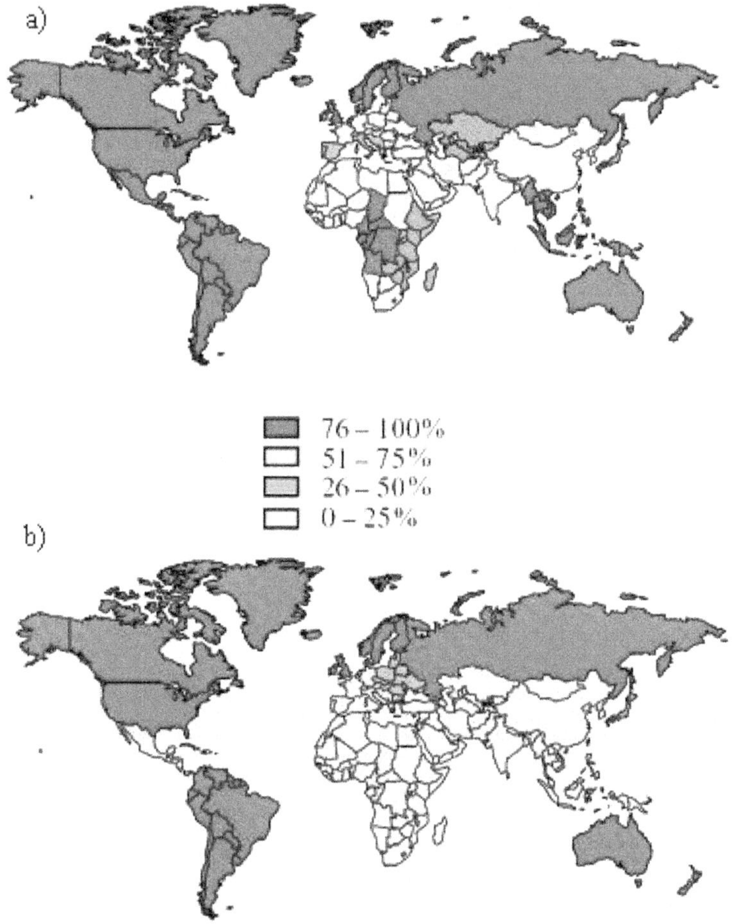

Fig. 15.9 Percentage of annual renewable freshwater left for aquatic ecosystems after meeting all human needs **a** now and **b** in 2050 (Reprinted from Wallace et al., 2003 with permission of The Royal Society. Copyright 2003)

15.3.2 Integrated Management Principles

Sustainable development and its political consequences are the foundations of integrated management (Vallega, 1999). Agenda 21 asserts that States are committed to integrated management and sustainable development and must undertake the necessary actions to deal with this task. In order to design integrated management, several principles should be established.

15.3.2.1 Maximization of Consistency Between Specific, Local, and Final Goals

The first principle of integrated management is to maximize the consistency between specific, local, and final goals (Vallega, 1999). By focusing on the objective, management is integrated through programs and measures designed to optimize sustainable development through correlated goals in pursuit of ecosystem integrity, economic efficiency, and social equity (Cicin-Sain and Knecht, 1995).

15.3.2.2 Holistic and Conservation-aimed Management of Individual or Group of Contiguous Ecosystems

The second principle of integrated management is to provide a holistic and conservation-aimed management of individual or group of contiguous ecosystems (Vallega, 1999). Sustainable development implies that the ecosystem is the primary and essential resources for development, therefore the management of resources is integrated when the whole ecosystem is embraced by comprehensive programs and measures.

15.3.2.3 Design of Proper Cross-issue and Cross-sector Programs

The third principle of integrated management is to design proper cross-issue and cross-sector programs (Vallega, 1999). Programs of resources management are generally perceived by the local community, adopted by local decision-makers and address a single issue. Therefore, the management is integrated as the single issue is considered jointly with other existing issues.

15.3.2.4 Benefits from Inter-sector and Inter-rank Coordination Between Decision-makers

The fourth principle of integrated management is to benefit from inter-sector and inter-rank coordination between decision-makers (Vallega, 1999). Integrated management is effectively undertaken when all the concerned decision-makers have the same understanding of the need for resources management and implement coordinated actions. Coordination is needed between decision-makers involved in different sectors and levels.

15.3.2.5 Coordination in Content and Time Between Decision-making Processes

The fifth principle of integrated management is to provide coordination in content and time between decision-making processes (Vallega, 1999). Sustainable resources development are optimized through measures adopted by decision-makers which are compatible with others and interact positively. Coordination is also required over time since the times sequences of provisions by the different decision-makers need be correlated.

15.3.2.6 Management Coordination of Resources Area with Surrounding Environment

The sixth principle of integrated management is the coordination of the management of the resources area with that of surrounding environment (Vallega, 1999). The concept of external environment has an essential function: as the resources ecosystem interacts with the surrounding environment, it is subject to inputs generated by its external environment and impacts upon the contiguous ecosystems with which it is linked. Also, settlements and human activities are influenced by factors acting, and processes developing, in other regions and countries. Consequently, the resources community is able to address impulses to other communities and to establish interactions with its social, economic and political external environment. The assessment of these interactions is critical to resources management.

15.3.3 The Four 1992 Dublin Principles

Integrated water resources management encompasses land- and water-related aspects and should be carried out at the watershed level. Integrated water resources management is based on the four 1992 Dublin principles (UNEP, 1992):

- "Fresh water is a finite and vulnerable resource, essential to sustain life, development and the environment;
- Water development and management should be based on a participatory approach, involving users, planners and policy-makers at all levels;

- Women play a central part in the provision, management and safeguarding of water;
- Water has an economic value in all its competing uses and should be recognized as an economic good".

15.3.4 Water Poverty Index (WPI)

A more holistic approach to water resources management can be achieved by using integrated water management tools such as the Water Poverty Index (WPI) developed by Sullivan et al. (2002). This approach incorporates environmental attributes into its framework and allows decision-makers to discriminate between different locations or communities based on water related indices. Sullivan et al. (2002) grouped these indices into five criteria (Figure 15.10):

- Water resources assessment, including the variability and quality of surface and groundwater resources;
- Access to water for domestic use and irrigation;
- A measure of how water is used for productive purposes;
- Values reflecting the water management capacity, based on education, health, membership of water user groups and access to finance;
- Environmental impact of water utilization, which currently serves as a proxy for the incorporation of ecological water needs.

Normalized scores (between 0 and 100) for each of these criteria can be identified from existing data. By applying a weighted average, an overall index value, WPI

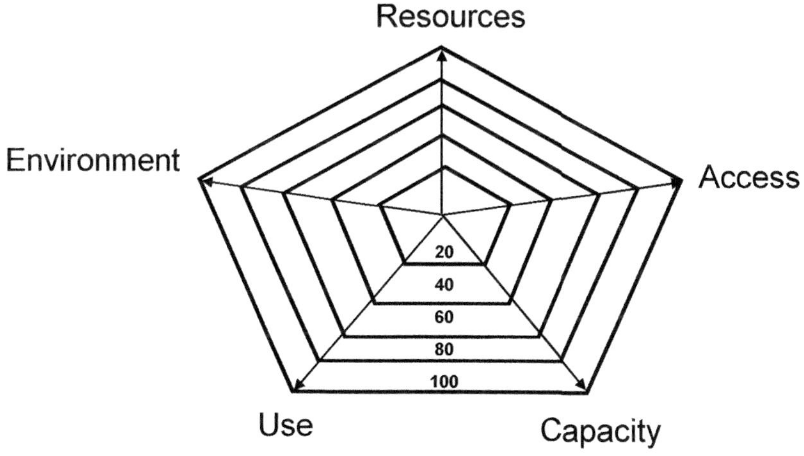

Fig. 15.10 Water Poverty Index (WPI) (Modified from Sullivan et al., 2002)

can be generated. The WPI allows comparison between locations and assesses changes over time. By plotting the components together on a pentagram, the nature and relative importance of the water resources management issues at each site and over time can be analyzed and compared (Sullivan et al., 2002).

15.4 Application to Transboundary Water Resources

Water resources sustainability is the ability to provide and manage water quantity and quality to meet the present needs of human and environmental ecosystems, while not impairing the needs of future generations to do the same (WCED, 1987). The 1992 UN Rio Declaration set the principle of cooperation by imposing solidarity concept in solving shared water resources problems in transboundary systems (UNCED, 1992a). The main elements of the 1997 UN Convention on International Waters are: equitable and reasonable use of transboundary water resources, information of activities that may have adverse transboundary environmental effects on shared water resources; obligation to consult between neighboring States, and obligation to immediately notify natural disasters and emergencies concerning shared water resources (UNCNUIW, 1997). Sustainable transboundary water resources management encompasses (UNESCO, 2003):

- Sharing water-related benefits among nations for regional economic integration rather than polarized claims for water;
- Balancing competing uses of basin and aquifer resources, especially upstream and downstream uses, in a transparent and participative way for local and regional sustainable development;
- Focusing on poverty reduction, public participation and gender balance to ensure equitable access to water for livelihoods;
- Recognizing the fundamental need of freshwater ecosystems for resource protection and natural risk prevention;
- Protecting watercourse during wars and conflict and post-conflict rehabilitation of water resources;
- Improving our knowledge about conflicts causes and potential policy responses to prevent conflicts triggered by competition for the resource among different uses and users, and environmental concerns (e.g., pollution);
- Developing capacity building on Integrated Water Resources Management.

Transboundary water resources and their use are of great importance to riparian States. Cooperation among those States may be desirable in conformity with existing agreements taking into account the interests of all riparian States concerned (UNCED, 1992b). In the case of transboundary water resources, riparian States need to formulate water resources strategies, prepare water resources action programs, and consider the harmonization of strategies and action programs (UNCED, 1992b). As water resources transcend political and administrative boundaries, water resources must be shared in a sustainable approach between individuals, economic sectors, intrastate jurisdictions, and sovereign nations. The integrated transboundary water resources management should include (INBO, 2004):

- Development of new national water legislation based on IWRM;
- Establishment of national and international aquifer and river basin organizations;
- Adoption of international conventions, treaties on freshwater management;
- Implementation of comprehensive monitoring systems and exchange of information and data;
- Elaboration and adoption of national and regional planning for water resources development;
- Creation of funding systems based on common causes and solidarity between freshwater resources systems.

Sustainable water resources development is a concept that emphasizes the need to consider the future as well as the present in the management of water resources under changing demands and environment over temporal and spatial scale without degradation of the ecosystem. Integrated water resources management provides methods and tools for an effective implementation of equitable water allocation, as we depends on water resources systems for our development, survival, and welfare. The global application of sustainable development and integrated management principles to water resources, especially in transboundary systems is critical for human and environmental security.

References

Capra, F., 1996, *The Web of Life: A New Scientific Understanding of Living Systems*, Anchors Books, New York.

Capra, F., 2002, *The Hidden Connections: Integrating the Biological, Cognitive, and Social Dimensions of Life into a Science of Sustainability*, Doublebay, New York.

Cicin-Sain, B., and Knecht, R. W., 1995, Measuring progress on UNCED implementation. *Ocean Coast. Manag.*,, special issue Earth summit implementation: Progress achieved on ocean and coasts, **29**(1–3):1–11.

Contanza, R., 2001, Visions, values, valuation, and the need for an ecological economics, BioScience, **51**(6):459–468.

Falkenmark, M., 1997, Society's interaction with the water cycle: a conceptual framework for a more holistic approach, *Hydrol. Sci.*, **42**:451–466.

Flint, R. W., 2003, Sustainable development: What does sustainability mean to the individual in the conduct of their life and business, in: *Handbook of Development Policy Strategies*, G. M. Mudacumura, ed., Marcel Dekker, Inc., New York.

Flint, R. W., 2004, The sustainable development of water resources, *Water Resour. Update*, **127**:41–51.

Flint, R. W., and Houser, W. L., 2001, *Living a Sustainable Lifestyle for Our Children's Children*, Campbell, CA: IUniverse.

Flint, R. W., Frick, F. C., Duffy, A., Brittingham, J., Stephens, K., Graham, P., and Borgmeyer, C., 2002, Characteristics of Sustainable Destination Resort Communities, Resort Municipality of Whistler, BC, Canada, (December 7, 2007); http://www.whistlerfuture.com/thinkit/pdfs/Whistler_Sustainit3.pdf.

Kranz, R., Gasteyer, S. P., Heintz, T., Shafer, R., and Steinam, A., 2004, Conceptual foundations for the sustainable water resources roundtable, *Water Resour. Update*, **127**:11–19.

International Network of Basin Organizations, (INBO), 2004, International organizations urge G8 world leaders to allocate funds to promote cooperation over transboundary water for development, security and peace, *The Network Newsletter*, **N°12**, December 2003–January 2004.

Lant, C., 2004, Water resources sustainability: An ecological economics perspective, *Water Resour. Update*, **127**:20–30.

Malthus, T. R., 1798, *An Essay on the Principle of Population as It Affects the Future Improvement of Society, with Remarks on the Speculations of Mr. Godwin, M. Condorcet, and Other Writers*, Published: London: J. Johnson, First edition, (September 21, 2007); http://www.econlib.org/library/Malthus/malPop.html.

Marchisio, S., 1999. Water management and general principles of international law, in: *Sustainable Development and Management of Water Resources: A Legal Framework for the Mediterranean*, S. Marchisio, G. Tamburelli, and L. Pecoraro, eds., Institute for Legal Studies on the International Community – CNR. Mediterranean Sustainable Development Law – MESDEL, Roma, Italy, pp. 103–118.

Sullivan, C. A., Meigh, J. R., and Fediw, T., 2002, *Derivation and Testing of the Water Porverty Index,* phase 1, final report. Wallingford, UK: Centre for Ecology and Hydrology.

United Nations Conference on Environment and Development, (UNCED), 1992a, *Rio Declaration on Environment and Development, Report of the United Nations Conference on Environment and Development, Agenda 21, Annex 1*, Rio de Janeiro, 3–14 June 1992b, (June 21, 2007); http://www.un.org/documents/ga/conf151/aconf15126-1annex1.htm.

United Nations Conference on Environment and Development, (UNCED), 1992b, *Rio Declaration on Environment and Development, Report of the United Nations Conference on Environment and Development, Agenda 21, Chapter 18: Protection of the quality and supply of freshwater resources: Application of integrated approaches to the development, management and use of water resources*, Rio de Janeiro, 3–14 June 1992, (June 21, 2007); http://www.un.org/esa/sustdev/documents/agenda21/english/agenda21chapter18.htm.

United Nations Convention on Non-navigational Uses of International Watercourses, New York, (UNCNUIW), 1997, *Report of the Sixth Committee convening as the Working Group of the Whole*, (June 21, 2007); http://www.un.org/law/cod/watere.htm.

United Nations Environmental Programme, (UNEP), 1992, *Final Report of the International Conference on Water and the Environment*, Dublin, UNEP, Nairobi.

UNESCO, 2003, Report on the Involvement of IHP in the 3rd World Water Forum, 35th Session of the IHP Bureau, UNESCO, Paris, June 2–4, 2004, (June 21, 2007); http://unesdoc.unesco.org/images/0013/001306/130672e.pdf.

Vallega, A., 1999, *Fundamentals of Integrated Coastal Management*, Kluwer Academic Publishers, pp. 1–264.

Wackernagel, M., and Rees, W., 1996, Our Ecological Footprint, New Society Publishers, Gabriola Island, Canada.

Wallace, J. S., Acreman, M. C., and Sullivan, C. A., 2003, The sharing of water between society and ecosystems: From conflict to catchment-based co-management, *Phil. Trans. R. Soc. Lond. B*, **358**:2011–2026.

Water Encyclopedia, 2007, Integrated Water Resources Management, (September 21, 2007); http://www.waterencyclopedia.com.

World Commission on Environment and Development, (WCED), 1987, *Our Common Future*, The Brundtland Report, Oxford University Press, Oxford, UK, 383 pp.

Young, M. D., 1992, *Sustainable Investment and Resource Use. Equity, Environmental Integrity* Publishing Group, Paris.

Chapter 16
Assessing Chemical Status of Shared Groundwater Resources: A Crucial Political, Regulatory and Management Issue

Ewa Kmiecik, Jadwiga Szczepańska, Irena Twardowską*, Sebastian Stefaniak

Abstract According to the EU Water Framework Directive 2000/60/EU (WFD, 2000) and the project of the Daughter Groundwater Directive (GWD, 2003), water chemical status should be assessed for all usable groundwater bodies (GWBs) or group of bodies to identify any significant and sustained upward trend of concentrations of pollutants found in these bodies of groundwater, and define the starting point for trend reversals. Groundwater is a resource at risk, thus correct assessment of its chemical status and identification of risk, in particular in shared groundwater bodies is of a vital importance and has significant political dimension as a potential source of conflicts. The presented studies carried out on two selected objects within the EU 5FP BASELINE and 6FP BRIDGE projects were aimed to contribute to evaluating natural baseline quality in European Aquifer and to establishing reliable common approach for assessing groundwater chemical status based on the criteria set by WFD (2000) and GWD (2003), and with using also national regulations and threshold values (RMS, 2004). Assessment of groundwater chemical status performed in parallel in the point and spatial mode based on aggregation with use of median values proved this procedure to be the most appropriate. Application of the point method enables visualization of areas (within the drainage basin or groundwater bodies) of different groundwater chemical status for a correct decision-making, early identification of a risk and timely undertaking interception/reversal actions for restoration of good groundwater status. Spatial assessment of groundwater chemical status on the basis of concentration data aggregated with use of either mean or median values was proved to cause over- or underestimation of risks that lead to wrong decisions resulting in considerable economic losses, possible damage to human health and the environment, and to unavoidable conflicts in case of shared groundwater resources. The spatial assessment was found to be more reliable if performed on

Ewa Kmiecik, Jadwiga Szczepańska, Irena Twardowską, Sebastian Stefaniak
Polish Academy of Sciences, Institute of Environmental Engineering, M. Sklodowska-Curie St. 34, 41-819 Zabrze, Poland

* To whom correspondence should be addressed. E-mail: irena@ipis.zabrze.pl

C.J.G. Darnault (ed.), *Overexploitation and Contamination of Shared Groundwater Resources*.

the basis of median values, and not of arithmetic mean, as recommended by WFD (2000) and GWD (2003). This results from the fact that chemical constituents of groundwater usually display log-normal distribution, and in this case median is not susceptible to a deformation to such extent as mean value. In GWBs of multilayer structure, assessing groundwater chemical status should be accomplished separately for groundwater of shallow and deeper circulation due to different exposure to contamination sources.

Keywords Groundwater bodies, EU Groundwater Directive, chemical status, threshold values, assessment methods

16.1 Introduction

16.1.1 Groundwater Chemical Status and EU Current Policy

Groundwater resource and its protection is of particular importance for sustainability of ecosystems and mankind. According to UNEP (2000), some 1.5 billion people rely on groundwater sources, and withdraw annually about 600–700 km^3 that constitutes about 20% of global water withdrawals. In arid regions or in areas with dense population, the proportion of groundwater in total water consumption is considerably higher. The same source indicates that over 40% of the world's population live in water-scarce river basins; by 2025, further increase of water scarcity regions to 50% is anticipated.

Another serious problem is groundwater depletion due to pollution mostly from anthropogenic sources, which is estimated for about 20% of global withdrawals. Besides reducing availability of good quality water for consumption, it poses a risk to associated aquatic and terrestrial ecosystems. The specific feature of shallow renewable groundwater resources (that are most widely used for different purposes and actively participate in water circulation), is their susceptibility to non-point contamination from the diffuse sources that due to its spatial character is extremely difficult and expensive to control and to intercept. There is a growing evidence of increasing groundwater contamination from diffuse sources caused by the extensive application of fertilizers, growing waste reuse as soil amendment or as common fill, or due to weathering of geochemically unstable waste in disposal facilities.

It is generally recognized that the prevention of groundwater pollution is a major and a crucial issue that also contributes to the good quality status of associated aquatic ecosystems and directly dependent terrestrial ecosystems (GWD, 2003). Remediation of contaminated groundwater basins is mostly unsuccessful and expensive.

In the complicated geopolitical conditions, significant part of groundwater bodies protection is within the responsibility of two or even more states, as it occurs in the European Union. With respect to groundwater protection policy and its implementation mechanisms, EU constitutes a complicated political organism and hydrogeological system that currently comprises 25 Member and 3 Associated States with thousands kilometers of borders that divide river drainage basins and groundwater bodies between different states, thus creating shared groundwater resources. Although groundwater protection issues fall within the Water Framework Directive 2000/60/EC (WFD, 2000), the specificity of groundwater resource protection and the growing pollution from diffuse sources call for specific measures to prevent and control groundwater pollution. This requirement resulted in the development of a proposal for the Daughter Groundwater Directive (GWD, 2003) that have not yet been adopted for several reasons, such as a delay in establishing the selected lists of pollutants and related thresholds values due to scarce scientific data. Another critical issue is a lack of criteria for reliable assessment of chemical quality of groundwater based on monitoring data.

The explanatory memorandum to the project of the EU Groundwater Directive (GWD, 2003) can be summarized as:

- Sustainable management of shared groundwater resources requires harmonized approach for defining and monitoring groundwater status throughout Europe;
- At present, there are no common criteria and no common references (selected pollutants list and related threshold values) for groundwater to avoid conflicts, mistaken decision-making and unacceptable impacts on the environment;
- There is also not enough monitoring data and consolidated knowledge for reliable assessment of groundwater chemical status.

Under the circumstances, this could result in considerable economic losses and risks caused by wrong estimation of groundwater chemical status: in case of overestimation, unnecessary expenses on restoration measures might be spent; while in case of underestimation, evidence of deterioration might be overlooked, thus posing threat to human health and the environment.

16.1.2 Criteria for Assessing Groundwater Chemical Status

16.1.2.1 Criteria Set by the Directives of the European Communities

According to the EU Water Framework Directive (WFD, 2000), groundwater bodies (GWBs) or groups of bodies should be evaluated with respect to the groundwater chemical status. Such evaluation shall be performed in two stages, with a particular consideration of shared groundwater resources:

1. Initial characterization and identification for all GWBs to assess their uses and the degree of being at risk.
2. Further characterization for GWBs which cross the boundary between two or more Member States, or identified as being at risk, or without sufficient data to identify their chemical status and that are used or intended to future use for the abstraction of water providing at least an average of 10 m^3/day.

If also this analysis confirms poor status of groundwater quality, such GWBs shall be subject to monitoring in order to identify significant and sustained upward trend of pollutant concentrations, and define starting points and measures for trend reversals through programs of progressive reduction of groundwater pollution. Monitoring and assessment of groundwater chemical status are being carried out in accordance with criteria laid down in WFD (2000) and in still pending draft Directive on groundwater protection against pollution (GWD, 2003). According to WFD (2000), "the monitoring network shall be designed so as to provide a coherent and comprehensive overview of groundwater chemical status within each river basin district and to detect the presence of long-term anthropogenically induced upward trends of pollutants". River basin is considered a major water management unit and comprise interdependent surface waters and groundwater bodies.

Both WFD and GWD distinguish two kinds of groundwater chemical status: good and poor (Figure 16.1).

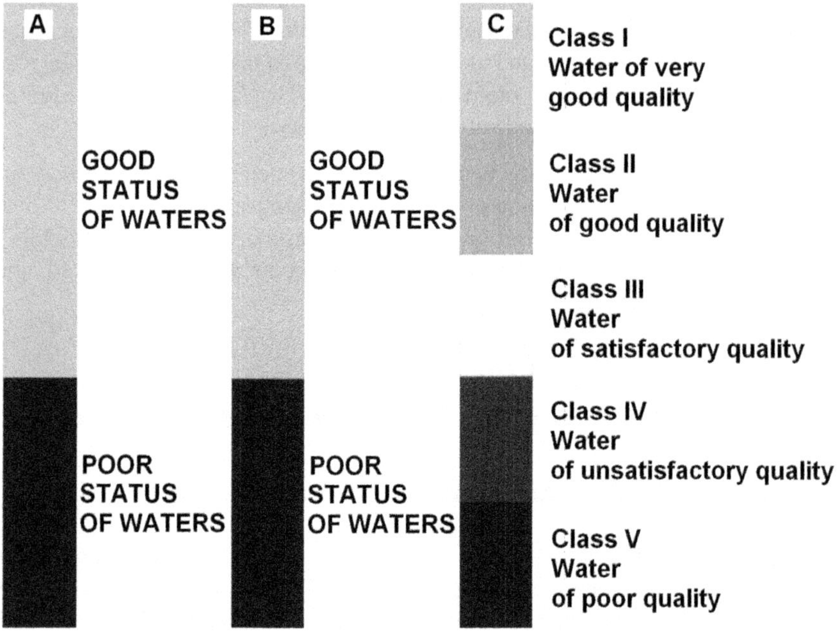

Fig. 16.1 Methods of presenting groundwater chemical status with use of color or shading code. **a** WFD—Water Framework Directive of 23.10.2000; **b** GWD—Groundwater Directive (GWD, 2003); **c** RMS—Directive of the Polish Ministry of Environment on the classification for presenting status of surface and groundwater of 11.02.2004

The classification is based on threshold values of concentrations for specific chemical constituents. Threshold values for assessing groundwater chemical status specified in GWD comprise only two constituents: nitrates (50 mg/L) and pesticides (0.1 μg/L of pure ingredient). Besides, GWD comprises a minimum list of nine substances and ions, for which the Member States should establish threshold values by the end of 2007 (amended term).

The WFD (2000) and GWD (2003) require the achievement of good groundwater status and provide for the criteria to be adopted for assessing good chemical status and criteria for identifying significant and sustained upward trends, as well as for defining starting points for trend reversals. According to these Directives, "in assessing status, the results of individual monitoring points within a groundwater body shall be aggregated for the body as a whole" on the basis of arithmetic mean values of the monitoring data at each point in the groundwater body or group of bodies. The aggregated mean values are to be used to demonstrate compliance with good groundwater chemical status by comparison with threshold values for those chemical parameters for which environmental quality standards have been set in the European Union legislation.

To summarize, the method of a chemical status assessment of a groundwater body or group of bodies within each river basin district set by WFD and GWD is based on a comparison of threshold values with the aggregated arithmetic mean values of monitored chemical parameters.

16.1.2.2 Criteria Set by the Polish Directive of the Ministry of Environment (2004)

In the context of problems with establishing threshold values and adopting GWD (2003), Polish Directive of the Ministry of Environment of 11.02.2004 (RMS, 2004) on the classification for presenting status of surface and groundwater (an amendment to this Directive is pending) might be mentioned as an example of regulations elaborated, enacted and implemented in one of the EU Member States at a national level, in compliance with quality criteria set by WFD (2000) and GWD (2003). According to this Directive, the assessment of groundwater chemical status has been based on 36 physico-chemical parameters, for which threshold values have been established and comprises an additional division of good and poor status in five classes. Good status comprises classes I, II, III, while poor status includes classes IV and V (Figure 16.1).

16.1.2.3 Studies Within the European Framework Programs

To support the efforts on establishing the principles of good governance with respect to groundwater resources, in particular on setting out criteria for assessing the chemical status of groundwater that would consider the natural variability of

European groundwater chemical composition, two projects under the European Framework Programs were undertaken, with broad participation of hydrogeologic institutions from about 30 European countries: the BASELINE project (Edmunds, 2003) carried out in 2001–2002 under the Fifth Framework Programme, and the BRIDGE project (2003–2006) performed within the Sixth Framework Programme by 28 European States including Poland. Both projects aimed at establishing the uniform criteria and methodology for the identification of threshold values of constituents, as well as the validation of criteria and methods for assessment of groundwater chemical status based on the monitoring data.

16.1.3 Aim and Objectives of the Presented Studies

The presented studies carried out in Poland within the aforementioned projects aimed at contributing to the consideration of a natural baseline quality in European aquifer and the establishment of reliable common criteria for assessing chemical status of a body or group of bodies of groundwater, which should result, according to the explanatory memorandum to GWD (2003), in a more harmonized approach for defining and monitoring groundwater status than the existing WFD specifications. Besides of improving a cost/benefit analysis and avoiding possible economic losses and risks resulting from wrong assessment of groundwater chemical status, this aim has a particularly significant political dimension for shared GWBs, which cross the boundary between two or more Member States, and would permit to avoid possible political conflicts.

16.2 Selected Objects and Monitoring Networks

For studies, two objects were selected, both located in the Southern Poland (Figure 16.2): (1) Major Groundwater Body MGWB 332 with a total area of 1540 km^2 and (2) The Koprzywianka River basin district with a total area of 707.4 km^2, and with a more complicated hydrogeology that comprise fragments of two GWBs. Both objects are typical for transboundary GWBs and thus, the derived conclusions could be generalized for shared groundwater resources.

Fig. 16.2 Location of the selected objects: **a** Major Groundwater Body 332 (MGWB 332) and **b** The Koprzywianka River basin district

16.2.1 MGWB 332

Major Groundwater Body MGWB 332 with a total area of 1540 km^2 is associated with continental sandy-gravel Tertiary formations (Sarmatian), from several meters to over 30 m thick. The body comprises also of a deep Quaternary aquifer in sandy-gravel sediments (Q_1) occurring in the prehistoric valley floors. This aquifer is hydraulically connected to the Sarmatian aquifer (Kleczkowski, 1990; Szklarczyk and Leśniak, 1990).

Hydrogeochemical studies of MGWB 332 carried out in 2001–2002 within the framework of a BASELINE project (Witczak et al., 2003a; Edmunds, 2003) covered the Eastern (Kędzierzyn) part of the body of an area 595 km^2 (Figure 16.3). For assessing the groundwater chemical status, the data for 37 points of a monitoring network that monitored the Tertiary (28 points) and the deep Quaternary aquifers (9 points) were utilized.

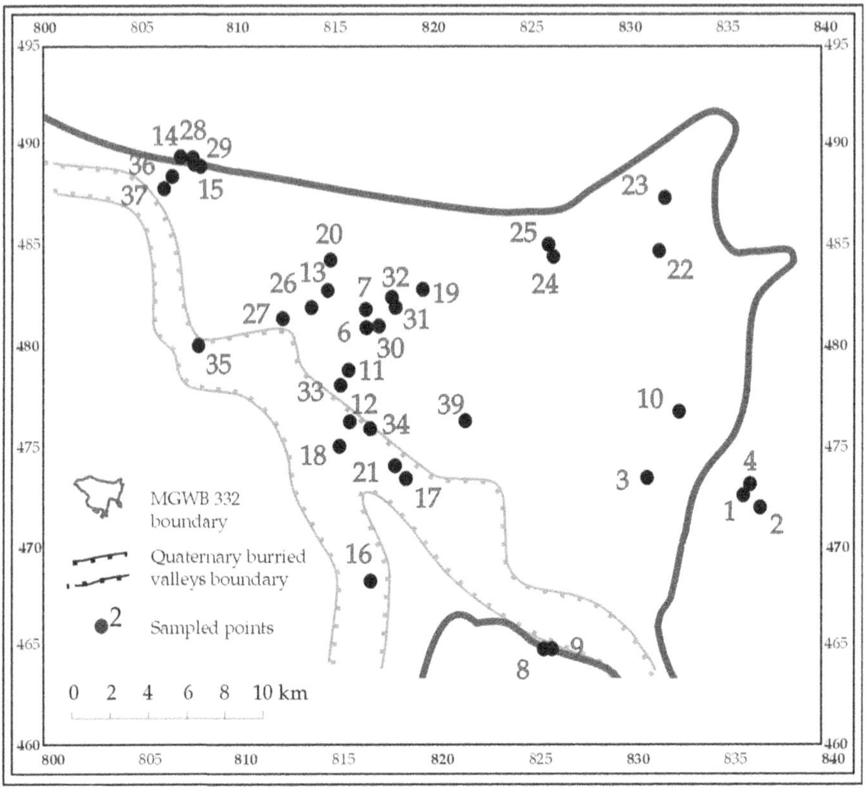

Fig. 16.3 MGWB 332. Observation points sampled within the framework of the BASELINE project (After Witczak et al., 2003a)

16.2.2 The Koprzywianka River Drainage Basin

The assessment of groundwater chemical status in the Koprzywianka River drainage basin was accomplished within the BRIDGE project (2005–2006).

The Koprzywianka River drainage basin with a total area of 707.4 km^2 is located at the border of two different hydrogeological regions. According to the division applied in the Hydrogeological Atlas of Poland (Paczyński, 1997), these regions are Holy Cross Subregion X_1 that belongs to Galicia Region X and Piedmont Carpathian Region XIII. In accordance with the Kleczkowski's regional division (Dowgiałło et al., 2002), the surveyed drainage basin is situated at the borderline of Paleozoic part of Holy Cross massif and Quaternary Piedmont Carpathian ridge.

In the current division of groundwater bodies (Herbich et al., 2005; Sadurski, 2005), the Koprzywianka River drainage basin is located at the boundary of two GWBs: 123 and 125. The boundaries of the Koprzywianka River drainage basin and the boundaries of hydrostructural units do not coincide. Within the boundaries

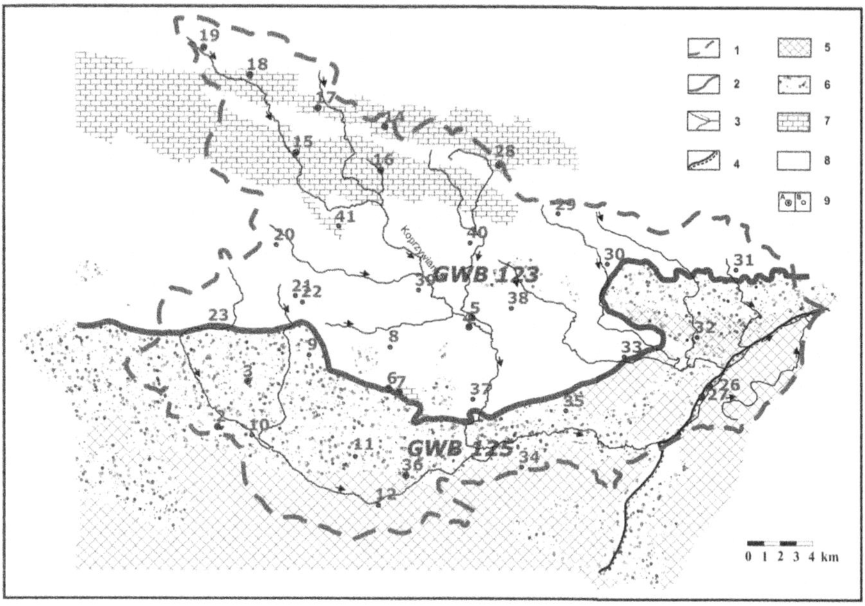

Fig. 16.4 Geological map of the Koprzywianka River drainage basin and localization of groundwater monitoring points. 1—Hydrologic drainage basin ; 2—Boundaries of GWBs 123 and 125; 3—Rivers; 4—Quaternary formations (porous); 5—Tertiary formations (low permeable); 6—Tertiary formations (fissured porous and porous); 7—Devonian formations (fissured karst); 8—Fissured formations of low water-bearing capacity (Paleozoic); 9—Groundwater monitoring points: A) Boreholes; B) Dug wells.

of the hydrological drainage basin occur just fragments of the aforementioned GWBs (Figure 16.4).

16.2.2.1 GBW 125

In GWB 125, water circulation comprises of two aquifers: Quaternary and Tertiary.

Quaternary aquifer is developed in the eastern part of the Koprzywianka River drainage basin within the Vistula River valley. The continuous sandy-gravel aqueous layer of the river origin, of mean thickness of 15 m and good hydraulic conductivity occurs here. Hydraulic conductivity coefficient is highly variable, from 5 to 85 m/d with a mean of 34 m/d. Potential well yields range from 30 to 70 m^3/h. Water table is free and occurs at depths from 1 to 4 m b.g.l. (Szklarczyk and Leśniak, 1990). Beyond the Vistula River valley, development of the Quaternary aquifer is less regular and limited to narrow strips along the rivers. Their mean thickness does not exceed 10 m, while the hydraulic conductivity ranges from 4 to 26 m/d (Prażak et al., 1981)

Tertiary aquifer occurs within the southern and south-eastern part of the Koprywianka River drainage basin. It is constituted of the Badenian formations of limestone falling in the south-east direction under the clayey-dusty neogene formations of insulating and semi-insulating character. Besides lithotamine limestone, also detritic limestone belonging to Sarmatian succession occurs here and forms the joint aquiferous layer in the zone not covered with clays. In the open area, it is represented by a layer of a free water-table occurring at the depth from several meters to over 20 m. It comprises porous-fissured-karst medium of variable parameters. Hydraulic conductivity varies here from 1.7 to 170 m/d. The average thickness of aquiferous layer is 20–30 m. Potential yields of wells range from 30 to 50 m^3/d (Kos, 1997a,b). In the eastern part of GWB 125, the lithological composition of Tertiary aquifer changes for sandy and sandy-clayey one. Tertiary aquiferous layer is deposited on the practically impermeable old Paleozoic formations and fed from atmospheric precipitation in outcrops, as well as by percolation from Quaternary stage.

16.2.3 GWB 123

In GWB 123, the major aquiferous structure consists of carbonate formations of Middle and Upper Devonian succession. Hydraulic conductivity coefficients in this aquifer are strongly variable, that is typical for fissure-karst medium and ranges from 0.2 to 0.5 m/d. The yields of wells are also strongly variable and can be up to 300 m^3/h. Devonian formations are fed from atmospheric precipitations in outcrops or through several meters to over 10 m thick Quaternary cover. Water table is free and occurs at low depths in the river drainage zones and at the depth of over 20–30 m in the watershed areas. Besides of feeding from the infiltration, side feeding from underground flow and partially surface runoff from the surrounding areas formed from old Paleozoic formations of a low hydraulic conductivity also play an important role. Surface watercourses flowing within an area of Devonian stage, to some extent feed the waters of this stage and then drain them at the discharge from the area (Witczak, 1996). Devonian aquifer is a basic source of water supply for communal needs and industry.

Old Paleozoic aquifer occurs in the whole area of GWB 123, where at the surface or under the Quaternary stage, Cambrian, Ordovician, Silurian and lower Devonian formations occur. In hydrogeology, this area is treated as waterless, i.e. depleted of the usable aquiferous horizons (Prażak et al., 1981; Wróblewska, 2000a,b). Potential yield of wells is up to 2 m^3/h. Despite of the limited water resources, this area is considered to play an important role in feeding surface watercourses and in side feeding a more affluent Devonian aquifer (Prażak et al., 2001; Witczak et al., 2003b).

16.2.3.1 Groundwater Monitoring Network in the Surveyed River Drainage Basin

A network for groundwater monitoring in the surveyed Koprzywianka River drainage basin comprises 37 points (boreholes and dug wells) for sampling water from Quaternary, Tertiary and Devonian aquifers. Their location is presented in Figure 16.4, while their assignment to the specific aquifers and groundwater bodies is given in Table 16.1.

16.3 Groundwater Sampling and Analysis

Both networks were sampled once, in MGWB 332 in 2002 and in the Koprzywianka River basin in May 2005. Sampling procedure was carried out in accordance with the guidelines for groundwater monitoring set out by the State Inspectorate for Environment Protection PIOS (Witczak and Adamczyk, 1994, 1995). Field analyses comprised unstable parameters: pH, specific electric conductivity, turbidity, color and temperature. Of-site chemical analysis was performed at the Chemical Laboratories of the Hydrogeology and Engineering Geology Department at the University of Science and Technology in Krakow with use of mass spectrometry with inductively coupled plasma (ICP MS Perkin Elmer SCIEX Elan) and comprised 42 constituents and parameters: total, carbonate and non-carbonate hardness, silica, sodium, potassium, lithium, beryllium, calcium, magnesium, barium, strontium, iron, manganese, silver, zinc, copper, nickel, cobalt, lead, mercury, cadmium, selenium, antimony, aluminum, chromium, molybdenum, vanadium, zirconium, arsenic, thallium, wolfram, chloride, brome, iodine, sulfate, bicarbonate, nitrate, nitrite, fluoride, phosphate and boron. At the Laboratories of the Regional Inspectorate of Environmental Protection WIOS in Krakow, six parameters were analyzed: total and dissolved organic carbon (TOC and DOC), Kjeldahl nitrogen, BOD_5, COD-Cr, COD-Mn, ammonium ions. Analyses of TOC and DOC were performed according to the European Standard EN 1484 (1999) with use of carbon analyzer TOC 5000 Shimadzu.

The scope of analysis comprised parameters specified in RMS (2004). Part of these substances has been laid down in the list given in the draft GWD (2003), therefore the assessment of groundwater chemical status in the MGWB 332 and the Koprzywianka River drainage basin could be accomplished on the basis of these analyses.

Table 16.1 Monitoring points in the Koprzywianka River drainage basin

No	Point Number	Point Location	Point Type	Stratigraphy of Aquiferous Horizons
		GWB 123		
1	4	Górki Klimontowskie, Klimontów community, well S1	Borehole	Q
2	5	Górki Klimontowskie, Klimontów community, well S3	Borehole	Q
3	8	Witowice, Klimontów community	Dug well	Q-Cm
4	19	Nieskurzów Stary 73a, Baćkowice community	Borehole	Q-Cm
5	20	Gryzikamień, Iwaniska community	Dug well	Q-Cm
6	21	Miłoszowie, Bogoria community	Dug well	Q-Cm
7	22	Miłoszowie, Bogoria community	Dug well	Q-Cm
8	23	Wola Małkowska, Bogoria community	Dug well	Q-Cm
9	29	Kurów, Lipnik community	Dug well	Q-Cm
10	30	Jugoszów, Obrazów community	Dug well	Q-Cm
11	31	Obrazów, Obrazów community	Dug well	Q-Cm
12	33	Chobrzany, Samborzec community	Dug well	Q
13	37	Nawodzice, Klimontów community	Dug well	Q-Cm
14	38	Pęchów, Klimontów community	Dug well	Q-Cm
15	39	Konary, Klimontów community	Dug well	Q-Cm
16	40	Łownica, Klimontów community	Dug well	Q-Cm
17	6	Jurkowice, Bogoria community	Borehole	
18	7	Budy, Bogoria community	Borehole	
19	14	Kobylany village, Opatów comm.	Borehole	
20	15	Piskrzyń mine, Iwaniska comm.	Borehole	
21	16	Wymysłów mine, Iwaniska comm.	Borehole	
22	17	Moliborzyce, well 2, Baćkowice community	Borehole	D
23	18	Baćkowice, well B1, Baćkowice community	Borehole	
24	28	Włostów, Lipnik community	Borehole	
25	41	Ujazd, Iwaniska community	Dug well	
		GWB 125		
26	9	Moszyny, Bogoria community	Dug well	
27	11	Smerdyna, Staszów community	Dug well	
28	26	Szewce, well 2, Samborzec comm.	Borehole	
29	27	Szewce, well 4, Samborzec comm.	Borehole	Q
30	32	Miechowice, Samborzec community	Dug well	
31	34	Sulisławice, Koprzywnica comm.	Dug well	
32	35	Gieraszowice, Koprzywnica comm.	Dug well	
33	2	Szombiergi, Staszów community	Borehole	
34	3	Zimnowoda, Bogoria community	Borehole	
35	10	Wiśniowa Poduchowa, Staszów community	Borehole	Tr
36	12	Wiązownica Duża, Osiek community	Dug well	
37	36	Wiązownica, well W2, Osiek community	Borehole	

16.4 Assessing Hydrogeochemical Background

Applied methods for assessing hydrogeochemical background is exemplified here in MGWB 332. For determining the variability of groundwater chemical parameters and the hydrogeochemical background of the studied groundwater body, descriptive statistics and graphical methods were applied. Data obtained below the detection limit were replaced by half of a value declared by the laboratory as the detection limit (<DL = 0.5 DL), in accordance with the principle adopted within the BASELINE project (Edmunds, 2003).

In Table 16.2, the descriptive statistics of analyzed macro- and micro-constituents in water samples taken from the MGWB 332 are presented, along with the corresponding values of the hydrogeochemical background (assessed by computation method for the percentile range 16–84%) and maximum permissible values of these parameters in water intended for human consumption (DWD, 1998; RMZ, 2002; WHO, 2004).

The variability range of the analyzed chemical parameters of groundwater in MGWB 332 is demonstrated in box-and-whisker plots (Figure 16.5) and probability plots (Figure 16.6).

These plots were used also for assessing background concentrations of the analyzed constituents by the graphical method. In most cases, the range of the actual hydrogeochemical background of the analyzed parameters was at least one order of magnitude lower than standard values for water intended for human consumption. Only iron, manganese and ammonium occurred in concentrations exceeding these values; the concentrations of these constituents were though controlled by the natural hydrogeochemical processes (Witczak et al., 2003a). Both descriptive statistics and graphical methods appeared to characterize adequately the background concentrations and the natural variability of chemical status of groundwater bodies.

Table 16.2 MGWB 332. Basic descriptive statistics for the analyzed physico-chemical parameters (Witczak et al., 2003a):

Lp.	Parameter	Unit	N	Mean	Median	Mode	Std. dev.	16%	84%	MPL
1	Temperature	(°C)	37	12.03	12.00	11.50	1.29	11.02	12.96	
2	pH		37	7.24	7.25	7.06	0.33	7.06	7.56	6.5–9.5
3	Eh	(mV)	37	81.62	54.70	-27.70	107.01	24.33	86.83	
4	DO	(mg/L)	35	0.28	0.02	0.01	1.07	0.01	0.04	
5	SEC	(μS/cm)	37	655.05	560.30	125.30	653.41	288.24	731.92	
6	Ca	(mg/L)	37	99.20	81.50	5.54	110.77	32.88	122.44	
7	Mg	(mg/L)	37	13.53	11.80	10.60	13.13	4.79	16.09	
8	Na	(mg/L)	37	18.59	14.20	8.24	20.27	4.97	27.14	200
9	K	(mg/L)	37	3.45	3.19	2.62	2.87	1.59	4.57	
10	Cl	(mg/L)	37	24.30	5.31	4.63	60.57	2.43	27.39	250

(continued)

Table 16.2 (continued)

Lp.	Parameter	Unit	N	Mean	Median	Mode	Std. dev.	16%	84%	MPL
11	SO4	(mg/L)	37	135.74	43.09	3.62	299.73	7.20	192.01	250
12	HCO3 field.	(mg/L)	37	217.18	210.00	205.07	76.73	150.97	292.13	
13	HCO3 lab.	(mg/L)	37	217.43	217.90	217.90	66.33	160.60	285.78	
14	NO3	(mg/L)	37	0.166	0.011	0.011	0.773	0.011	0.011	50
15	NO2	(mg/L)	37	0.006	0.006	0.006	0.000	0.006	0.006	0.5
16	NH4	(mg/L)	37	0.395	0.350	0.311	0.235	0.156	0.583	0.5
17	P	(μg/L)	37	249.709	127.19	115.80	673.49	66.981	218.931	
18	TOC	(mg/L)	37	1.652	1.582	0.335	0.862	0.484	2.644	
19	DOC	(mg/L)	36	1.511	1.301	0.349	0.799	0.522	2.516	
20	F	(mg/L)	36	0.047	0.040	0.040	0.033	0.040	0.040	1.5
21	Br	(mg/L)	37	0.102	0.020	0.035	0.282	0.013	0.077	0.01
22	I	(mg/L)	37	0.203	0.010	0.007	0.747	0.005	0.037	
23	Si	(mg/L)	37	7.829	7.740	7.306	0.804	7.152	8.520	
24	Ag	(μg/L)	37	0.473	0.010	0.007	2.261	0.007	0.026	
25	As	(μg/L)	37	0.685	0.138	0.036	2.265	0.036	0.500	10
26	B	(μg/L)	37	55.614	48.900	3.230	41.841	14.536	93.440	1000
27	Ba	(μg/L)	37	68.178	65.100	38.800	45.651	23.872	107.760	
28	Be	(μg/L)	37	0.026	0.018	0.018	0.017	0.018	0.049	
29	Bi	(μg/L)	37	0.160	0.013	0.003	0.411	0.005	0.284	
30	Cd	(μg/L)	37	0.063	0.018	0.018	0.247	0.018	0.018	5
31	Co	(μg/L)	37	0.277	0.158	0.149	0.525	0.069	0.224	
32	Cu	(μg/L)	37	0.327	0.236	0.018	0.297	0.122	0.549	2000
33	Fe	(mg/L)	37	2.209	1.330	0.030	3.292	0.488	3.257	
34	Hg	(μg/L)	37	0.044	0.024	0.024	0.049	0.024	0.057	
35	Li	(μg/L)	37	21.175	19.600	12.700	16.129	9.066	26.092	
36	Mn	(mg/L)	37	0.284	0.210	0.230	0.422	0.092	0.324	0.05
37	Mo	(μg/L)	37	0.350	0.291	0.251	0.222	0.215	0.543	
38	Ni	(μg/L)	35	1.769	1.170	0.398	2.670	0.418	2.264	20
39	Pb	(μg/L)	37	0.275	0.074	0.006	0.909	0.022	0.243	10
40	Rb	(μg/L)	37	2.051	1.660	1.660	2.380	0.892	2.429	
41	Sb	(μg/L)	37	0.007	0.006	0.006	0.001	0.006	0.007	5
42	Se	(μg/L)	37	1.097	0.572	0.360	1.986	0.360	1.016	10
43	Sr	(μg/L)	37	855.408	622.000	1840.000	749.947	139.560	1840.000	
44	U	(μg/L)	37	0.039	0.017	0.013	0.075	0.005	0.042	
45	V	(μg/L)	37	0.039	0.012	0.012	0.084	0.012	0.012	
46	Y	(μg/L)	36	0.044	0.029	0.005	0.089	0.008	0.051	
47	Zn	(μg/L)	37	25.217	15.200	2.740	47.146	6.928	30.272	

N—number of observations; Values: mean, median, mode, standard deviation; 84%—percentile range; MPL—Maximum Permissible Levels of parameters in drinking water (DWD, 1998; RMZ, 2002; WHO, 2004)

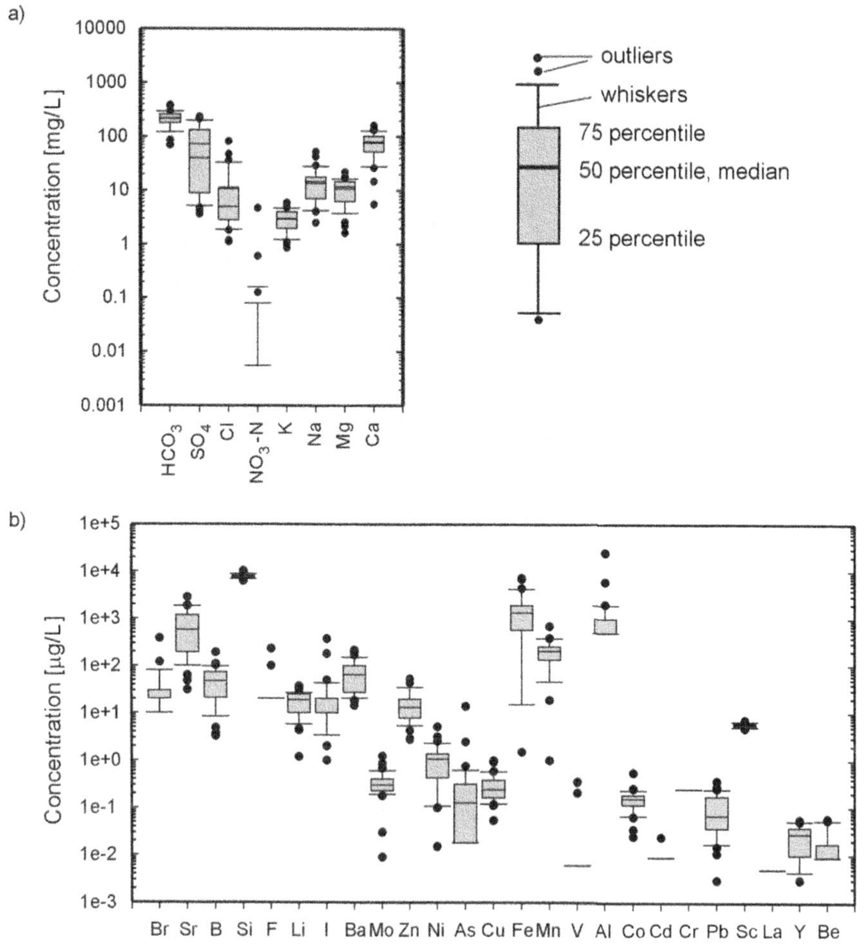

Fig. 16.5 Box-and-whisker plots for: **a** major ion concentrations; **b** minor and trace elements determined in groundwater. Data for MGWB 332 studied within the BASELINE project (Witczak et al., 2003a; Edmunds, 2003)

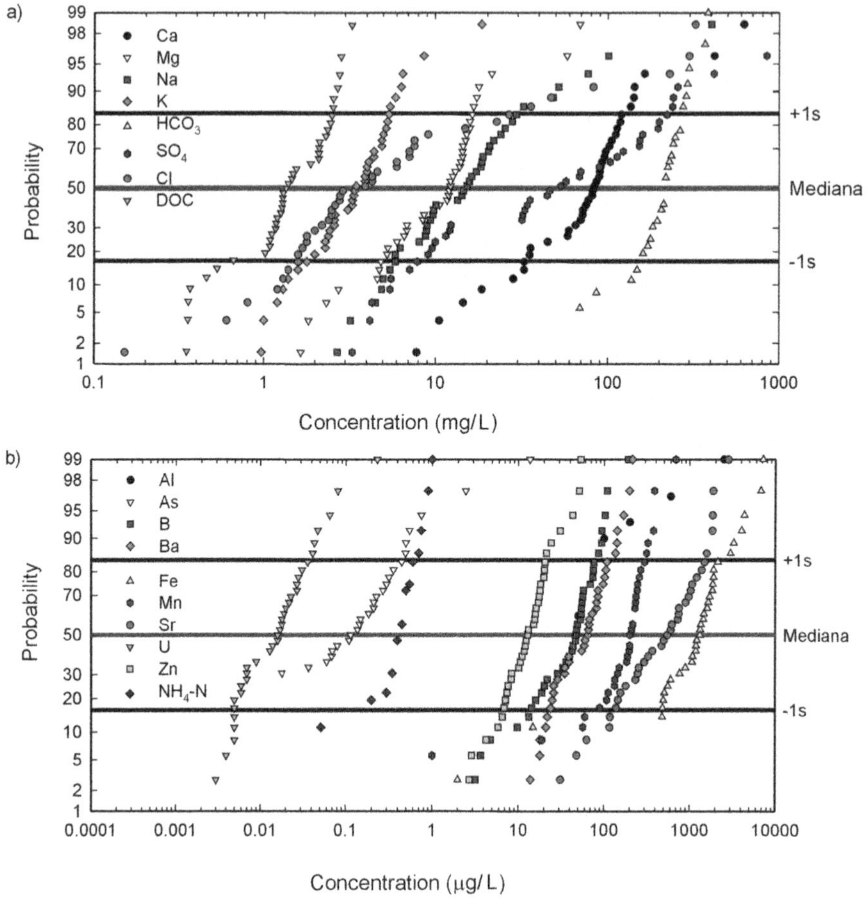

Fig. 16.6 Probability plots for the assessment of hydrogeochemical background for: **a** major ion concentrations; **b** minor and trace elements determined in groundwater. Data for MGWB 332 studied within the BASELINE project (Witczak et al., 2003a; Edmunds, 2003).

16.5 Assessing Groundwater Chemical Status of the Studied Groundwater Bodies

16.5.1 General Principles

Groundwater chemical status of the studied bodies or group of bodies of groundwater was assessed in a point mode in each particular monitoring point and in a spatial mode by the aggregation of data obtained for specific monitoring points in accordance with the European Water Framework Directive and a project of the Daughter Groundwater Directive (WFD, 2000; GWD, 2003). The data were aggregated in two ways: on the basis of mean and median values.

General principles of groundwater classification with respect to its chemical status are presented in Figure 16.1. The results were used for a comparative analysis of the data.

16.5.2 MGWB 332

16.5.2.1 Point Assessment

Among the 37 monitoring points of MGWB 332, 31 (83.8%) had water representing III class of water quality (waters of satisfactory quality); although MPL was exceeded for iron (RMS, 2004). In two cases (5.4%), water quality was classified as unsatisfactory (IV class) and in two other cases (5.4%) as water of poor quality (V class). Only in one case (point 332/17), water fulfilled the I class quality criteria, and in another case (point 332/25) water was classified as II class (water of good quality) (Table 16.3).

Table 16.3 MGWB 332. Groundwater chemical status. Point assessment

Water Quality Class	Number of Points	% of Total
I—Water of very good quality	1	2.71
II—Water of good quality	1	2.71
III—Water of satisfactory quality	31	83.78
IV—Water of unsatisfactory quality	2	5.40
V—Water of poor quality	2	5.40
Total	37	100.00

In Figure 16.7, the groundwater status is presented with the use of color (shading) codes in each particular monitoring point, according to the Directive of the Ministry of the Environment (RMS, 2004).

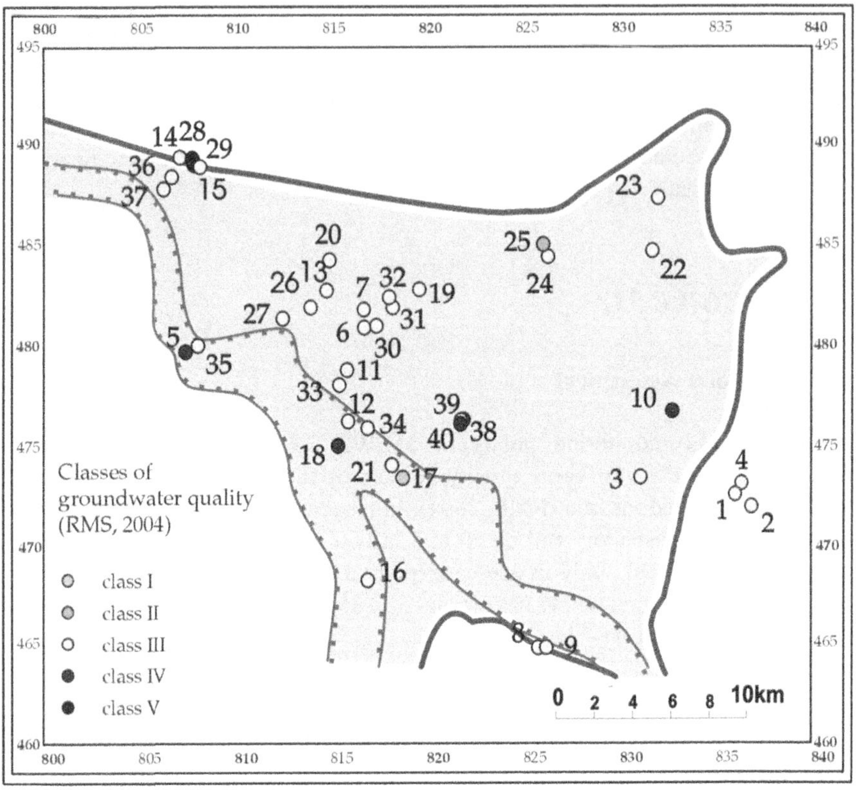

Fig. 16.7 MGWB 332. Groundwater chemical status. Point assessment (RMS, 2004)

16.5.2.2 Spatial Assessment

According to the European Water Directives (WFD, 2000; GWD, 2003), for spatial evaluation of groundwater chemical status, data obtained for specific monitoring points should be aggregated. Therefore, for the spatial assessment of the chemical status of groundwater in the MGWB 332, data obtained for the particular monitoring points were aggregated in two ways: with use of mean and median values.

For the data aggregated by the method of arithmetic mean, groundwater appeared to belong to IV class, i.e. to waters of unsatisfactory quality that means that chemical status of waters is low. In turn, if median values were used for aggregation, groundwater in the studied MGWB 332 could be reckoned to be of III class, i.e. of satisfactory quality that corresponds to good groundwater chemical status (Figure 16.8).

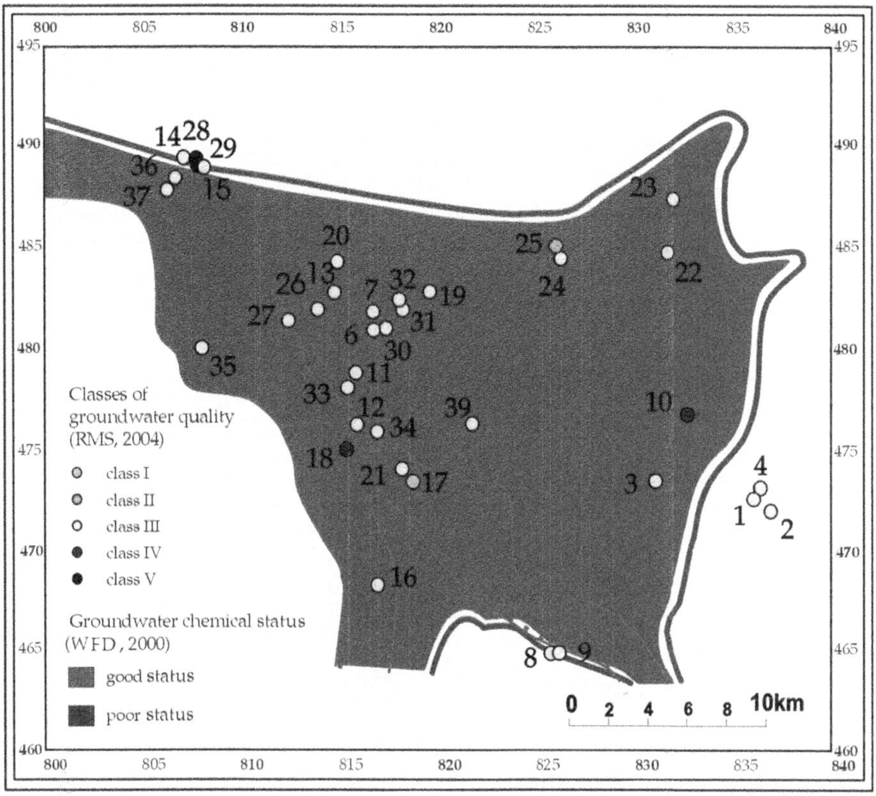

Fig. 16.8 MGWB 332. Groundwater chemical status. Spatial assessment with use of median values for data aggregation (WFD, 2000; GWD, 2003)

The comparison of the groundwater chemical status assessment in MGWB 332 accomplished by point and spatial method with the use of mean and median concentration values showed significant differences that made the evaluation vague and uncertain.

16.5.3 The Koprzywianka River Drainage Basin

Groundwater chemical status in the Koprzywianka River drainage basin was assessed in a point mode in each particular monitoring point and in a spatial mode:

- For the whole Koprzywianka River drainage basin;
- In the areas of each groundwater body within the Koprzywianka River drainage basin: GWB 123 and GWB 125;

- For each particular GWB (GWB 123 and GWB 125) with the division on waters of a shallow and a deep circulation.

16.5.3.1 Point Assessment

For assessing groundwater chemical status in each particular monitoring point of the Koprzywianka River drainage basin in accordance with guidelines of the Polish RMS—Directive of the Ministry of Environment (2004), the computer program by Klasa was used (Figure 16.9).

Groundwater quality

well	location	2005	well
2	KP2	IV (Ni)	2
3	KP3	IV (Ni)	3
4	KP4	IV (Ni)	4
5	KP5	IV (NO3,Ni)	5
6	KP6	IV (NO2)	6
7	KP7	IV (NO3,HCO3)	7
8	KP8	IV (NO3,PO4,HCO3,K)	8
9	KP9	IV (NO3,K)	9
10	KP10	IV (NO3)	10
11	KP11	V (NO3,NO2,PO4,K)	11
12	KP12	IV (NO3,HCO3)	12
14	KP14	IV (NH4,HCO3,Fe_og)	14
15	KP15	III (Fe_og)	15
16	KP16	III	16
17	KP17	III (HCO3)	17
18	KP18	IV (Fe_og)	18
19	KP19	III (pH)	19
20	KP20	III	20
21	KP21	V (NO2,PO4)	21
22	KP22	IV (K)	22
23	KP23	IV (NO3)	23
26	KP26	IV (NH4,NO2,Fe_og)	26
27	KP27	IV (NH4)	27
28	KP28	III (HCO3)	28
29	KP29	IV (HCO3,K)	29
30	KP30	IV (NO3,HCO3)	30
31	KP31	IV (NO2)	31
32	KP32	V (NO3,HCO3)	32
33	KP33	III (HCO3)	33
34	KP34	V (NO3,NO2,HCO3,K)	34
35	KP35	V (PEW,NO3,NO2,HCO3,S	35
36	KP36	II	36
37	KP37	IV (HCO3,K)	37
38	KP38	V (NO3,HCO3,Ca)	38
39	KP39	V (NO3,HCO3,Ca)	39
40	KP40	III	40
41	KP41	V (F)	41

Fig. 16.9 Point assessment of groundwater chemical status in the Koprzywianka River drainage basin (screen projection from the computer program by Klasa (Janecka-Styrcz, 2004; Szczepańska and Kmiecik, 2005)

The detailed program description after Janecka-Styrcz (2004) can be found elsewhere, also in publications of these authors (Szczepanska and Kmiecik, 2005).

In Figure 16.10, the groundwater chemical status in the Koprzywianka River drainage basin has been presented for monitoring points that take water from all the aquiferous horizons. Among 37 points located in the Koprzywianka River drainage basin (Figure 16.10), no water of class I was found, and only in one point (KP36) water quality fulfilled criteria of class II (2.7%), while in 8 points, water of class III (21.6%) occurred. In 20 points, water quality represented class IV (54.1%), and in 8 points class V (21.6%). Thus, in general, water quality was classified as unsatisfactory (class IV) and poor (class V). This classification of water was mainly due to high concentrations of nitrates and bicarbonate hardness.

16.5.3.2 Spatial Assessment

As indicated above, spatial assessment of groundwater chemical status for the studied river drainage basin was performed for several variants on the basis of data aggregated by using either mean or median values. The data obtained are presented in Figures 16.11 and 16.12, and in Table 16.4.

On the basis of the data for 37 monitoring points, groundwater chemical status was assessed for the whole surveyed river drainage basin (Figure 16.11, Table 16.4). By using mean values as an aggregation method, these waters were assigned to class V (Figure 16.11a), which indicates poor chemical status. When using median values, these waters were assigned to class III, indicating good chemical status (Figure 16.11b).

If chemical status is assessed for each GWB located in the Koprzywianka River drainage basin separately (Figure 16.12, Table 16.4), GWB 123 displays good chemical status independently of the data aggregation method (in both cases, water quality of class III), while GWB 125 demonstrates poor chemical status (aggregation using mean value results in water quality of class V, while aggregation based on median results in class IV in term of water quality).

Table 16.4 Assessment of groundwater chemical status in the Koprzywianka River drainage basin on the basis of aggregated data (with use of mean and median values) for the whole drainage basin, groundwater bodies GWB 123 and 125, and aqueous horizons in the individual GWBs

No	Area for which data Aggregation was Done	Points Total	Number of Points in the Water Quality Class					Aggregation with Use of Mean Values		Aggregation with Use of Median Values	
			I	II	III	IV	V	Class	Status	Class	Status
1	Koprzywianka River drainage basin	37	–	1	8	20	8	V	Poor	III	Good
2	GWB 123	25	–	–	8	13	4	III	Good	III	Good
3	Horizon Q	16	–	–	4	9	3	IV	Poor	III	Good
4	Horizon D	9	–	–	4	4	1	III	Good	III	Good
5	GWB 125	12	–	1	–	7	4	V	Poor	IV	Poor
6	Horizon Q	7	–	–	–	3	4	V	Poor	IV	Poor
7	Horizon Tr	5	–	1	–	4	–	III	Good	III	Good

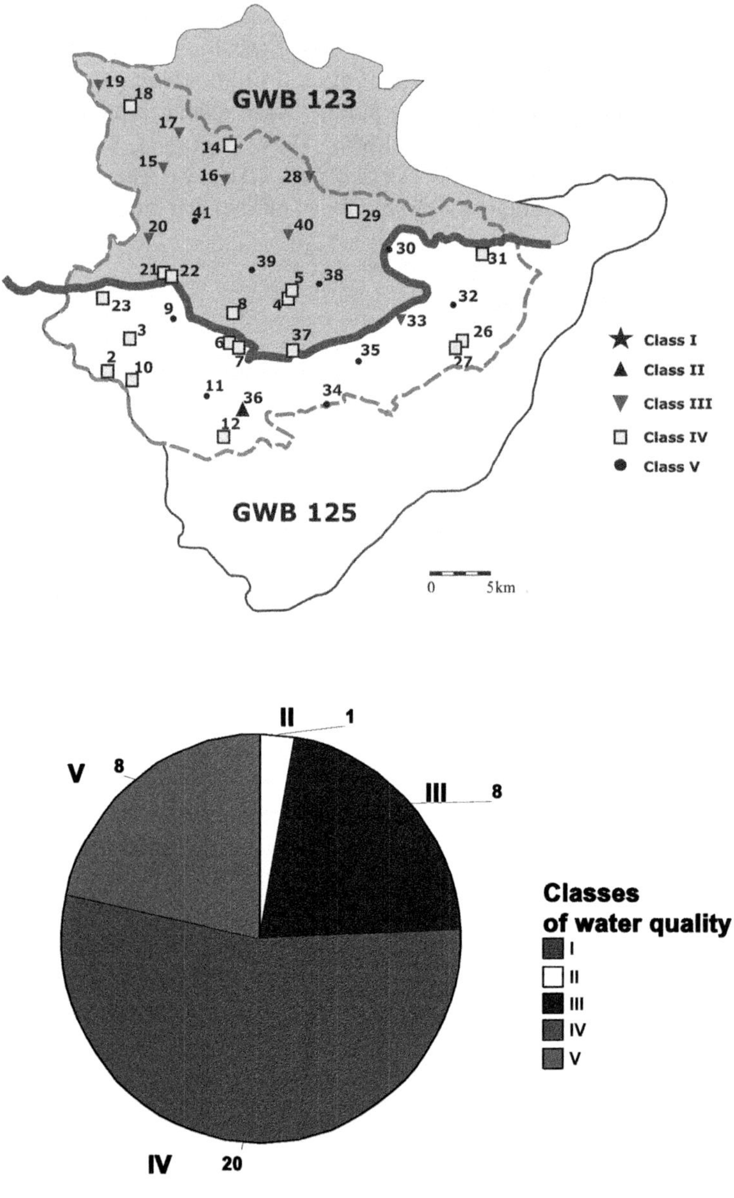

Fig. 16.10 Point assessment of groundwater chemical status in the Koprzywianka River drainage basin. A border of the drainage basin is marked with broken line

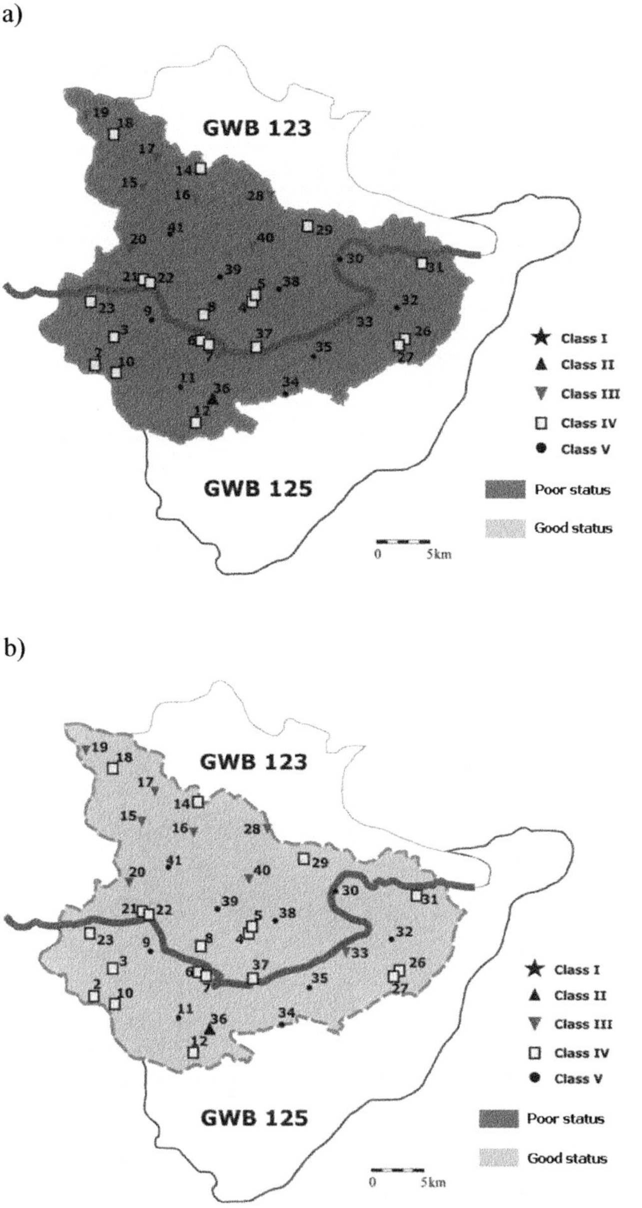

Fig. 16.11 Spatial assessment of groundwater chemical status in the Koprzywianka River drainage basin: **a** aggregation with use of mean values; **b** aggregation with use of median values

a)

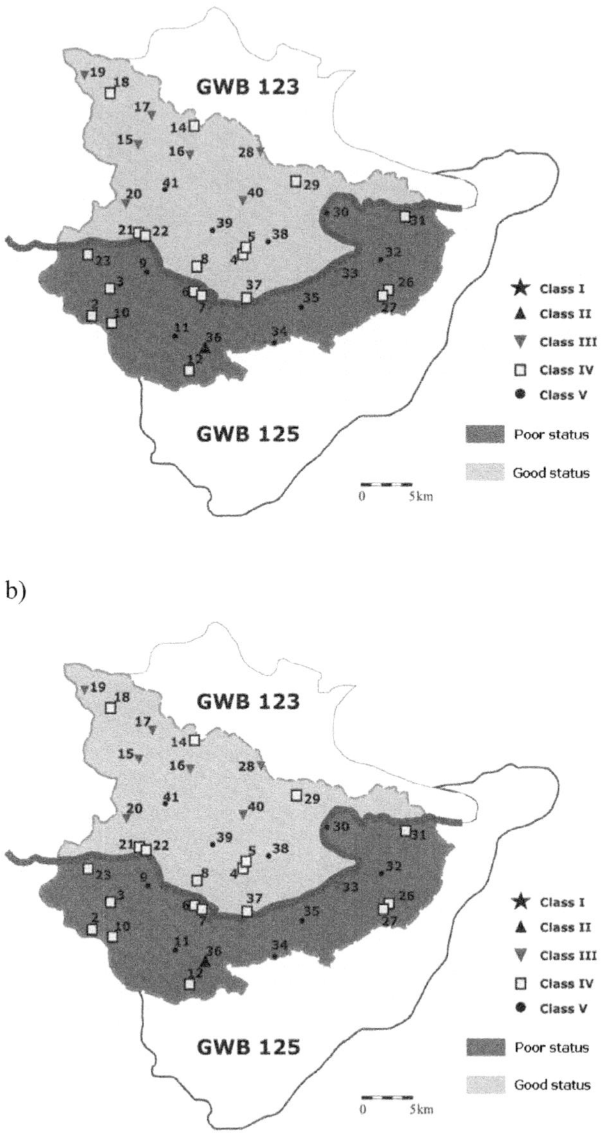

Fig. 16.12 Assessment of a spatial groundwater chemical status in the Koprzywianka River drainage basin divided in two GWBs: **a** aggregation with use of mean value; **b** aggregation with use of median value

Again, different results will be obtained if data aggregation is accomplished for each GWB with division in waters of shallow (lines 3 and 6 in Table 16.4) and deeper circulation (lines 4 and 7 in Table 16.4). In GBW of multilayer structure, assessing water chemical status should thus be accomplished separately for the waters of shallow and deeper circulation. The presented analysis shows that groundwater of shallow circulation susceptible to anthropogenic contamination displays poor chemical status, while deeper, better protected groundwater has good chemical status.

16.6 Discussion

Adoption of correct criteria for assessing actual groundwater chemical status and for identifying significant and sustained upward trends, in particular in shared groundwater resources, is of crucial importance and has a substantial political dimension since the groundwater quality might be a basis of serious arguments and conflicts. Chemical status of groundwater bodies determines their current use, influences the management strategy and potential protection/ remediation actions, and establishes the baseline conditions for planning future sustainable development with respect to this vital natural resource.

Parallel assessment of groundwater chemical status in the Major Groundwater Body MGWB 332 and in the Koprzywianka River drainage basin district with use of existing EU and national methods and guidelines, as well as these recommended in the draft proposal for the EU Directive on the protection of groundwater against pollution revealed significant differences, depending on the applied methods of data interpretation (WFD, 2000; GWD, 2003; RMS, 2004).

16.6.1 Mean or Median Value?

With respect to the spatial groundwater chemical status, assessment based on the aggregated concentration data as recommended by WFD (2000) and GWD (2004) has significant differences due to the fact that the physico-chemical parameters of groundwater show in most cases asymmetric, usually log-normal distribution. In such cases, mean value is not an adequate average measure.

Hydrogeochemical parameters of groundwater in MGWB 332 display generally an asymmetrical distribution, with a right-side asymmetry (Figure 16.13). In such case, the median-based method of aggregation appears to be better.

Also, with regard to the groundwater bodies occurring in the other studied object, the Koprzywianka River drainage basin district, the concentration data for all studied variants and parameters showed in the most cases similar distribution pattern. Figure 16.14 presents example of histograms of chemical parameters (silica

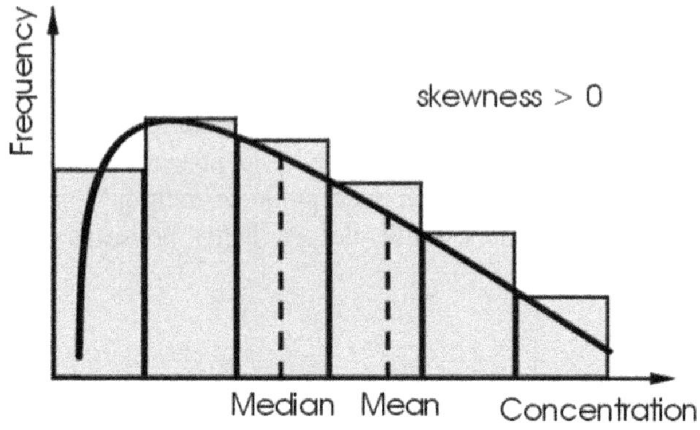

Fig. 16.13 Example of a distribution histogram with the right-side asymmetry

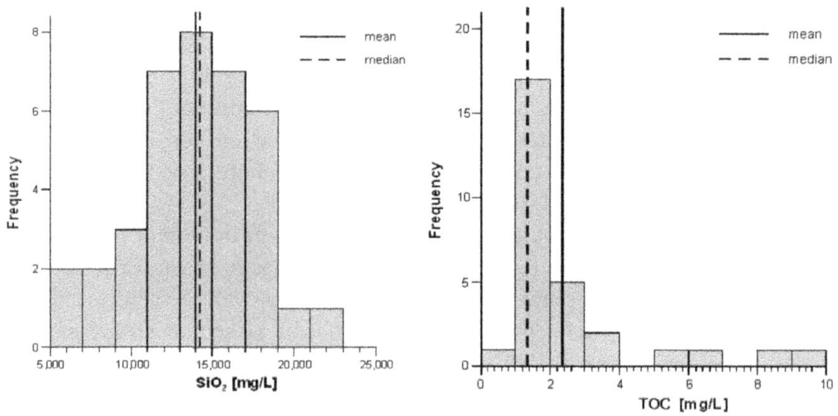

Fig. 16.14 Example of histograms of physico-chemical parameters (silica and TOC) distribution in the groundwater of the Koprzywianka River drainage basin

and TOC) distribution in groundwaters of the Koprzywianka River drainage basin, drawn by using SPSS v. 12.0 computer program.

Silica concentrations display approximately a symmetric (normal) distribution, thus mean value (14.04 mg/L) is close to the median (14.13 mg/L). With respect to TOC, the distribution shows right-side asymmetry, thus mean value (2.39 mg/L) significantly differs from the median (1.34 mg/L).

In general, application of mean values as a method of data aggregation results in taking into consideration for assessing groundwater chemical status a higher concentration of a given parameter, which is associated with assignment of groundwater to the higher/worse quality class.

Although mean value might be computed precisely on the basis of all observations, it inadequately describes the data set in case of one or more outlying results or in particular in the case of the asymmetric data distribution. In case of MGWB 332, the application for aggregation of mean values caused unjustified assessment of the groundwater chemical status as poor (Figure 16.8).

Median value is not susceptible to deformation so easy as mean value; besides, in case of a symmetric distribution, it is close to average (Figure 16.14). On the other hand, the chemical status assessment based on median concentrations of chemical constituents for all monitoring points causes camouflaging zones of unsatisfactory (class IV) and poor quality waters (class V) in the averaged picture. In practice, this may lead to absolve from the timely remedial action for restoring good quality waters in the groundwater bodies being at risk. This indicates the need of application of additional measures for early identification of pollution sources.

16.6.2 Point Assessment as Early Warning Tool

As it was shown in Section 16.5.2.1., point assessment of groundwater chemical status in the particular monitoring points of MGWB 332 in compliance with the national regulations by comparison of the analyzed constituent concentrations with the maximum permissible levels for water quality classes (I–V) resulted in the general evaluation of groundwater in the MGWB 332 as belonging to class III (satisfactory quality), but locally to class IV and class V (unsatisfactory and poor quality, respectively) (RMS, 2004) (Figure 16.7). This picture indicates clearly the points at risk, possibly the sources of MGBW contamination that, if timely intercepted, may prevent the groundwater deterioration and further costly restoration. Therefore, point assessment may serve as an early warning system.

As it was shown earlier, spatial assessment of groundwater chemical status in the same MGWB 332 conducted according to the guidelines specified in the EU Water Framework Directive and in the proposal for the EC Directive on the protection of groundwater against pollution based on the aggregated mean values of chemical parameters for the whole basin, resulted in the evaluation of the chemical status of waters in the MGWB 332 as generally poor (WFD, 2000; GWD, 2003) (Figure 16.8). Consequently, unnecessary restoration measures might be taken, resulting in wasting considerable amounts of money. On the contrary, data aggregation based on the median values leads to assessing chemical status of this water body as generally good (Figure 16.7). In such case, evidence of pollution might be overlooked along with a risk to human health and the environment, and even heavier economic burden of late deterioration reversal.

Either distorting or camouflaging the diversified risk in the averaged chemical status picture is more visible by comparing point and spatial assessments of the chemical status in the GWBs in the Koprzywianka River basin (Table 16.4). While at the point assessment, for 55.5–100% of the monitoring points, the chemical status of the evaluated groundwater bodies were assessed as poor (mostly class IV and in some cases class V); spatial assessment which uses aggregation mean values of the measured concentrations, resulted for four out of seven studied variants in classification of groundwater chemical status as poor (three variants had the lowest quality, class V), and for the remaining three variants, it was ranked good (satisfactory quality, class III). In turn, spatial assessment using median values for aggregation resulted in evaluation of groundwater chemical status as good (satisfactory quality, class III) for five of seven studied variants, and as poor (unsatisfactory quality, class IV) for the other two variants.

These results confirm that the spatial mode of chemical status assessment based on the median concentrations of chemical constituents may result in overlooking evidence of pollution. Thus, the above analysis of spatial assessment of groundwater chemical status based on the aggregated data with using either mean or median values arouses doubts. As it has been stated in an explanatory memorandum to the project of GWD (2003), any such doubts and wrong assessment that cause either under- or overestimation of a risk will not only have adverse effects on decision-making and could result in considerable economic losses and risks, but also will result in a loss of public confidence.

The presented results indicate suitability of point assessment of groundwater chemical status in parallel with its spatial evaluation in the whole groundwater body or group of bodies based on the data aggregation using median values. This approach would assure early indication of a risk and avoid wrong decisions and conflicts, particularly dangerous in such sensitive areas as shared groundwater resources.

16.7 Conclusion

The conclusions derived from the presented studies may be briefly summarized as follows:

- Both descriptive statistics and graphical methods appeared to be adequate tools for characterizing the background concentrations and the natural variability of chemical status of groundwater bodies;
- Spatial assessment of groundwater chemical status on the basis of concentration data aggregated with using either mean or median values might lead to over- or underestimation of risks resulting in considerable economic losses, possible damage to human health and the environment, and to unavoidable conflicts in the case of shared groundwater resources;
- Analysis of distribution of physico-chemical parameters variability and multilayer assessment of groundwater chemical status based on the aggregated data show that the spatial assessment should be conducted on the basis of median values, and not on arithmetic mean, as it is recommended by WFD (2000) and GWD (2003). It results from the fact that chemical constituents of waters usually

display log-normal distribution, and in this case median is not susceptible to a deformation to such extent as mean value;

- Assessment of groundwater chemical status carried out in the point and spatial mode, suggests that conducting in parallel a such two-stage procedure for water status analysis is the most appropriate. Application of the point method enables visualization of areas (within a drainage basin or a groundwater body or a group of bodies) of different groundwater chemical status, early identification of a risk and timely undertaking interception/reversal actions for restoration of good groundwater status;
- For proper assessment of groundwater chemical status, a selection of representative monitoring points is of a particular importance. In GBWs of multilayer structure, assessing water chemical status should be accomplished separately for the waters of shallow and deeper circulation that have different exposures regarding contamination.

References

BRIDGE project, 2006, *Background Criteria for the Identification of Groundwater Thresholds*, 2005–2006, Contract No. 006538 (SSPI), EU 6PR, Brussels.

Dowgiałło, J., Kleczkowski, A. S., Macioszczyk T., and Różkowski, A., eds., 2002, *Dictionary of Hydrogeology*, Department of Geology of the Ministry of Environment, Polish Geological Institute, Warsaw (in Polish).

DWD—Drinking Water Directive, 1998, 80/778/EEC (OJ L 229, 30.8.1980) as amended by Council Directive 98/83/EC of 3 November 1998 on the quality of water intended for human consumption (OJ L 330, 5.12.1998).

Edmunds, M., 2003, Final report on the studies carried out within the 5th framework program UE BASELINE, *Natural Baseline Quality in European Aquifers: A Basis for Aquifer Management*, Brussels, (unpublished).

GWD—The Groundwater Directive, 2003, *Proposal for a Directive of the European Parliament and the Council on the Protection of Groundwater Against Pollution*, COM (2003) 550 final, Brussels, 19.9.2003.

Herbich, P., Hordejuk, T., Kazimierski, B., Nowicki, Z., Sadurski, A., and Skrzypczyk, L., 2005, Groundwater bodies in Poland, in: *Contemporary Problems of Hydrogeology* XII, A. Sadurski, and A. Krawiec, eds., Toruń (in Polish).

Janecka-Styrcz, K., 2004, *Classification and Assessment of Ground Waters*, Computer program, PIG—Polish Geological Institute, Kielce (in Polish).

Kleczkowski, A. S., ed., 1990, *Map of Major Groundwater Basins (MGWB) and Aquifers in Poland Requiring Particular Protection*, Scale 1: 500000, AGH Publisher, Krakow.

Kos, M., 1997a, *Hydrogeological Map of Poland 1: 50000 with Comments Sheet Staszów (886)*, Polish Geological Institute, Warsaw (in Polish).

Kos, M., 1997b, *Hydrogeological Map of Poland 1: 50000 with Comments Sheet Klimontów (887)*, Polish Geological Institute, Warsaw (in Polish).

Paczyński, B., ed., 1997, *Hydrogeological Atlas of Poland 1:500000*, Polish Geological Institute, Warsaw (in Polish).

Prażak, J., Tracz, M., and Sideł, G., 1981, *Hydrogeological Documentation of Ground water Resources of the Western Part of the Tarnobrzeg Region along with the Framework Project of Geological Studies for a Category "C" of Regions Prospective for Construction of Large*

Groundwater Intakes, "South" Geological Combine ZPiDG, Kielce Department (unpublished, in Polish).
Prażak, J., Witczak, S., and Żurek, A., 2001, Problems with assessment of disposable groundwater resources in the drainage basins of rivers with water flow limited by the uderground discharge, in: *Contemporary Problems of Hydrogeology* X, part.1, Wrocław (in Polish).
RMS, 2004, *Directive of the Minister of Environment of 11 February 2004*, on the defining classification for the presentation of surface water and groundwater status, on the method of monitoring and the method of data interpretation and presentation of the state of these waters. Dz.U. 32.284.2004 (in Polish, pending amendments).
RMZ, 2002, Directive of the Minister of Health of 19 November 2002 on the requirements concerning drinking water quality, Dz.U. 203.1718.2002 (in Polish).
Sadurski, A., 2005, Transboundary groundwater bodies as a task of the Polish State Hydrogeological Survey, *Special Papers* Vol. 18, Polish Geological Institute, Warsaw.
Szczepańska, J., and Kmiecik, E., 2005, *Assessment of Groundwater Chemical Status on the Basis of Monitoring Data*, Wyd., AGH, Krakow (in Polish).
Szklarczyk, T., and Leśniak, J., 1990, *Detailed Conception of Groundwater Protection for Separated Hydrogeologic Regions*, MGWB 332. PG, Krakow (unpublished, in Polish).
UNEP, 2000, *World Resources 2000-2001. People and Ecosystems*, Elsevier, Amsterdam.
WFD—The EU Water Framework Directive, The Directive 2000/60/EC of the European Parliament and the Council establishing a framework for the Community action in the field of water policy, OJ L 327, 22 December 2000.
WHO, 2004, *Guidelines for Drinking Water Quality*, 3rd ed., WHO Guidelines, Brussels.
Witczak, S., 1996, *Groundwater Balance of the Koprzywianka River Drainage Basin*, Archive Materials of the University of Science and Tachnology, Krakow (unpublished, in Polish).
Witczak, S., and Adamczyk, A., 1994, 1995, *Catalogue of the Selected Physical and Chemical Indicators of Groundwater Contamination and Methods of Their Determination, Vol. 1, 2*, PIOŚ, The Library of the Environmental Monitoring, Warsaw (in Polish).
Witczak, S., Duda, R., Duliński, M., Jackowski-Korczyński, M., Karlikowska, J., Kmiecik, E., Kuc, T., Opoka, M., Różański, K., Szczepańska, J., Śliwka, I., Zdechlik, R., and Zuber, A., 2003a, Baseline Quality of PL2 Groundwater 332 Kędzierzyn-Głubczyce Subthrough, Final Report on the Studies Carried out within the 5th Framework Program UE BASELINE in: *Natural Baseline Quality in European Aquifers: A Basis for Aquifer Management*, Archives of ZHiOW, AGH, Krakow, (unpublished).
Witczak, S., Duda, R., Żurek A., and Szklarczyk, T., 2003b, Groundwater flow model of different hydrogeological systems, in: *Contemporary Problems of Hydrogeology* XI, vol. 1, H. Piekarek-Jankowska and B. Jaworska-Szulc, eds., Gdańsk.
Wróblewska, E., 2000a, *Hydrogeological Map of Poland 1: 50000 with Comments Sheet Opatów (854)*, Polish Geological Institute, Warsaw (in Polish).
Wróblewska, E., 2000b, *Hydrogeological Map of Poland 1: 50000 with Comments Sheet Łagów (853)*, Polish Geological Institute, Warsaw (in Polish).

Chapter 17
A Risk-based Integrated Approach for Managing Transboundary Groundwater Resources

Jacques Ganoulis*

Abstract The increasing demand for water as a consequence of population increase, socio-economic growth and climate variations together with the deterioration in water quality from various pollution sources, has resulted in upgrading the role and importance of transboundary waters including transboundary aquifer resources. In this chapter, the UNESCO-ISARM (Internationally Shared Aquifer Resources Management) approach is formulated in terms of a risk-based methodology for integrated groundwater management.

Keywords Transboundary groundwater resources, risk, integrated transboundary aquifer resources management

17.1 Introduction

Although it is very difficult to evaluate the exact quantity of available groundwater resources, it is widely acknowledged that groundwater constitutes the most important and most precious freshwater resource on Earth. Even though estimations in the literature on the quantity of available groundwater resources at a global level vary by some orders of magnitude, it is generally accepted that among all other sources of freshwater this quantity is the most important, except for vast quantities of freshwater blocked in icecaps and glaciers. As shown in Figure 17.1, freshwater resources on Earth represent only 2.5% of the total water available, because the majority of water is salt water held in the oceans and seas (Shiklomanov, 2005).

Jacques Ganoulis
UNESCO Chair and Network INWEB, Department of Civil Engineering, Aristotle University of Thessaloniki, 54124 Thessaloniki, Greece;
* To whom correspondence should be addressed. E-mail: iganouli@civil.auth.gr

C.J.G. Darnault (ed.), *Overexploitation and Contamination of Shared Groundwater Resources.*

Except for freshwater trapped in glaciers and ice sheets, the majority of freshwater resources, i.e. about 30% of the total, are in the form of groundwater. Only 0.3% of freshwater is surface water in lakes (87%) and rivers (2%).

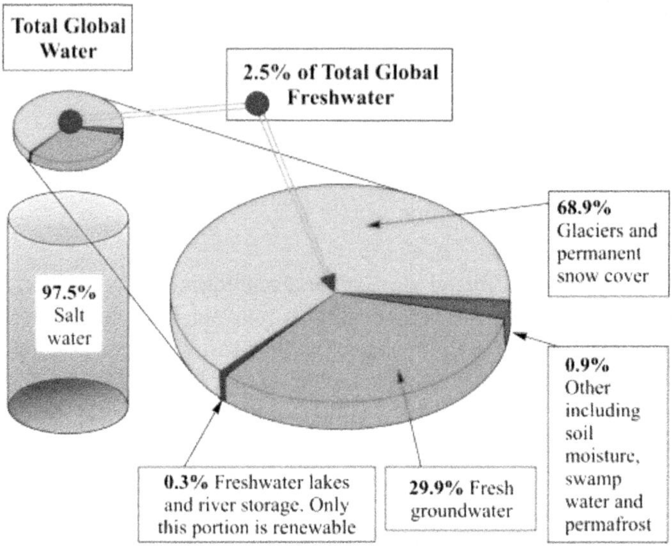

Fig. 17.1 Distribution of water in the hydrosphere

Groundwater is a key source of drinking water, particularly in rural and coastal areas. Table 17.1 shows the importance of groundwater for municipal water supply in southern European countries (Llamas, 2004).

Table 17.1 Groundwater Uses in some Mediterranean Countries

Country	Groundwater Use by Sector (Percentage of Total Abstractions)			Demand Covered by Groundwater (Percentage by Sector)		
	Water Supply	Agriculture	Industry	Water Supply	Agriculture	Industry
Spain	17	80	3	26	21	–
France	63	6	31	71	4	55
Italy	39	57,5	3,5	91	25	7
Greece	37	58	5	–	20,5	–
Israel	20	75	5	45	60	20
Turkey	64	36	–	64	36	–

Transboundary water resources are far from negligible. The 2003 UN Report (UN WWDR, 2003) entitled "Water for Life Water for People", listed 263 international transboundary basins. Apart from their significance in terms of area and conflict potential, it should be noted that these basins

- Cover 45% of the land surface of the Earth;
- Affect 40% of the world's population;

- Account for approximately 80% of global river flow;
- Cross the political boundaries of 145 nations.

The distribution of transboundary basins among the continents in terms of number and as a proportion of the total surface is shown in Figure 17.2. It can be seen that Europe comes first with 73 basins, followed by Africa and Asia.

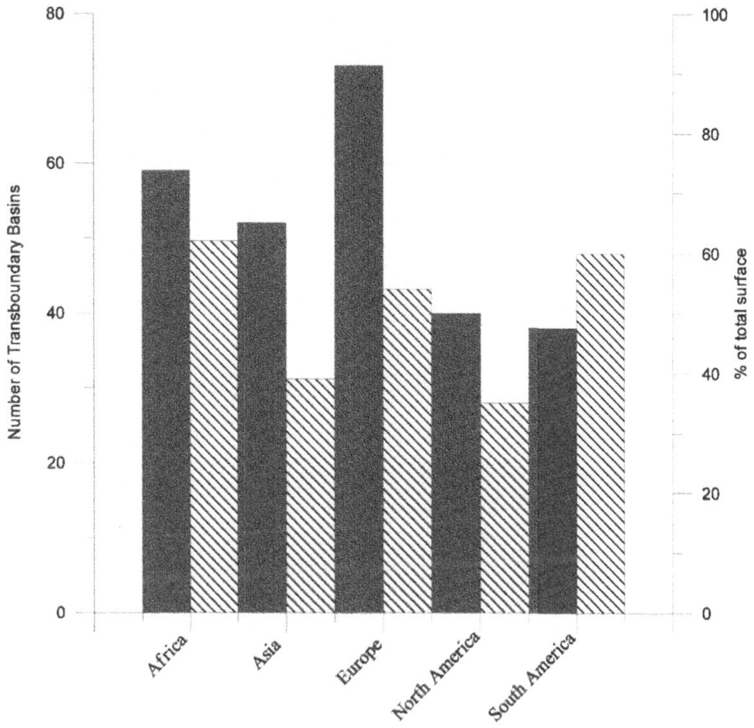

Fig. 17.2 Distribution of transboundary basins among the continents

In the case of internationally shared surface waters much progress has been made on how to determine what type of water resources problems need or will need to be resolved by bilateral or multilateral interstate agreements. A large number of such international agreements (more than 200 during the last 50 years) have been signed to resolve various types of interstate surface water resources problems. These are available for reference and act as precedents.

The situation is quite different in the case of internationally shared groundwater aquifer resources. Difficulties arise in scientific and technical matters (groundwater monitoring, data interpretation and modeling), and because of a lack of political willingness for cooperation and the weakness of the institutions involved. Major difficulties arise in designing groundwater development plans because groundwater flow and groundwater quality are subject to several types of uncertainties and are affected by these to a much greater degree than in surface water hydrology. These uncertainties are related to the high variability in space and time of the hydrogeological, chemical and biological processes. The principal challenge is to set up a cooperative framework between countries involved, so that institutions from both sides can work together effectively (Ganoulis et al., 1996).

In many real situations, interactions between surface and groundwaters on both sides of the international border may create international disputes. As shown in Figure 17.3 (UNESCO-ISARM, 2001) groundwater overexploitation on one side of the boundary may lower the water level of a shared surface lake or river or accelerate the sea water intrusion in a coastal zone located in the other country.

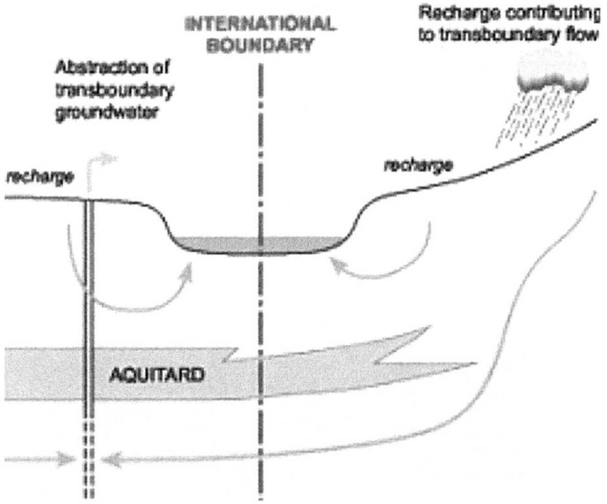

Fig. 17.3 Interaction between transboundary surface and groundwater flows

A very characteristic case of groundwater–surface water interdependencies can be found in the southern Balkans, in the region of the Doirani/Dojran Lake, which is internationally shared by Greece and the Former Yugoslav Republic of Macedonia (FYROM). In the many years of drought during the last decade, extensive evaporation together with overexploitation for irrigation purposes on the Greek side may have contributed to substantially lowering the lake's water level. In all cases, cooperation between countries is of primary importance in order to understand the problems, to agree about the underlying causes and to try to develop reliable solutions (INWEB, 2007).

17.2 Typology of Transboundary Aquifers

Water infiltrating into the soil circulates through various geological formations. Depending on the boundary conditions (impermeable or semi-permeable layers of soil, atmospheric pressure, rivers and lakes) the groundwater forms various types of subsurface reservoirs, called *aquifers*. These are extensive permeable rock formations in which water partially accumulates and through which water partially flows. Figure 17.4 gives an overview of different types of groundwater aquifers in various geological formations. According to their geological formation characteristics, aquifers may be classified into three main groups:

1. Alluvial and sedimentary aquifers.
2. Limestone and karstic aquifers.
3. Crystalline fractured rock aquifers.

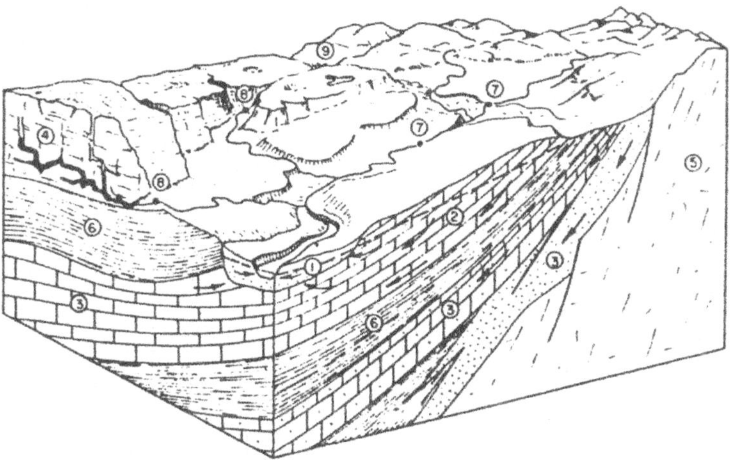

1 Alluvial aquifer in connection with a river
2 Regional aquifer with free surface in limestone
3 Regional aquifer between impermeable layers
4 Karstic aquifer
5 Crystalline fractured rock
6 Sedimentary impervious layer
7 Spring from phreatic aquifer
8 Spring from karstic aquifer
9 Leakage to a river

Fig. 17.4 Groundwater in various geological formations (Bodelle and Margat, 1980)

(1) *Alluvial and sedimentary aquifers*: This category of aquifers is characterized by successive layers of different hydrogeological properties: permeable, semi-permeable or impermeable. The water circulates in the successive layers, which consist mostly of gravel, sand, clay and silt. Phreatic, confined or semi-confined (leaky) aquifers are formed.

(2) *Limestone and karstic aquifers*: Solution processes caused by acidified rainwater increase the permeability of limestones and dolostones forming secondary aquifers. Karstic phenomena are extreme cases of such processes, creating subterranean fractures and water conduits of high permeability. In karstic regions surface runoff is almost nil and large volumes of groundwater can be found at various depths.

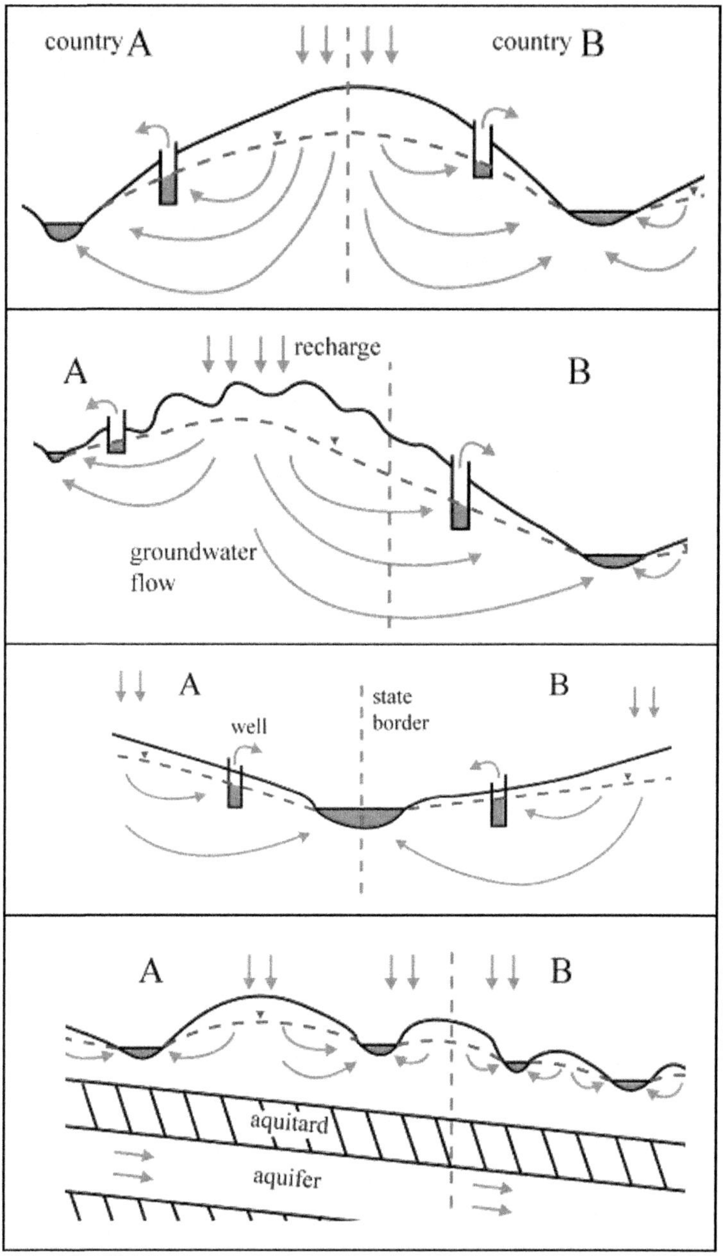

Fig. 17.5 Types of transboundary sedimentary aquifers (Chilton, 2007)

(3) *Crystalline fractured rock aquifers*: The importance of groundwater resources in these rocks depends on two factors (i) the rate of fracturing and (ii) the chemical weathering of the surface layer, through which precipitation water percolates into the rock. This geological formation is divided in several blocks by secondary and primary fractures.

In transboundary situations, depending on the location of the international border, sedimentary and alluvial aquifers may be classified in four different types, as follows (Figure 17.5, Chilton, 2007):

- *Type (a):* the state border follows the basin and groundwater divide. Very limited discharge occurs across the border;
- *Type (b):* the state border is separate from the basin and groundwater divide. Recharge occurs in one country and discharge in the other one;
- *Type (c):* the state border follows a transboundary river or lake. Little transboundary groundwater flow occurs in the alluvial aquifer connected to the river;
- *Type (d):* Large deep aquifer recharged far from the border. Transboundary groundwater flow not connected to the surface may be important.

In the case of deep karst aquifers covered by sediments, recharge of groundwater may occur in one country and water can appear at the surface in the form of a spring in another country (Figure 17.6). This occurs frequently in the Dinaric karst (Western Balkans), between Bosnia and Herzegovina (upstream) and Croatia (downstream, near the Adriatic coast).

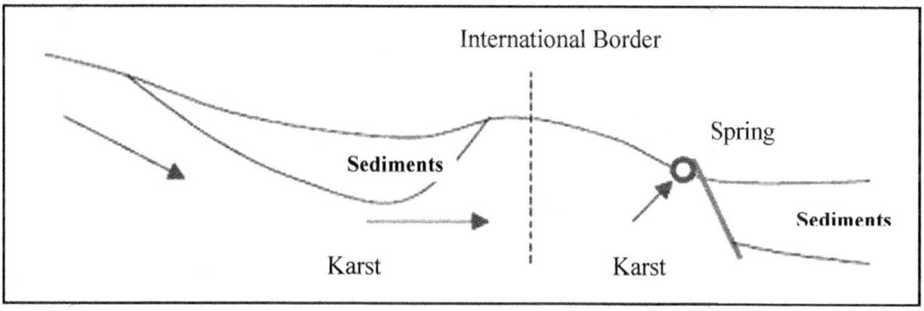

Fig. 17.6 Large deep karst aquifer recharged in one country and forming a spring in a neighboring country

17.3 ISARM's Methodology

The UNESCO-ISARM Programme (UNESCO-ISARM, 2001) has identified the following five key focus areas for the sound management of transboundary aquifer water resources:

- Scientific-hydrogeological approaches;
- Legal aspects;

- Socio-economic issues;
- Institutional considerations;
- Environmental protection issues.

17.3.1 Scientific–Hydrogeological Approaches

The management of groundwater quantity and quality is a complicated multidisciplinary scientific field requiring good cooperation between various disciplines, such as:

- *Hydrogeology*: geophysical and geological prospecting, drilling techniques, mapping;
- *Groundwater hydrodynamics*: quantitative aspects of flows, mathematical modeling, calibration and prediction scenarios;
- *Groundwater management*: systems analysis, optimization techniques, risk analysis and multi-objective decision-making methods;
- *Hydrochemistry*: chemical composition of the soil and water;
- *Hydrobiology*: biological properties of groundwater systems.

Modern tools for groundwater development extensively use new information technologies, databases, computer software, mathematical modeling and remote sensing.

17.3.2 Legal Aspects

International conventions on transboundary waters should include provisions for the monitoring and assessment of transboundary waters, including measurement systems and devices and analytical techniques for data processing and evaluation. Guidelines on how to effectively exchange information and monitoring data and undertake measures to reduce impacts from transboundary water pollution are also very important. As surface and groundwaters are interconnected, measures to protect ecosystems and drinking water supply should also include the monitoring and assessment of transboundary groundwaters.

An international convention has already been agreed upon for the monitoring and assessment of transboundary rivers and lakes (UNECE, 2000). No such international treaty yet exists for transboundary aquifers. The monitoring and assessment of surface waters are also part of the 1999 Protocol on Water and Health to the Convention on the Protection and Use of Transboundary Watercourses and International Lakes. This Protocol contains provisions regarding the establishment of joint or coordinated systems for surveillance and early-warning systems to identify issues related to water pollution and public health, including extreme weather conditions. It also includes the development of integrated information systems and databases, the exchange of information and the sharing of technical and legal knowledge and experience.

The complexities of groundwater law have been described by many authors in the technical literature. Overexploitation can cause groundwater quality to deteriorate through salinity problems, either by seawater intrusion (Chapter 9) or evaporation–deposition. Overexploitation of groundwater in one country can endanger the future freshwater supplies of another country. The Bellagio Draft Treaty, developed in 1989, attempts to provide a legal framework for groundwater negotiations. The treaty describes principles based on mutual respect, good neighborliness and reciprocity for the joint management of shared aquifers. Although the draft is only a model treaty and not the result of accommodating actual state practice, and accepts that collecting groundwater data may be difficult and expensive and should rely on cooperation; it does provide a general framework for groundwater negotiations.

Only three bilateral agreements are known to deal with groundwater supply (the 1910 convention between Great Britain and the Sultan of Abdali, the 1994 Jordan–Israel peace treaty and the Palestinian–Israeli accords (Oslo II). In addition, the 1977 Geneva Aquifer Convention is also an important reference for the internationalization of shared aquifer management and regulation by intra-State authorities for transboundary cooperation. Treaties that focus on pollution usually mention groundwater but do not quantitatively address the issue. In August 2005, the third report on shared groundwater resources was presented in Geneva to the United Nations International Law Commission (UNILC, 2005). In this report a set of articles for a draft international convention on the law of transboundary aquifers is proposed (Chapter 3).

17.3.3 Socio-economic Issues

It is widely accepted today that the use of water resources, the protection of the environment and economic development are not separate challenges. Development cannot take place when water and environmental resources are deteriorating, and similarly the environment cannot be protected and enhanced when growth plans consistently fail to consider the costs of environmental destruction. Nowadays, it is clear that most environmental problems arise as "negative externalities" of an economic system that takes for granted—and thus undervalues—many aspects of the environment. The integration of environmental and economic issues is a key requirement in the concept of sustainability, not only for the protection of the environment, but also for the promotion of sustainable long-term economic development, especially in areas where water is scarce.

The ISARM Framework Document (UNESCO-ISARM, 2001) makes a preliminary overview of different socio-economic aspects of transboundary aquifer management. The main driving forces behind the overexploitation of groundwater resources resulting in negative impacts are: population growth, concentration of people in big cities and inefficient use of water for agricultural irrigation. The agricultural sector is most often mainly responsible for groundwater overexploitation. The situation becomes particularly difficult when neighboring countries share common transboundary groundwater resources, as a number of differences arise in:

- Socio-economic level;
- Political, social and institutional structures, including strict region-specific positions on national sovereignty;
- Objectives, benefits and economic instruments;
- International relations, national legislation and regulation.

Competition for the use of groundwater for different purposes on one or both sides of the border may generate potential conflicts. Effective governance should consider specific hydrogeological conditions, aquifer recharge rates and multi-objective use of renewable groundwater resources involving multidisciplinary regional working groups.

17.3.4 Institutional Considerations

International commissions have proved to be the most effective institutional settings for transboundary surface water resources management for transboundary watercourses and lakes. No such common institutions exist for transboundary groundwaters. Whether transboundary groundwater management should be a specific task of one or more specialized committees belonging to the same international river or lake committee, or whether a separate common institutional body should be created for this purpose, remains a question unanswered. In view of the physical interactions between surface and groundwaters, coordination between different specialized institutions is necessary for the overall sustainable management of water resources.

In the present situation, national institutions dealing with groundwater are not sufficiently or effectively prepared to be able to undertake the joint management of transboundary groundwaters. Groundwater management units, when they exist, are often a mere side-line or even invisible in surface water dominated water administrations and groundwater is not explicitly addressed in national water legislations. Capacity building is essential, especially the development of joint capacity and consultation mechanisms at decision-maker level, including the harmonization of domestic groundwater law supported by common monitoring systems and the sharing of information and data. The role of regional partnerships between different decision-makers, scientists from different disciplines and other water stakeholders is also important for preventing conflicts and enhancing cooperation. It is important to link and reconcile transboundary aquifer management with land management, and with regional political, social and economic regional cooperation and development policy.

17.3.5 Environmental Protection Issues

Preservation of groundwater quality and ecosystem biodiversity should be an important objective for sustainability. Environmental protection should be realistically based on Environmental Risk Analysis (ERA) rather than on some precautionary principles, which may not lead to any action. ERA is a general and very useful approach for studying risks related to overexploitation or pollution of water in sensitive areas.

The application of ERA consists of two main phases:

1. Risk assessment;
2. Risk management.

17.4 Risk-Based Integrated Transboundary Aquifer Management (RITAM)

Sustainable management of transboundary groundwater resources should be based on current best practices, which are grouped under the term of Integrated Water Resources Management (IWRM). The term was first coined in 1977 at the UN Conference in Mar del Plata and according to the Global Water Partnership (GWP)—an NGO based in Stockholm—IWRM is defined as "a process which promotes the coordinated development and management of water, land and related resources to maximize the resultant economic and social welfare in an equitable manner without compromising the sustainability of vital ecosystems". According to IWRM, groundwater should be considered in relation to surface waters and should be studied at the river basin scale, which is the most appropriate unit of water management. Scientific–technical, environmental, economic and social issues should be taken into account, as explained in UNESO-ISARM's approach.

In what follows, the UNESCO-ISARM approach is formulated in terms of a risk-based multidisciplinary methodology called RITAM. For the integrated management of shared groundwater resources, four important risk indices are defined: technical, environmental, economic and social. It is explained how under alternative socio-economic and climate scenarios, different modeling techniques or expert judgments may be used in order to quantify risks. Risk quantification is an important step for initiating the process of risk management and sustainable use of transboundary aquifer resources.

Furthermore, in Chapter 18, Risk-based Multicriterion Decision Analysis (RMCDA) is presented as a tool for risk management and conflict resolution in internationally shared groundwater resources.

17.4.1 Definition of the Engineering Risk

In a typical problem of technical failure under conditions of uncertainty, there are three main questions, which may be addressed in three successive steps:

1. When should the system fail?
2. How often is failure expected?
3. What are the likely consequences?

The first two steps are part of the uncertainty analysis of the system. The answer to question 1 is given by the formulation of a critical condition, producing the failure of the system. To find an adequate answer to question 2, it is necessary to consider the frequency or the likelihood of failure. This can be done by use of the probability calculus. Consequences from failure (question 3) may be accounted for in terms of economic losses or profits.

As explained in the book by Ganoulis (1994), a variable reflecting certain external conditions of stress or loading may be defined as *load* ℓ. There is also a characteristic variable describing the capacity of the system to overcome this external load. This system variable may be called *resistance* r. A *failure* or an *incident* occurs when the load exceeds this resistance, i.e. *Failure or Incident*: $\ell > r$. Otherwise we have: *Safety or Reliability*: $\ell \leq r$.

In a probabilistic framework, ℓ and r are taken as random or stochastic variables and the chance of failure occurring is defined as the *engineering risk*. In this case we have: *Risk* = *Probability of failure* = $P(\ell > r)$.

This simple definition of engineering risk as the probability of exceeding a certain value of load is not unique (Duckstein and Plate, 1987). Generally speaking, risk is a complex function of the probability of failure and its consequences. In the literature, the product of the probability and its consequences are often taken as the risk function. However, different risk indices may be found in the literature for describing economic and social risks.

17.4.2 Technical, Environmental, Economic and Social Risks

The RITAM approach for transboundary groundwater resources planning and operation aims to reduce not only technical and economic but also environmental and social risks in order to achieve four main objectives (Figure 17.7): (1) Technical reliability, (2) Economic effectiveness, (3) Environmental safety, and (4) Social equity.

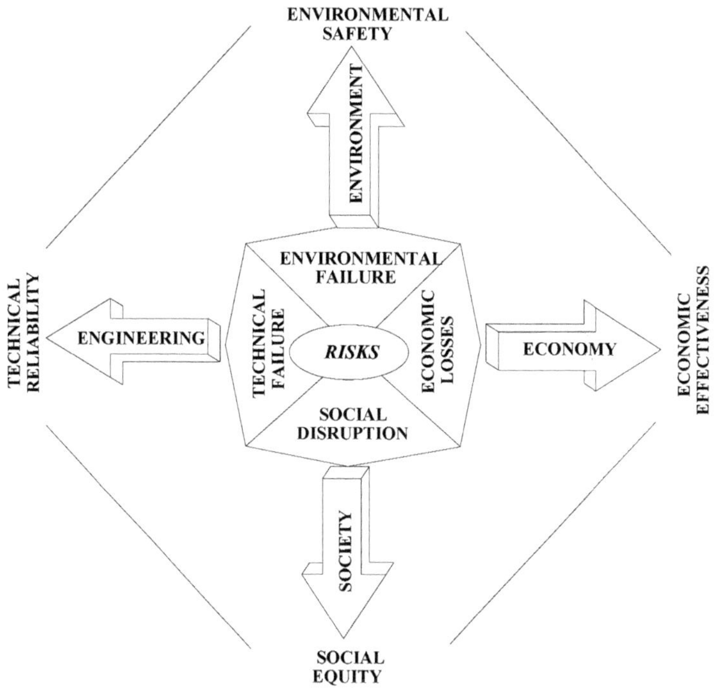

Fig. 17.7 Technical, environmental, economic and social objectives for RITAM

17.5 Risk Quantification in Aquifer Resources Management

Aquifer formations are complex hydrogeological systems with properties and hydrodynamic characteristics varying both in space and time. Any planning strategy for groundwater resources development and protection depends upon two main conditions:

1. The ability to predict the multiple risks and consequences from alternative strategies or operational policies under different socio-economic and climate scenarios.
2. The ability to analyze and rank the reliability of various strategies or operational policies by use of multiple quantitative criteria.

As shown schematically in Figure 17.8, the previous two conditions reflect the duality of modeling the physical and decision-making or societal parts of the process. In fact, the first condition may be fulfilled through various modeling techniques of

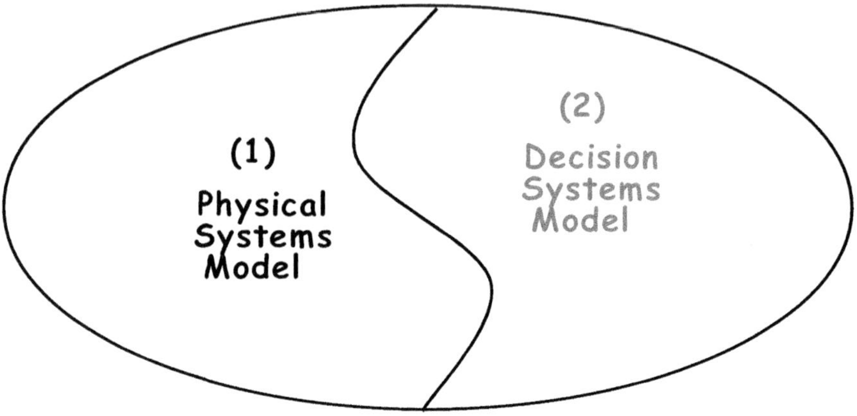

Fig. 17.8 The duality of modeling physical and decision systems for RITAM

groundwater flows, environmental impacts (like groundwater pollution and ecosystem analysis) and also modeling socio-economic risks. The second condition may be based on different decision analysis tools using multiple criteria under risk. Although important progress has been made in developing sophisticated modeling techniques, final judgments are actually based on experts' opinions or intuitive political considerations. However, physical modeling, optimization and application of risk and reliability techniques may be found to be useful tools for decision-makers.

17.5.1 Modeling Groundwater Flow

For groundwater hydrodynamics, conceptual models were developed as idealizations of natural aquifer systems (form, areal extension, physical properties of the aquifers) and their constituent processes (flow conditions, boundary conditions). Vertically integrated equations are usually used to represent flow in regional aquifers. These equations are obtained in the horizontal plane x–y by application of two basic laws:

- The law of mass conservation;
- Darcy's law.

Introducing the *piezometric head* or *hydraulic head* h as the sum of the *pressure head* p/ρg and the *elevation* z, i.e. $h = p/(\rho g)+z$ and using the definition of the storage coefficient S, the general mass balance differential equation for confined or unconfined groundwater flow takes the following form (Ganoulis, 1994):

$$S(\frac{\partial h}{\partial t}) = \nabla(KC\nabla h) - \sum_i q_i \delta_i \tag{1}$$

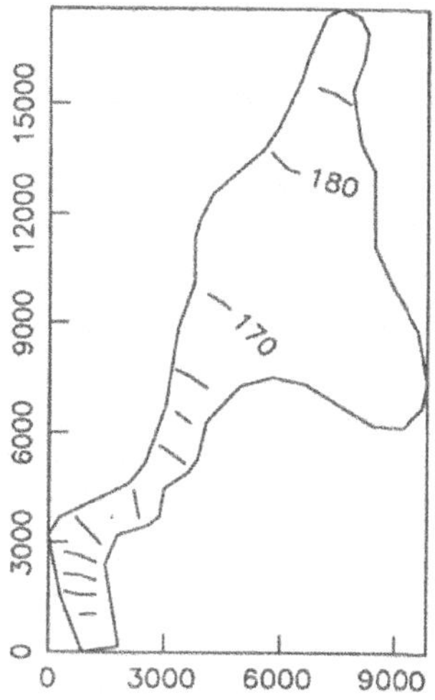

(a)

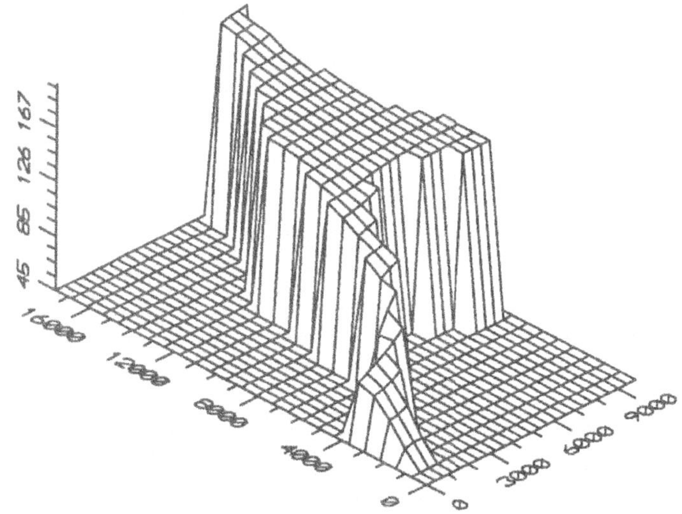

Fig. 17.9 Modeling the water table elevation of the Gallikos aquifer (in meters)

where K is Darcy's permeability coefficient, C = h for unconfined aquifer or C = b = thickness of a confined aquifer, $q_i > 0$ for pumping wells (in $m^3/s/m^2$), $q_i < 0$ for recharging wells (in $m^3/s/m^2$) and δ_i is the Dirac delta function for point i.

For modeling groundwater flows in regional aquifers, several analytical methods, finite difference and finite element numerical algorithms and, more recently, stochastic approaches of various levels of sophistication have been developed. These techniques for simulating the aquifer's hydrodynamics have been validated using physical models in the laboratory and in-situ measurements in real, relatively homogeneous aquifers of limited extent.

The results of a numerical simulation representing the distribution of the water table elevation of the Gallikos aquifer in two and three dimensions are given in Figure 17.9. This is an almost homogeneous, unconfined alluvial aquifer, located near the Axios River in Macedonia, Greece, which partly supplies the city of Thessaloniki with water (Ganoulis, 1994).

Because of the natural variability in space and time, the main problem for evaluating risks in groundwater flow and aquifer contamination is the fact that physical parameters and variables of the aquifer show random deviations in space. To this randomness, one must add various other uncertainties due to the scarcity of the information concerning the inputs (flow rates and pollutant loads), the value of parameters (measurement and sampling uncertainties) and also the imperfection of models (modeling uncertainties).

The natural variability of aquifer parameters and uncertainties in boundary conditions can be simulated using stochastic modeling and fuzzy logic approximation techniques (Ganoulis, 1994). In fact, during the last years, there has been an increasing number of publications on the application of stochastic and fuzzy logic-based methods to groundwater flow in aquifers. This indicates that more and more scientists are engaged in this area and the stochastic modeling and management of groundwater resources is an active subject of research.

17.5.2 Modeling Groundwater Pollution

For conservative pollutants, such as saline waters, this interaction is negligible and for regional 2-D groundwater flows, the following dispersive convection equation may be used:

$$\frac{\partial C}{\partial t} + u\frac{\partial C}{\partial x} + v\frac{\partial C}{\partial y} = D_x\frac{\partial^2 C}{\partial x^2} + D_y\frac{\partial^2 C}{\partial y^2} \qquad (2)$$

where C(x, y, t) is the pollutant concentration (M/L^3), u(x,y,t), v(x,y,t) are the groundwater velocity components (L/T), and D_x, D_y are the dispersion coefficients (L^2/T).

In fact, Eq. (2) is a random partial differential equation. The causes of randomness and variability are (i) the random variation of the velocity components (u, v) owing to the spatial variability of the aquifer parameters (porosity, permeability), and (ii) the variation of the dispersion coefficient D as a result of the random fluctuations of the velocity components. In general, stochastic simulation and risk analysis techniques can be used to quantify the effect of various uncertainties in the dispersion process (Ganoulis, 1994).

Several particle-oriented models in hydrological applications have been developed in the past (Bear and Verruijt, 1992). It seems that particle methods based on random walks are more flexible and easier to use and lead to relatively accurate results.

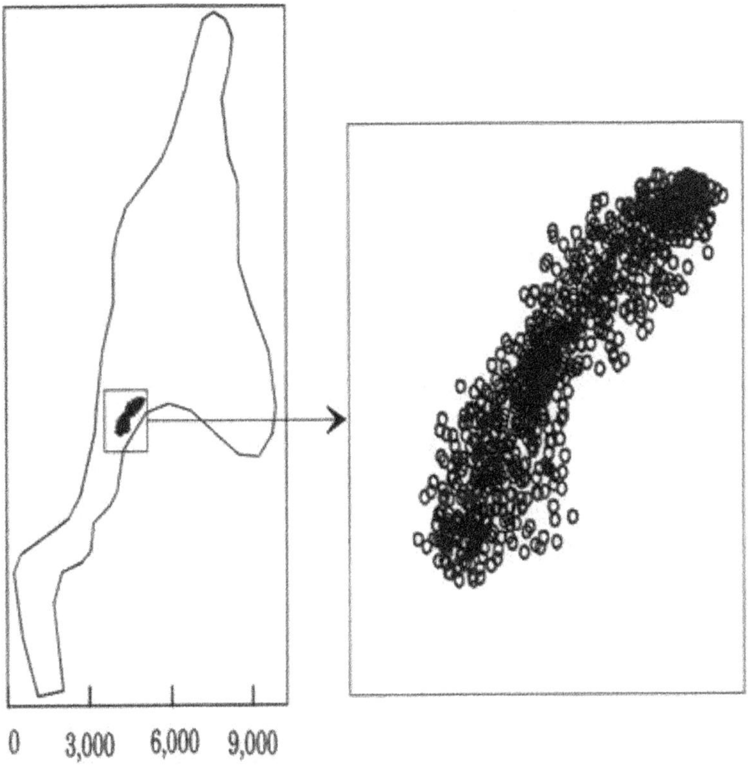

Fig. 17.10 Random walk simulation in the Gallikos aquifer (dimensions are in meters)

Consider at time t = nΔt, a large number of particles N located at the positions:

$$\vec{r}_{n,p} = (x_{n,p}, y_{n,p}) \qquad p = 1, 2, ..., N \tag{3}$$

According to the random walk principle the probability of finding a particle at a given position after time Δt follows a Gaussian law of mean value 0 and variance $s^2 = 2\Delta tD$, where D is the dispersion coefficient. The particles move from time t = nΔt to time t+Δt = (n+1)Δt according to the relations:

$$x_{n+1,p} = x_{n,p} + u\Delta t + \xi \tag{4}$$

$$y_{n+1,p} = y_{n,p} + v\Delta t + \eta \quad (5)$$

where u, v are the velocity components, and ξ, η random variables following a normal distribution of mean value 0 and variance $s^2 = 2\Delta tD$.

This is illustrated in the case of the Gallikos aquifer, where vulnerability of the groundwater from pollutant sources has been investigated using random walks. Introducing the corresponding velocity field, the groundwater pollution from a point source is obtained, as shown in Figure 17.10.

17.6 Conclusion

Risk-based Integrated Transboundary Aquifer Management (RITAM) methodology is proposed in order to integrate multiple risk indices into a multi-objective planning and decision-making process for sustainable transboundary groundwater use. Modeling techniques or expert judgments may be used to evaluate not only technical reliability and cost effectiveness but also environmental safety and social equity.

In order to achieve sustainability of transboundary aquifer resources, multiple risk indices are defined, such as technical, environmental, economic and social. These are used in Chapter 18 in order to rank alternative strategies for transboundary groundwater resources management and conflict resolution.

Using particle tracking and random walks, the risk quantification methodology is illustrated for evaluating risk of groundwater pollution in the Gallikos aquifer, Macedonia, Greece.

References

Bear, J., and Verruijt, A., 1992, *Modeling Groundwater Flow and Pollution,* Reidel, Dordrecht, 414 pp.

Bodelle, J., and Margat J., 1980, *L' Eau Souterraine en France*, Masson, Paris.

Chilton, J., 2007, Personal Communication.

Duckstein, L., and Plate, E., eds., 1987, *Engineering Reliability and Risk in Water Resources*, E. M. Nijhoff, ed., Dordrecht, The Netherlands, 565 pp.

Ganoulis, J., 1994, *Risk Analysis of Water Pollution: Probabilities and Fuzzy Sets*, Wiley-VCH, Weinheim, Oxford, NY, 306 pp.

Ganoulis, J., Duckstein, L., Literathy, P., and Bogardi, I., eds., 1996, *Transboundary Water Resources Management: Institutional and Engineering Approaches*, NATO ASI Series, Partner Sub-Series 2. Environment, Vol.7, Springer-Verlag, Heidelberg, Germany, 478 pp.

INWEB, 2007, (August 8, 2007); http://www.inweb.gr.

Llamas, M. R., 2004, Sustainable groundwater use and over-exploitation in groundwater, in: *Encyclopedia of Life Support Systems* (EOLSS), L. Silveira, S. Wohnlich, and E. J. Usunoff, eds., Developed under the auspices of the UNESCO, Eolss Publishers, Oxford, UK, (January 16, 2007); http://www.eolss.net.

Shiklomanov, I. A., 2005, The hydrological cycle, in: *Encyclopedia of Life Support Systems* (EOLSS), I. A., Shiklomanov, ed., Developed under the auspices of the UNESCO, Eolss Publishers, Oxford ,UK, (January 16, 2007); http://www.eolss.net.

UNILC, 2005, *Third Report on Shared Natural Resources: Transboundary Groundwaters,* by Mr. Chusei Yamada, Special Rapporteur, Geneva, (August 8, 2007); http://untreaty.un.org/ilc/sessions/57/57docs.htm.

UN WWDR, 2003, *Water for People, Water for Life*, UNESCO Publishing, Paris, (August 8, 2007); http://www.unesco.org/publishing.

UNECE, 2000, *Guidelines on Monitoring and Assessment of Transboundary Groundwaters,* Lelystad, UNECE Task force on monitoring and assessment, under the Convention on the Protection and Use of Transboundary Watercourses and International Lakes (Helsinki 1992).

UNESCO-ISARM, 2001, *A Framework Document*, Paris, UNESCO, Non-Serial Documents in Hydrology.

Chapter 18
Multicriterion Decision Analysis (MCDA) for Conflict Resolution in Sharing Groundwater Resources

Jacques Ganoulis*

Abstract Different socio-economic attributes and goals of countries sharing transboundary aquifer resources may lead to potential conflicts and political tensions. The MCDA methodology is adapted in order to compromise different management strategies suggested by adjacent countries. The methodology incurporates the results of a Risk-based Integrated Transboundary Aquifer Resources Management (RITAM) approach, which is presented in Chapter 17, in order to suggest common acceptable policies. An example of its application is given for the case of Mesta/Nestos River flowing between Bulgaria and Greece.

Keywords Attributes, criteria, risk, transboundary groundwater resources management

18.1 Introduction

There are many examples where potential conflicts over the use of internationally shared groundwaters could arise. In South Eastern Europe (SEE) for example, since the collapse of the Yugoslav Federation, about 90% of the region lies within international basins, as compared to a world average of 50%. More than half of these transboundary basins belong to three or more riparian states. Transboundary groundwater resources are the most important source for drinking water in the region and competition over the use of this water is constantly increasing, not only between different sectors within each country but also between countries (Ganoulis et al., 1996). The case of the Dinaric karst aquifer system situated along the Adriatic

Jacques Ganoulis
UNESCO Chair and Network INWEB, Department of Civil Engineering, Aristotle University of Thessaloniki, 54124 Thessaloniki, Greece

*To whom correspondence should be addressed. E-mail: iganouli@civil.auth.gr

C.J.G. Darnault (ed.), *Overexploitation and Contamination of Shared Groundwater Resources.*

coast is very characteristic of the natural complexity of transboundary groundwater resources and the involvement of two or three countries (INWEB, 2007). In this area, vast amounts of water are stored in karst aquifers originating in high mountains and distributed between two or three countries. Waters may recharge and circulate through deep karst formations located in one country (inner Dinarides) and appear in the form of karst springs in a neighboring country, usually along the Adriatic coast (outer Dinarides).

The case of sharing groundwaters in the Middle East is also very acute, especially since 1967 when Israel occupied the West Bank where strategic aquifers are located. Although the Palestinian-Israeli accords (Oslo II) were concluded in 1994, Palestinians still protest over the sharing of groundwater in the region.

Potential conflicts in sharing transboundary groundwaters may arise at two different scales:

- National or internal scale;
- International or external scale.

Internal conflicts are often due to competition over sharing water quantities among various sectors, like agriculture, urban water supply and industry. International conflicts may occur between neighbouring countries for different reasons:

- Sovereignty and other rights;
- National jurisdiction;
- Historical reasons;
- Competition over resources;
- Complexity of regional issues;
- Lack of participation of involved stakeholders.

To deal with potential water-related problems, UNESCO developed a special educational training project called PCCP: from Potential Conflict to Cooperation Potential. PCCP is a programme component within UNESCO's World Water Assessment Programme (WWAP) (UN WWDR, 2003). The WWAP was conceived to respond to the seven key challenges formulated in the Declaration of the Ministerial Conference held in The Hague during the IInd World Water Forum in March 2000. One of the key challenges identified was "Shared Water Resources Management." Within WWAP, UNESCO was given the task of elaborating the response to this challenge. The objectives of PCCP, moreover, are consistent with achieving the Millennium Development Goals (MDGs) agreed at the World Summit in Johannesburg in 2002, since PCCP aims at strengthening man's ability to cope with water related problems and to govern wisely in water related issues. This is vital if increased water security is to be achieved, extreme poverty to be eradicated and environmental sustainability to be ensured. Through PCCP, UNESCO has produced and published an extremely valuable and comprehensive knowledge base on conflict resolution in the water context, which was first presented at the 3rd World Water Forum in March 2003. This knowledge base consists of

- 19 papers and reports reviewing the legal, technical and diplomatic tools available for the anticipation and resolution of water conflicts;
- 9 case studies from around the world drawing lessons from both root causes of conflicts and successful cooperation in water resources management;
- 5 educational modules addressed to a large target audience with an interest in water management, ranging from post-graduate students to high-ranking decision-makers.

Many alternate negotiation strategies are available to modify a complex framework of transboundary groundwater management issues. The best policy is that which provides benefits to both sides. This is a "win–win" solution or a "positive–sum" policy. On the contrary, the worst overall policy is a "zero–sum" or "win–lose" solution, in which one country wins and the other loses. In all cases, potential water-related conflicts may worsen when there is water scarcity in the region (Ohlsson, 2004).

Since it is very difficult to increase the actual amount of water available, the best way to reverse a "win–lose" situation is through developing cooperation between riparian countries and implementing common management policies. In fact a study conducted by Wolf (1998) concluded that around the world there are more agreements for cooperation on sharing waters than conflicts between countries on the same issue. In the Transboundary Freshwater Dispute Database developed by Wolf, the full text of 140 water-related treaties is available, as well as negotiating notes from 14 basins and files on water-related agreements. Good examples of cooperation along large international river catchments are cited in the literature, such as the Rhine and the Danube rivers in Europe and the Mekong River in South Eastern Asia.

The main problem is the implementation of existing agreements by local institutions and decision-makers. In this context, MCDA methodology for conflict resolution may be a helpful tool in order to develop trust and initiate a compromise strategy based on a "win–win" policy.

18.2 Optimization Versus Compromise Solutions

In the past, traditional engineering approaches for water resources management emphasized the effective use of economic resources in planning and operation. Whilst still providing a reliable framework, investment and maintenance costs were to be minimized. As shown schematically in Figure 18.1, the main objective was to minimize total costs under a given degree of technical reliability. If only one objective is taken into account, an optimization problem can be formulated.

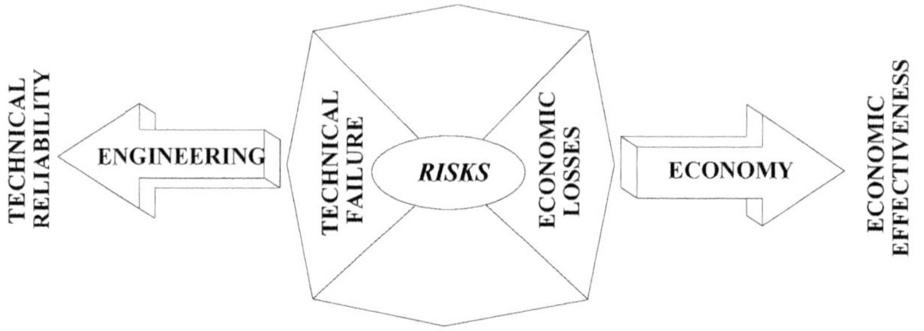

Fig. 18.1 Economic effectiveness versus technical reliability

18.2.1 Economic Optimization Under Risk

When using engineering modeling for the design of a water management plan, a number of options or alternative solutions usually emerge. The selection of any one particular solution depends upon the criteria used and is part of the decision process. In some simple cases, the particular objectives can be formulated as functional relationships between the problem variables. In cases where there is only one objective, analytical or numerical optimization techniques can be applied (Ang and Tang, 1984; Mays and Tung, 1992). Using such techniques, maximization or minimization of the objective function and the choice of an "optimum" solution may be achieved either under conditions of certainty or risk (Ganoulis, 1994).

To clarify this, let us first consider a simple, one-dimensional decision problem. As shown in Figure 18.2, a flood levee is to be constructed having a crest height h above the mean water level h_0 (free board). To determine one value of the variable h, which ensures an acceptable protection from possible floods, first the uncertainty conditions and the objectives of the project should be defined.

If sufficient experience from other cases is available, then we can assume that the levee operates under deterministic or certainty conditions, although overtopping of the levee is still possible. Apart from the investment costs of building the levee, we should also consider the costs arising from the consequences of a flood, when the water overtops the levee. Different kinds of damage behind the levee can be considered: damage to property, loss of life, environmental consequences, decrease in aesthetic values, etc. One reasonable objective should be to minimize the sum of both investment and damage costs.

For example, let us assume that investment costs C_I increase proportionally to the free board height h (Figure 18.3). The function C_I (h) has the form:

$$C_I = C_o + A\,h \tag{1}$$

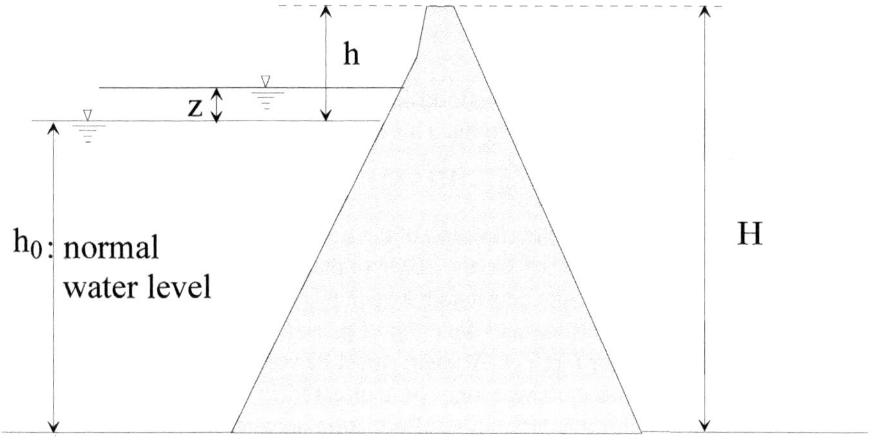

Fig. 18.2 The flood levee optimization problem

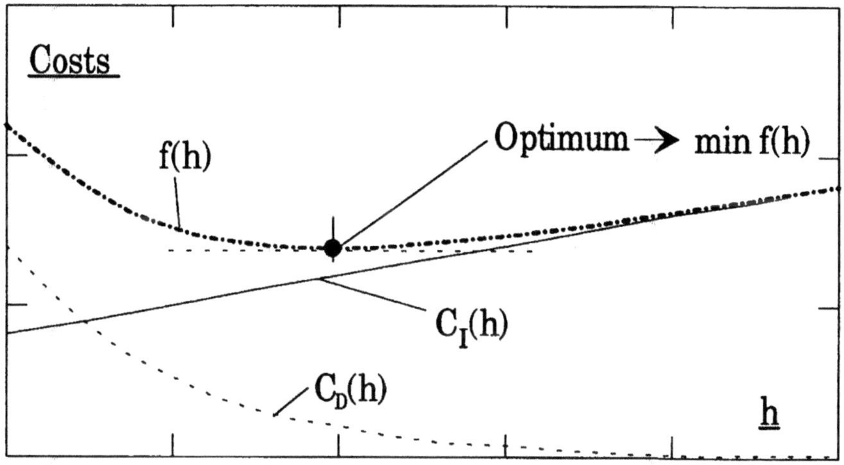

Fig. 18.3 Optimization of total costs under certainty

Damage costs C_D may decrease exponentially with h (Figure 18.3), i.e.

$$C_D = Be^{-\lambda h} \tag{2}$$

The *objective function* f(h) is written as

$$f(h) = C_I(h) + C_D(h) = C_0 + Ah + Be^{-\lambda h} \tag{3}$$

and the optimal solution (Figure 18.3) is at the minimum f(h), i.e.

$$f_{opt} = \min f(h) \tag{4}$$

In reality, the decision problem of flood protection usually involves uncertainties. These uncertainties may be quantified in terms of risk, which may be taken as a decision variable in optimization.

In a simple example such as the flood levee, let us consider the hydrological risk p_F as the probability of overtopping. This may be expressed as

$$p_F = P(z + h_o > H) = P(z > H - h_o) = F(h) \tag{5}$$

where P is the probability, z is the elevation of the flood above the normal water level h_O, and $h= H - h_O$ is the free board, i.e. the height of the levee above h_O (Figure 18.2).

From Eq. (5), a relation may be found between p_F and h. The objective function given by Eq. (3) may be written as a function of p_F and the optimum solution may be found in terms of p_F or ($-\ln p_F$). At every level of risk, there are consequences implying potential damages. These may be expressed in terms of *damage costs* having monetary or non-monetary values. Protection against damage should imply some other costs, called *protection costs.*

For low risk, the damage costs are low and they increase as risk increases. The opposite is true for the protection costs: high investment is necessary to keep the risk as low as possible. As risk increases so protection costs decrease. Generally speaking, we can state that

- *Damage costs* increase as risk increases and decrease as safety increases;
- *Protection costs* decrease as risk increases and increase as safety decreases.

To illustrate these statements let us consider a simple example in which the probability of overtopping is known. It is assumed that the probability density distribution of the flood elevation above the normal water height is exponential (Ang and Tang, 1984), with a mean value 2 m above h_O. In order to find the risk corresponding to the economically optimum design and the corresponding height h of the water level above h_O, it will be assumed that only one overtopping is expected with damage cost (C_D / overtopping) = 70,000 US dollars. The construction costs have the functional form (1), with C_O = 20,000 US dollars and A = 7,500 US dollars.

It is given that the probability density function of the flood elevation z above the normal water level is known. It can be expressed as an exponential distribution with a mean value 2 m above h_O. We have

$$f(z) = \lambda\, e^{-\lambda z} \tag{6}$$

$$E(z) = \langle z \rangle = 1/\lambda = 2 \tag{7}$$

$$\begin{aligned} P\ (h_0 + z > H) &= \int_{z=H-h_0}^{\infty} f(z)dz = \int_{z=H-h_0}^{\infty} \lambda e^{-\lambda x} dx = \\ &= -e^{-\lambda z}\Big|_{H-h_0}^{\infty} = e^{-\lambda(H-h_0)} = e^{-(H-h_0)/2} \end{aligned} \tag{8}$$

The probability of overtopping, i.e. the probability of having z > h (Figure 18.2) may be calculated as the probability of overtopping defined by the *engineering risk* or *probability of failure* p_F. From Eq. (8) it follows that

$$p_F = e^{-h/2} \qquad \text{or} \qquad h = -2 \ln p_F \tag{9}$$

- Protection Costs: C_p. These are proportional to h. The general expression is

$$C_p = C_0 + A\ h = C_0 - 2\ A \ln p_F \tag{10}$$

From Eq. (10), C_p decreases as p_F increases.

- *Damage Costs:* C_D. Suppose that B represents the expected costs for every overtopping. Then the total damage costs are

$$C_D = E\,(\text{less/overtopping})\ P\,(\text{overtopping}) = B\ p_F \tag{11}$$

The total costs are

$$C_T = C_P + C_D = C_0 - 2\ A \ln p_F + B\ p_F \tag{12}$$

It can be seen from Figure 18.4 that if safety ($-\ln p_F$) is chosen as a variable, investment costs are an increasing function of safety, whereas damage costs decrease with increasing safety. The risk corresponding to the optimum (minimum) cost is shown in Figure 18.4.

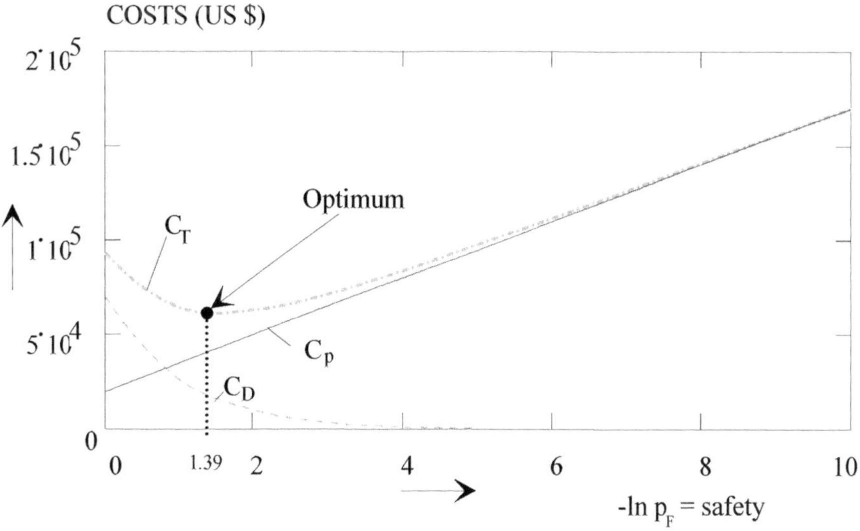

Fig. 18.4 Economic effectiveness versus technical reliability or safety

18.2.2 Multicriteria Compromise Methodologies

To obtain sustainable water resources management the four pillars of sustainability should be respected, which, as shown in Figure 18.5, are

- Technical reliability;
- Environmental safety;
- Economic effectiveness;
- Social equity.

For every specific case of a given river basin, the above four objectives can be hierarchically structured in attributes and goals. This is the hierarchical MCDA approach, shown in Figure 18.6 (Bogardi and Nachtnebel, 1994; Vincke, 1989).

MCDA techniques are gaining importance as potential tools for solving complex real world problems because of their inherent ability to consider different alternative scenarios, the best of which may then be analyzed in depth before being finally implemented. (Goicoechea et al., 1982; Szidarovszky et al., 1986; Pomerol and Romero, 2000).

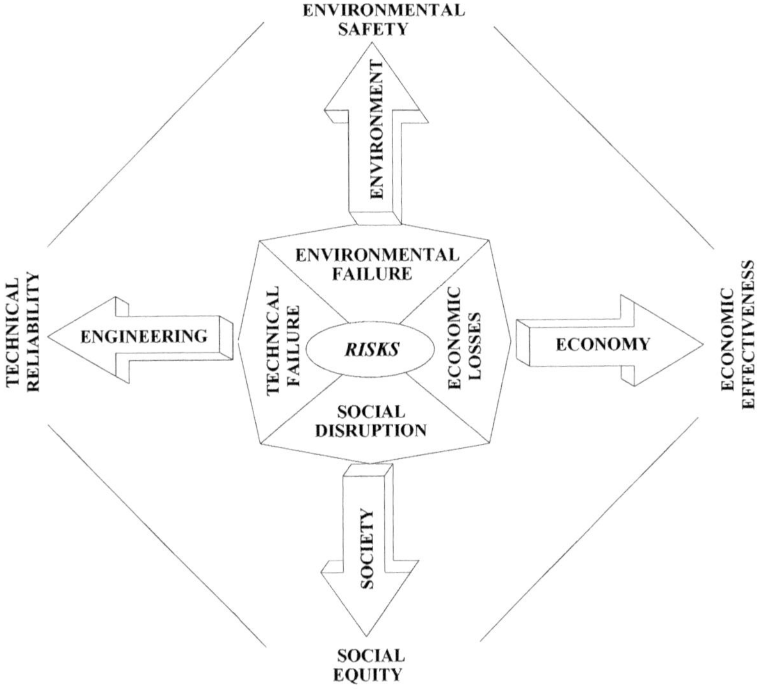

Fig. 18.5 The four pillars for sustainable water resources management

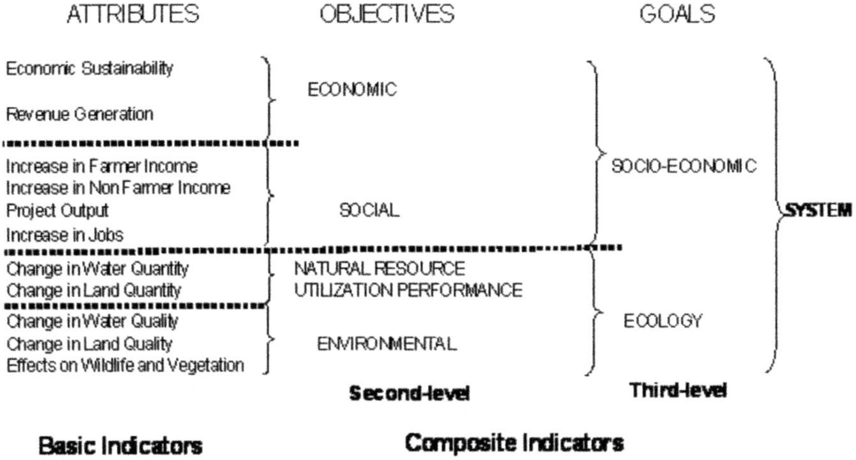

Fig. 18.6 Attributes, objectives and goals for sustainable water resources management

In order to apply MCDA techniques, it is important to specify the following:

- *The attributes*: which refer to the characteristics, factors and indices of the alternative management scenarios. An attribute should provide the means for evaluating the attainment level of an objective;
- *The objectives*: which indicate the directions of state change of the system under examination, and which need to be maximized, minimized or maintained in the same position;
- *The criteria*: which can be expressed either as attributes or objectives;
- *The constraints*: which are restrictions on attributes and decision variables that can or cannot be expressed mathematically.

A multi-criterion programming problem can be represented in a vector notation as

$$\text{“Satisfy”} \qquad f(x) = (f_1(x), f_2(x), \ldots, f_I(x)) \tag{13}$$

$$\text{Subject to} \qquad g_k(x) < 0, \ k = 1,2,\ldots, K \tag{14}$$

$$x_j \geq 0, \ j = 1,2,\ldots, J \tag{15}$$

Here there are I objective functions each of which is to be “*satisfied*” subject to the constraint sets (14) and (15). The region defined by this constraint set is referred to as the feasible region in the J-dimensional decision space. In this expression, the set of all J-tuples of the decision variable x, denoted by X, forms a subset of a finite J-dimensional Euclidean space; in many other applications, X is defined to be discrete. In the further special case when X is finite, then the most satisfying alternative plan has to be selected from that finite set X. It is important to note at this point that the word “optimum” which includes both the maximization of desired outcomes and minimization of adverse criteria is replaced by the word "satisfactum" and "optimize" is replaced by "satisfy" in this discussion. The reason is that when dealing with two or more conflicting objectives, one cannot, in general, optimize all the objectives simultaneously (Simon, 1957) as an increase in one objective usually results in a deterioration of some other(s). In such circumstances, trade offs between the objectives are made in order to reach solutions that are not simultaneously optimum but still acceptable to the decision-maker with respect to each objective (Goicoechea et al., 1982; Roy, 1996).

In a mathematical programming problem such as the one defined by Eqs. (13)–(15), the vector of decision variables and the vector of the objective functions **f**(x) define two different Euclidean spaces. These are (1) the J-dimensional space of the decision variables in which each coordinate axis corresponds to a component of vector X, and (2) the I-dimensional space F of the objective functions in which each coordinate axis corresponds to a component of vector **f**(x). Every point in the first space represents a solution and gives a certain point in the second space that determines the quality of that solution in terms of the values of the objective functions. This is made possible through a mapping of the feasible region in the decision space X into the feasible region in the objective space F, using the I-dimensional objective function.

18.2.2.1 Feasible, Non-dominated and Efficient Solutions

In Multi-Criterion Decision Analysis (MCDA), the question is not to obtain an optimal solution as in the case of one objective. Instead of an optimum solution, we speak about a "non-inferior" or "non-dominated" solution. This is a solution for which no improvement in a single objective can be achieved without causing a degradation of at least another objective.

Let us consider, for example, the problem of "maximizing" two conflicting objectives Y_1 and Y_2 subject to a set of constraints:

$$g_j(x_1, x_2, \ldots, x_n) \leq = \geq 0 \quad j = 1, 2, .., m \tag{16}$$

As shown in Figure 18.7, each couple of values Y_1 and Y_2 that satisfy the constraints lies within the *feasible region or feasible space*. This region is limited by a curve ABCD called a *feasibility frontier*. All points of this frontier form the set of "*non-inferior*" or "*non-dominated*" solutions. Every decision vector on this curve is defined by a maximum value of the objective Y_2 given a value of the objective Y_1. This particular solution is "optimal" in the sense that there can be no increase in one objective without a decrease in the value of the other objective.

A selection of one particular solution from a set of non-inferior solutions depends on the preferences of the decision-maker. This may be indicated by a family of *iso-preference* or *indifference curves* (Figure 18.7). In this figure, the *efficient solution* is defined by the point B on the feasibility frontier that has the maximum level of preference.

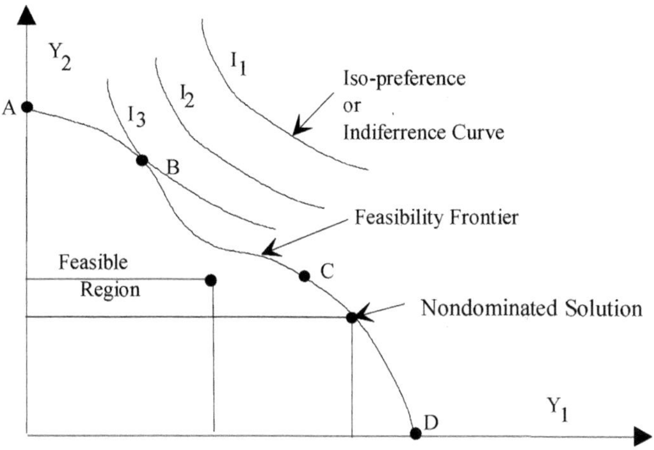

Fig. 18.7 Non-dominated solutions for a two-objective problem

18.2.2.2 Solution Procedures and Typology of MCDA Techniques

Finding the set of efficient solutions of a mathematical programming problem is usually determined using a generating procedure, in which an objective function vector is used to identify the non-dominated subset of feasible decisions. This procedure deals mostly with the objective realities of the problem (e.g., the set of constraints) without necessarily taking into consideration the preference structure of the decision-maker.

In order to clarify the technique choice procedure, the classification of MCDA models given in Tecle and Duckstein (1994) is now summarized. Five types are distinguished:

1. *Value or utility-type*: which essentially coalesce the multiple objectives into a one-dimensional "multi-attribute" function. It can be a value function that is deterministic or a utility function that includes a measure of risk.
2. *Distance-based techniques*: which seek to find a solution as "close" as possible to an ideal point, such as *compromise* and *composite* programming or else, a solution as "far" as possible from a "bad" solution, such as the Nash cooperative game concept.
3. *Outranking techniques*: which compare alternatives pair wise, and reflect the imperfection of most decision-makers' ranking process (Roy, 1996) namely, alternative A(j) is preferred to alternative A(k) if a majority of the criteria C(i) are better for A(j) than for A(k) and the discomfort resulting from those criteria for which A(k) is preferred to A(j) is acceptable. As a result, non-comparability of certain pairs of alternatives is an acceptable outcome; this is in contrast with the previous two types of approaches where a complete ordering of alternatives is obtained. Techniques such as ELECTRE and PROMETHEE are recommended.

4. *Direction-based*: interactive or dynamic techniques where a so-called progressive articulation of preferences is undertaken.
5. *Mixed techniques*: which utilize aspects of two or more of the above four types. In planning problems, a general class of methodology has been developed to rank different alternatives with various conflicting objectives under risk, (Goicoechea et al., 1982).

One of the promising methods is the *Composite* or *Compromise* Programming. First, trade-offs between objectives may be made in different levels to obtain some composite economic or ecological indicators. Then, ranking between different strategies or options may be done using different techniques, such as the one based on the minimum composite distance from the ideal solution (Figure 18.8), (Duckstein and Szidarovszky, 1994).

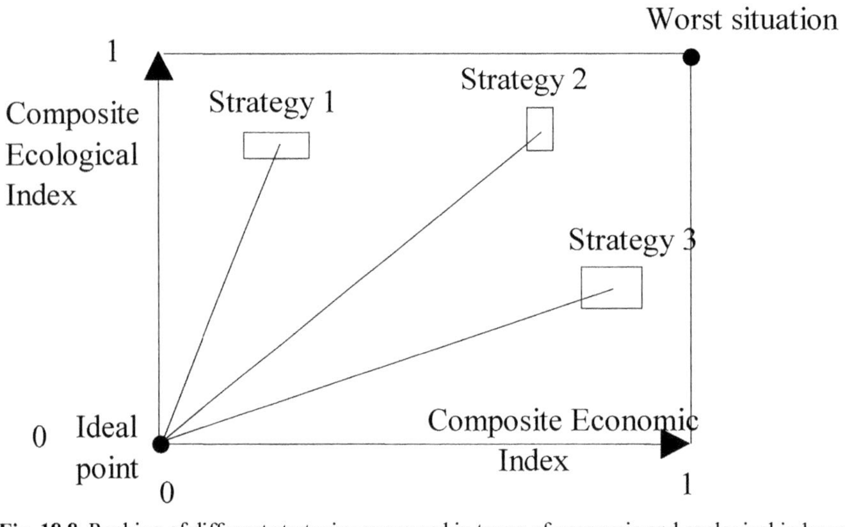

Fig. 18.8 Ranking of different strategies expressed in terms of economic and ecological indexes

18.3 Modeling Transboundary Conflicts

Conflict situations in transboundary groundwater resources management may occur on at least two levels:

- Conflict among specific attributes, in particular economic, environmental and social ones;
- Conflicts of goals or general interests between countries and among groups of actors involved.

Broadly speaking, every state has social, economic and political goals linked to water resources development, conservation, and control and protection of the river basin. Economic goals may be to obtain new water resources in order to increase food production, conservation goals may be to control water pollution, and control and protection goals may concern defense against floods or drought control. These goals may be achievable by jointly building water reservoirs. This would entail the states involved cooperating together and solving possible areas of conflict.

Goals are accomplished by various water resources developments, transfers of water from the water-surplus adjacent river basins, water conservation, control and protection. Each particular goal means satisfying some particular purpose, which may have to do with irrigation, drainage, hydropower production, navigation, water supply, water pollution control, flood defence, drought control, or other.

Finally, to satisfy the purposes of state goals in water resources development, one must define and then maximize or minimize particular economic, social, monetary and political attributes. The particular purposes, attributes and interests in water resources development of the river basin should be strictly taken into consideration in any future cooperation on conflict resolution between the states.

18.3.1 MCDA for Conflict Resolution

Three different approaches are suggested for conflict resolution. In the first approach, each country proceeds separately and evaluates alternatives according to its own objectives (Figure 18.9).

In the second approach, the different attributes used by the two countries are first traded-off and then alternatives are ranked according to the composite objectives (Figure 18.10).

The third approach is based on the aggregation of the countries' different alternatives in order to obtain a consensus between them (Figure 18.11).

As an extension of the present methodology, two different types of uncertainties can be taken into consideration:

1. Uncertainties in attribute and goal values.
2. Uncertainties in the preferences of the decision-makers and other interest groups.

The methodology can be applied either for internationally shared surface or groundwaters. As an example, the case of the transboundary Nestos/Mesta River, flowing between Greece and Bulgaria is presented.

18.4 A Case Study: The Mesta/Nestos Transboundary Waters

Different management alternatives and different projects were suggested from both countries in order to address the following regional problems:

1. *Water availability*: water supply for urban and rural settlements, agriculture, recreational activities and hydropower generation are competing for more water especially in summer and in periods of drought.
2. *Water quality*: the lack of landfills and wastewater treatment facilities upstream, the unsystematic breeding of cattle and the overexploitation of groundwater resources for irrigation and drinking water downstream has caused water quality problems and salinization of coastal areas near the river's delta.
3. *Environmental*: the upper part of the basin is part of the Pirin national park and the delta region is a RAMSAR convention protected area. Water quality degradation created negative impacts on fauna and flora and loss of biodiversity.
4. *Development problems*: poor infrastructure and lack of facilities have resulted in a very low level of tourism, aquaculture and industry in the area.

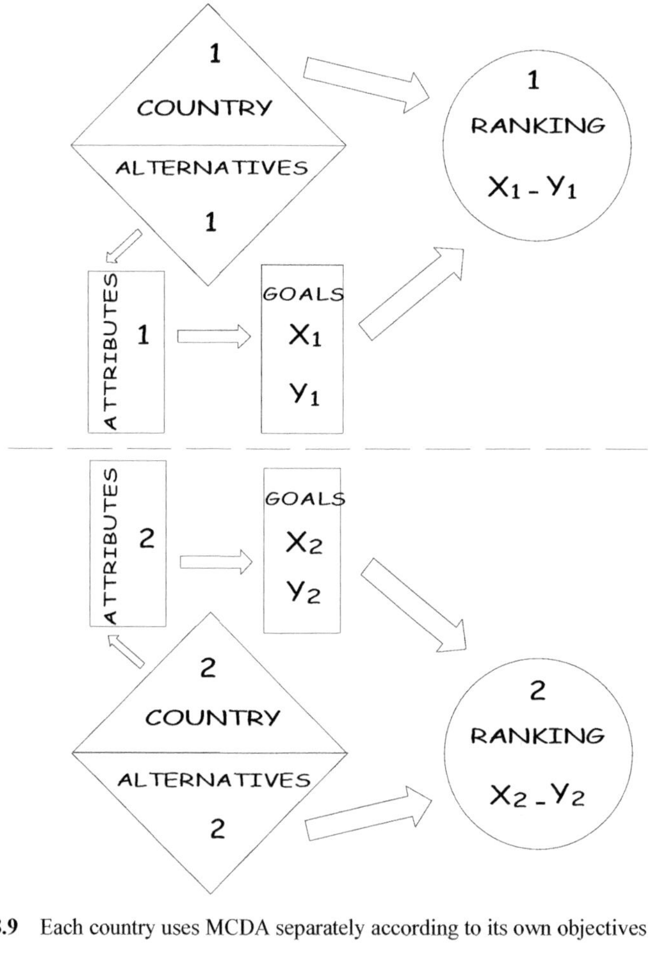

Fig. 18.9 Each country uses MCDA separately according to its own objectives

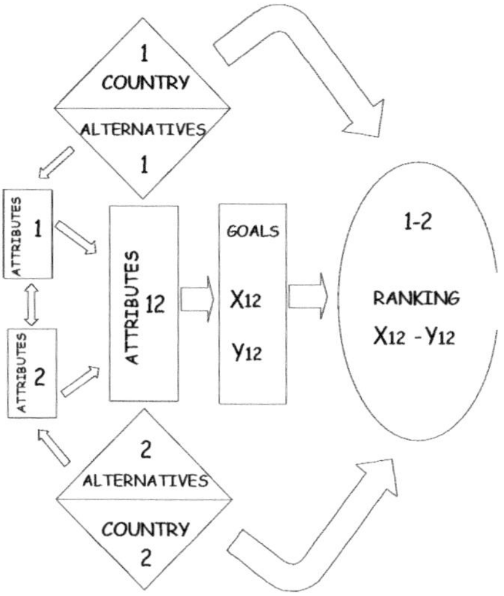

Fig. 18.10 Compromising countries' different attributes

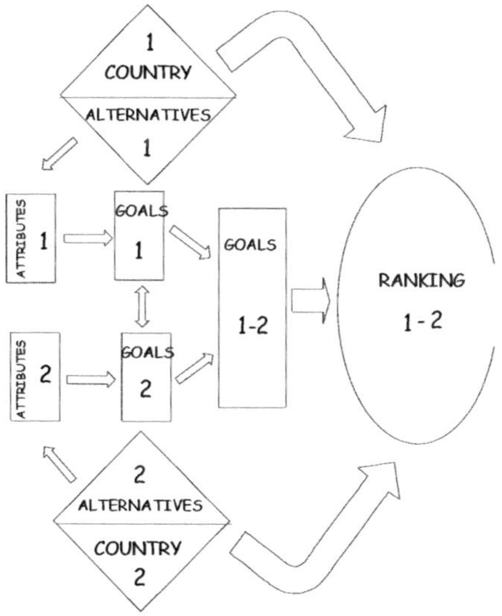

Fig. 18.11 Compromising countries' different goals

For this case study, four different management options (1–4) were suggested by the country A and four other options (5–8) by the country B. Because of different attributes and goals, every country gives preference to their options. Individual rankings by country give the following results: Country A 3,2,6,8 and country B 6,8,5,7. By compromising the different countries' attributes and goals, and using non-dimensional aggregated *socio-economic* and *ecological indexes* varying from 0 (worst) to 1 (ideal) the obtained results are shown in Figures 18.12 and 18.13.

Both methods suggest that management options 3 and 6 have higher priority over the others because they are located closer to the ideal point. This is consistent with the countries' individual preferences.

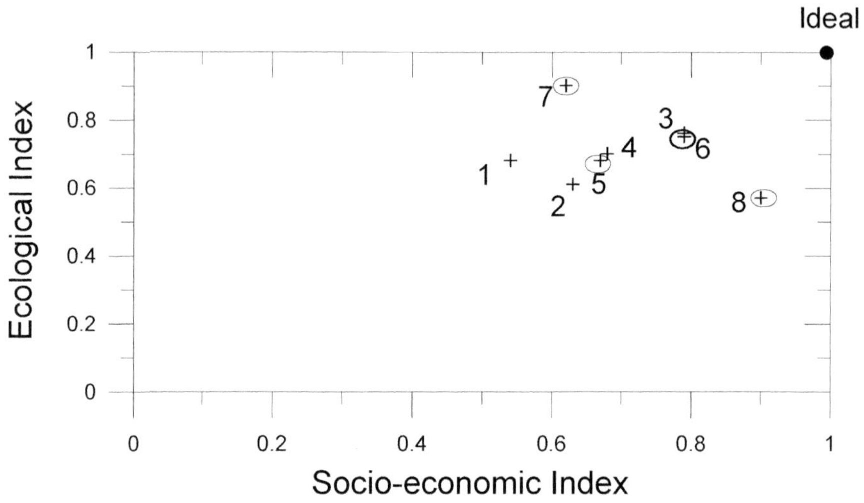

Fig. 18.12 Conflict resolution by trading off countries' different attributes

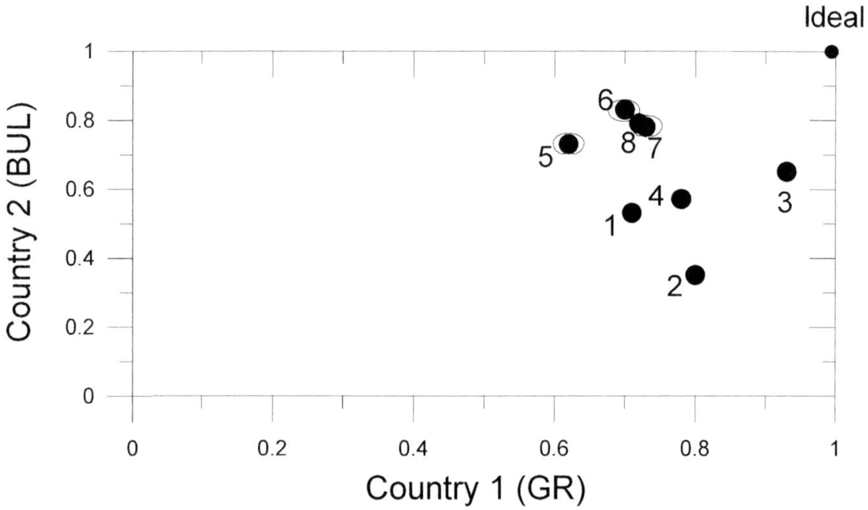

Fig. 18.13 Conflict resolution by trading off countries' different goals

18.5 Conclusion

The Risk-based Integrated Transboundary Aquifer Management (RITAM) methodology presented in Chapter 17 is based on mathematical modeling techniques or expert judgments in order to evaluate for every specific management project risk indices for technical reliability, cost effectiveness, environmental safety and social equity.

In this chapter, the MCDA methodology was adapted in order to rank alternative strategies for transboundary groundwater resources management and conflict resolution. The technique in based on aggregating countries' different attributes or goals deriving by application of the RITAM multiple risk indices.

The methodology is illustrated by a case study, where trade-offs made either at the level of countries' different attributes or countries' different goals lead to similar compromise results.

References

Ang, A. H., and Tang, W. H., 1984, *Probability Concepts in Engineering Planning and Design* Vol. 2: *Decision, Risk and Reliability*, J. Wiley, New York, 562 pp.

Bogardi, I., and Nachtnebel, H. P., 1994, *Multicriteria Decision Analysis in Water Resources Management*, IHP, UNESCO, Paris, 469 pp.

Duckstein, L., and Szidarovszky, F., 1994, Distance based techniques in multicriterion decision making, in: *Multicriterion Decision Analysis in Water Resources Management,* J. Bogardi and H. P. Nachtnebel, eds., UNESCO, Paris, pp. 86–112.

Ganoulis, J., 1994, *Risk Analysis of Water Pollution: Probabilities and Fuzzy Sets*, Wiley-VCH, Weinheim, Oxford, New York, 306 pp.

Ganoulis, J., Duckstein, L., Literathy, P., and Bogardi, I., eds., 1996, *Transboundary Water Resources Management: Institutional and Engineering Approaches*, NATO ASI SERIES, Partnership Sub-Series 2, Environment, Vol. 7, Springer-Verlag, Heidelberg, Germany, 478 pp.

Goicoechea, A., Hansen, D. R., and Duckstein, L., 1982, *Multiobjective Decision Analysis with Engineering and Business Applications*, J. Wiley, New York, 519 pp.

INWEB, 2007, *Inventories of Transboundary Groundwater Aquifers in the Balkans,* UNESCO Chair and Network INWEB, Thessaloniki, Greece, (to be published), (August 8, 2007); http://www.inweb.gr.

Mays L., and Tung, Y. K., 1992, *Hydrosystems Engineering and Management*, McGraw Hill, New York, 529 pp.

Ohlsson, L., 2004, Water and conflict, in: *Water and Development*, C. M. Marquette, ed., in: *Encyclopedia of Life Support Systems (EOLSS),* Developed under the Auspices of the UNESCO, Eolss Publishers, Oxford ,UK, (February 7, 2007); http://www.eolss.net.

Pomerol, J. C., and Romero, S. B., 2000, *Multicriterion Decision in Management: Principles and Practice*, Kluwer Academic Publishers, the Netherlands.

Roy, B., 1996, *Multi-criteria Methodology for Decision Aiding,* Kluwer, Dordrecht, the Netherlands.

Simon, H. A., 1957, *Administrative Behaviour*, Free Press, New York, (2nd edition, 1976).

Szidarovszky, F., Gershon, M., and Duckstein, L., 1986, *Techniques for Mutli-objective Decision Making in Systems Management*, Elsevier, Amsterdam.
Tecle, A., and Duckstein L., 1994, Distance based techniques in multicriterion decision making, in: *Multicriterion Decision Analysis in Water Resources Management,* J. Ganoulis, ed., pp. 86–112.
UN WWDR, 2003, *Water for People, Water for Life*, UNESCO Publishing, Paris, (August 8, 2007); http://www.unesco.org/publishing.
Vincke, P., 1989, *L'Aide Multicritère à la Décision*, Editions de l'Université de Bruxelles.
Wolf, A., 1998, Conflict and cooperation along international waterways, *Water Policy* **1**(2):251–265.

Chapter 19
Strategies for Groundwater Resources Conflict Resolution and Management

W. Todd Jarvis*

Abstract This chapter surveys contemporary thinking on the state of conflict and cooperation over transboundary groundwater resources and the evolution of approaches to conflict management to address potential conflicts over the use of transboundary groundwater resources. The findings of this overview reveal that groundwater professionals often overlook their role in both causing and resolving conflicts over groundwater. This chapter will be of interest to groundwater professionals will little to no formal training in dispute resolution and conflict management.

Keywords Conflict resolution, groundwater governance, groundwater management, groundwater domains

19.1 Introduction

Over 40% of the world's population relies on transboundary water resources for their secure and stable livelihoods (Wolf and Giordano, 2002). Worldwide there are over 263 transboundary river basins that cover over 45% of the global landmass (Wolf et al., 1999; Wolf and Giordano, 2002; Bakker, 2006). Intranational and international conflicts over the right to use water traditionally focus on surface water resources. Conflicts over these resources are predicted to increase over the next 15 years as many countries press against the limits of available water (Postel and Wolf, 2001).

W. Todd Jarvis
Senior Researcher in Water Resources, Institute for Water and Watersheds, 210 Strand Hall, Oregon State University, Corvallis, Oregon, USA 97331-5506

* To whom correspondence should be addressed. E-mail: todd.jarvis@oregonstate.edu

C.J.G. Darnault (ed.), *Overexploitation and Contamination of Shared Groundwater Resources*.

Less widely recognized are the nearly 240 transboundary groundwater systems or "aquifers" identified by the World-wide Hydrogeological Mapping and Assessment Program (WHYMAP) as described by Struckmeir et al. (2006). Analogically, a large percentage of the world population also resides in lands underlain by transboundary aquifers. Many of these underground aquifers are regional in extent with flow paths ranging from meters to hundreds of kilometers across continents, and consequently shared by several countries. Shiklomanov and Rodda (2003) estimated that almost 90% of all accessible freshwater is found in aquifers. According to Puri et al. (2001), the hidden nature of groundwater and the lack of international law governing shared aquifers invite misunderstandings leading to conflict.

Pumping of groundwater is among the most intensive human-induced changes in the hydrologic cycle. With dramatic changes in drilling technology, pumping technology and the availability of electrical power over the past 60 years, the number of wells has increased exponentially in many parts of the world (Moench, 2004). According to Zekster and Everett (2004), groundwater is the world's most extracted raw material, with withdrawal rates approaching 600–700 km^3 per year. With global abstractions of freshwater estimated by Shiklomanov and Rodda (2003) approaching between 4,600 and 5,800 km^3 per year by the year 2025 and comparing this to the estimated volume of groundwater stored in the Earth's crust approaching 23.4 million km^3, there also exists a perception that humans are using a miniscule proportion of the potential global groundwater resources.

Until recently, conflicts over groundwater have generally focused on contamination of wells (Gleick, 2006). Yet concerns over access to water in drought prone regions such as Somalia to have a new generation of conflict over groundwater. *The Washington Post* reported a "War of the Well" in 2006. Workers of the International Medical Corps reported that "It's like the start of the water wars right here in Somalia" (Wax, 2006). A "Silent Revolution" is occurring where millions of farmers pursue short-term benefits associated with the intensive use of groundwater for agricultural use in India, China, Mexico, and Spain and the need for proactive governmental action is needed to avert water conflicts (Llamas and Martinez-Santos, 2005). These events, along with the "Silent Trade" of groundwater contaminated with hazardous waste flowing across international boundaries into Lebanon reported by Jurdi (2002), have accelerated interest in managing and governing transboundary groundwater resources.

Thomasson (2005) indicates that while the water situation in many poor countries is dire, most sources on conflict and conflict practice are from developed countries, and that this unbalance is unlikely to change in the near future. Thomasson (2005:6) points out that "There is no lack of societal conflicts in rich countries, only there they rarely turn violent." This chapter speaks to how people have, and have not, dealt with conflict over groundwater and the development of management schemes and governance models to avert conflict.

19.2 What is Conflict?

Thomasson (2005) cites Wallensteen (2002) as providing the best working definition of conflict: "conflict [...] (is) a social situation in which a minimum of two actors (parties) strive to acquire at the same moment in time an available set of scarce resources". Resources can mean any kind of resource, including material as well as political. Whereas the classic disputes over resources are over the territorial integrity of a state and the extent of government control as described by Anderson (1999), Thomasson (2005) posits that it is incompatibilities over resources that create grievances or conflicts and that "scarce" does not have to mean that the resource is limited. Incompatibilities often arise over the use and equitable, or inequitable, distribution of a resource.

19.3 The Debate on Conflict and Water

In water-related conflicts, sometimes there is a physical scarcity, but the conflicts over distribution are more common (Thomasson, 2005). Public discourse about the role of resource scarcity or abundance is the common thread of discussions regarding conflict at the intrastate as well as international scale. For example, Thomasson (2005) describes one foundation for conflict as the neo-Malthusians pointing to the role of population growth and finite resources as creating scarcities of resources. Other scholars place more emphasis on the distributional issues such as when a large part of the resources are being consumed by a certain group or a minority of the population—the situation where the United States consumes nearly 25% of the global production of oil yet only has about 4% of the global population serves as a good example of this argument.

Yet Thomasson (2005) argues that resource scarcity serving as a foundation for conflict has been challenged from at least three different perspectives. The "Cornucopians" argue that natural resources are abundant and can be traded, substituted through technological innovation, recycled, or rationed through market mechanisms when there is a shortfall of the resource. Desalinization of seawater is often referenced as an example. The "curse of resources" school of thought posits that resource abundance is more important than resource scarcity in creating conflict. Actors fight over material resources, with emphasis on resources that require a moderate investment in extraction and transportation, such as diamonds, minerals, oil, biomass, and timber. The trade disputes between the United States and Canada over lumber serve as a good example. The "liberal institutionalists" offer research on international shared water resources revealing that there are a larger number of cooperative treaties than of conflicts over water. The history of international water treaties regarding surface water is robust; over 400 treaties have been inventoried with the earliest dating back to 2500 BCE following the last documented war over water in Mesopotamia along the Tigris River (McCaffrey 1997, 2001; Wolf and Giordano, 2002).

19.4 What Causes Conflict over Water?

Moore (2003) and Rothman (1997) provide a conceptual model of the causes of conflict as depicted on Figure 19.1. While Moore's (2003) "Circle of Conflict" was developed primarily for assessing disputes between individuals, it can be related to larger groups (clans, villages, towns, cities) as well between nation–states. For example, Zeitoun and Warner (2006) provide examples of structural conflicts in their assessment of "hydro-hegemony" at the river basin level in the Nile River Basin, the Jordan River Basin, and the Tigris-Euphrates River Basin.

It is clear from chronologies of conflict compiled by Gleick (2006) that many conflicts commence at a local level, eventually escalating into an international one involving more than one nation as described by Trondalen and Munasinghe (2004). They describe the traditional types of international conflicts over water as arising from:

- Incompatible goals related to access to, control over and unsustainable use of international water systems through water diversions, dams and reservoirs;
- Problems created by utilizing the international water systems such as soil salination from irrigation, a change in water flow as a result of regulation, pollution from industry using the water, sewage from cities and communities;
- Effects from other activities affecting the river systems such as eutrophication, pollution from industries that do not use the water resource in the production process, soil erosion and silting of watercourses following deforestation or overgrazing.

For the basins at risk study described by Wolf et al. (2003), they investigated the data-based conflicts by testing some of the traditional indicators for future tensions over water including, climate, water stress, population, level of development, dependence on hydropower, dams or development *per se*, and "creeping" changes such as general degradation of quality and climate change-induced hydrologic variability. Their analysis determined that these parameters are only weakly linked to disputes.

Regardless if the conflict initially appears to be an interest- or resource-based conflict, it is important to acknowledge that identity is one of the foundations for nearly all conflicts (Rothman, 1997). Rothman indicates that many conflicts are poorly diagnosed since identity conflicts are usually misrepresented as disputes over tangible resources. Rothman proposes that identity-based disputes have foundations in people's need for "…dignity, recognition, safety, control, purpose, and efficacy…" He offers that identity-based conflicts are destructive, but once identified and processed with the right approach, these disputes can evolve into creative outcomes with significant opportunities for "…dynamism and growth…". Fitzhugh and Dozier (2001) indicate that the lack of respect afforded to technical professionals working on water-related issues also led to problems in negotiating a water use dispute in the state of Vermont within the United States. And one has to look no further than the situation of "dueling experts" as described by Wade (2004) to better appreciate the intensity and destructiveness of an identity-based dispute in a water conflict.

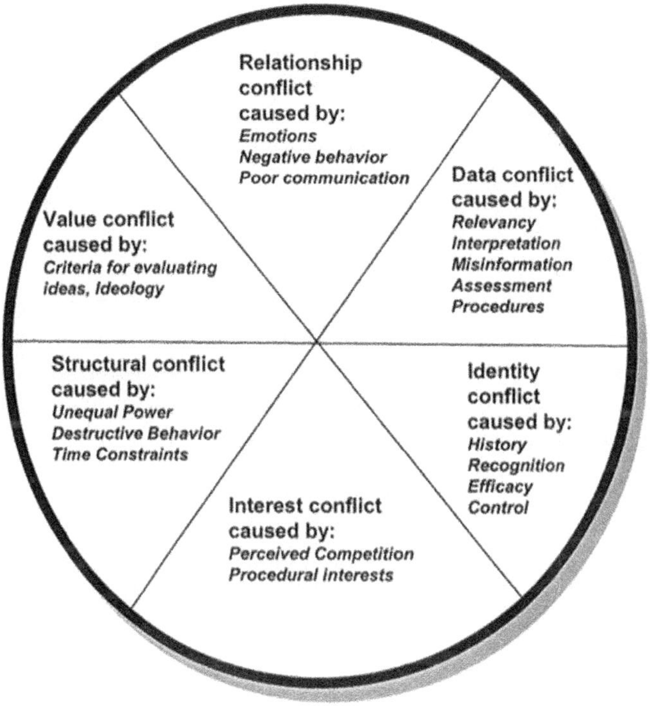

Fig. 19.1 Circle of conflict (Adapted from Moore, 2003; Rothman, 1997)

So why should water professionals care about conflict? Science is at the core of water issues because the interests and options are not easily defined without the assistance of specialists who can interpret causal chains (Renevier and Henderson, 2002). Reflecting on professional experience spanning over 20 years after publishing the landmark textbook *Groundwater* (Freeze and Cherry, 1979), hydrogeologic scholar Allen Freeze (2000) published *The Environmental Pendulum* where he tacitly implies that water professionals have both the opportunity and obligation to affect societal change. However, scientists and engineers typically do not receive training in the "soft" sciences such as conflict resolution in their technical curriculums at colleges and universities. This deficiency leads to misperceptions in the roles of the technical professionals in the discussion and solution of a wide range of societal problems, particularly those focusing on water. But this deficiency also leads to conflicts between technical professionals regarding the discussion and solution of water-related problems.

Conflicts over water require a holistic approach to address multidisciplinary and multimedia issues (Renevier and Henderson, 2002). Adler (2000) indicates that disputes over water are "often large in scale, broad in impacts, and laden with values that are at odds with each other. They are emotional to both "conscience" and "beneficiary constituents". At issue in many cases are matters of culture, economics, justice, health, risk, power, uncertainty, and professional, bureaucratic, and electoral politics. At local scales, conflicts may arise between parties because of the land–water nexus and the large investments required to purchase and develop the land while trying to weigh the value of maintaining a quality of life through open space initiatives and preserving the local water quality. In both developing and developed countries, conflicts also arise due to the plethora of beliefs surrounding the occurrence of water under the land held by the various parties.

Hydrologic professionals often develop conflicting conceptual models on how water is stored and flows in the subsurface. These models reflect the biases of the water professionals' institutional, disciplinary, and personal interests (Renevier and Henderson, 2002). Let us look at an example of a routine task in the development of conceptual models by groundwater professionals to better explore an emerging arena of value-based conflicts—the designation of boundaries for groundwater resource and user domains.

19.5 The Boundary Conundrum and Conflicts over Groundwater

Political geographers indicate that the likelihood and significance of boundary disputes over the territorial inte grity of a state and the extent of government control are greater now than at any time since the Second World War, especially with respect to transboundary movements where institutional capacity and international law are in the initial stage of formulation (Anderson, 1999). The political status and *de facto* contractual concept of boundaries between two entities sometimes serve as sources of conflict and obstacles for sharing information regarding water resources (Robertson, 2004).

Boundaries, either political or defining a resource or user domain, are obviously related to the control or the distribution of groundwater. Boundaries are used to exclude some users while at the same time providing the appropriators an opportunity to develop information and capture the benefits of organizing within the boundaries (Schlager 2004, 2007). Likewise, Blomquist and Ingram (2003) report that transboundary groundwater conflicts are often aggravated by the lack of information about the boundaries of the resource domain, resource capacity, and conditions suggestive of the water quality.

Resource domains define the fixed spatial dimensions or "boundaries" of resources (Buck, 1998). Fish stocks, for example, are natural resources found in the ocean resource domain. The spatial dimensions are used to define property rights which may be held by individuals, groups of individuals, communities, corporations, or nation-states. Rights to natural resource property are not a single right, but are rather composed of a "bundle of rights" such as rights of access, exclusion, extraction, or sale of the captured resource; the right to transfer rights between individuals, communities, corporations or nation-states; and the right of inheritance (Buck, 1998). Each "right" has an implied boundary.

Yet with the assumptions associated with the bundle of rights and implied boundaries comes the fact that the assumptions, knowledge and understandings that underlie the definition of the rights and associated boundaries are uncertain and often contested (Adams et al., 2003). For example, the question of identity and its relation to the domains of natural resources is often overlooked (Dietz et al., 2002). Choices about water resources are *value* choices that involve distinct local communities of interest (Blomquist and Schlager, 2005; emphasis added). Defining boundaries around water resource domains is "a supremely political act" because they represent different interpretations of key issues such as water quality, water quantity, nature, economics and history (Adams et al., 2003; Blomquist and Schlager, 2005). The resulting boundaries may range from the international scale, to the national, regional, local, or even the individual scale. These come from the fact that water resources are coupled with the larger reality of a region, including its environmental, social, legal, and economic characteristics. And to complicate matters even more, boundaries have no horizontal dimension: the crucial dimension of boundaries lies in the vertical plane or subsurface beneath the boundary (Anderson, 1999).

The broad spectrum of boundaries encompassing groundwater resources is diverse and fragmented. For example, Jarvis (2006) developed a typology of groundwater resource and user domains commonly mapped by groundwater professionals (Figure 19.2). The *bona-fide* boundaries reflect naturally occurring boundaries such as rivers, coastlines, and rock outcrops. Note that an "aquifer" is not listed as a resource or user domain. An emerging problem with the definition of resource and user domains in international agreements is the definition of an aquifer. The definition of an aquifer historically has relied on technical attributes of a permeable rock formation capable of transmitting and yielding usable quantities of water to a well or spring. Yet the International Law Commission (ILC) in their efforts to consider the international law applicable to transboundary aquifers defined an aquifer as "...a permeable [water-bearing] geological formation underlain by a less permeable layer and the water contained in the saturated zone of the formation". Eckstein (2004) indicates that the differences between the two definitions are significant in that the legal definition excludes the recharge and discharge areas and restricts an aquifer as only a formation that is water bearing.

Boundaries for groundwater resource domains evolve from a "static" or what may be considered a predevelopment condition, referred to as a *bona-fide* "commons" boundary, to a "dynamic" where there is a meshing of hydrology and hydraulics associated with development referred to as the *fiat* "hydrocommons" boundary. A good example is the capture area of a wellfield or the drainage areas of a qanat or mine.

The recognition of preserving groundwater resources for their social and cultural values as part of the common heritage of humankind is referred herein as the *fiat* "commons heritage" boundary. In order to protect deep confined aquifers, as well as spring waters and mineral waters used as therapeutic waters as part of a national or common heritage, de Marsily (1994) calls for the creation of "Hydrogeological Nature Reserves".

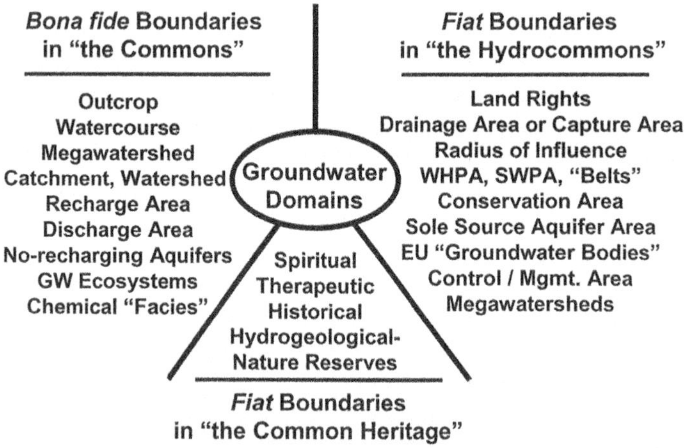

Fig. 19.2 Groundwater resource and user domain typology developed by Jarvis (2006)

Regardless of the physical setting of surface water or groundwater resources, boundaries are political in the traditional sense of the concept of a "state" or a sovereign spatial unit that defines who or what is "in" and who or what is "out", be it access to water, what can be located near water, or not using the water to preserve cultural and natural reserves (Blomquist and Schlager 2005). And while the geological and geographical areal extent of *bona-fide* groundwater resource domains do contain hydrological causes and effects, they do not necessarily include the social, economic, or other causes and effects which leads to conflict.

Clearly, boundaries matter when it comes to predicting conflicts over groundwater resources. The boundaries of groundwater resource and user domains are not only based on hydrogeology, but also based on values placed on the groundwater resources by the users. These boundaries are not clearly defined as a catchment or watershed; rather these boundaries are "shadows" on the land surface. The boundaries of the resource and user domains are transient, changing with continued stress imposed by intensive use and societal values.

19.6 Conflict Management Strategies

Given the scope of the problems and the resources available to address them, avoiding water conflict is vital. Conflict is expensive, disruptive, and interferes with efforts to relieve human suffering, reduce environmental degradation, and achieve economic growth. Developing the institutional capacity to monitor, predict, and preempt transboundary water conflicts, particularly in developing countries, is key to promoting human and environmental security in international river basins and for shared groundwater resources, regardless of the scale at which they occur.

19.6.1 Water and Conflict Resolution Strategies

It is beyond the scope of this chapter to discuss the myriad forms of conflict resolution. The reader is referred to Moore (2003) and Thomasson (2005) for theoretical and applied discussions.

Resolving conflicts over water provides unique challenges to groundwater professionals. Adler (2000) states that "Excellence in conflict resolution for water cases will derive from the way we meet the challenge of achieving powerful 'substantive' solutions to tough problems. Good process and improved relationships - the traditional measures of good mediation in other arenas, are necessary *but insufficient* for greater use of this method in water cases… In water cases, we must do better. We must be able to show outcomes that are Pareto-optimal, better than what can be achieved in litigation, better than expectations, or better than some other party-established baseline." (emphasis added). The tacit implication of Adler's statement is that it reinforces the premise that water professionals are at the core of negotiations over water as suggested by Renevier and Henderson (2002).

Effective conflict management strategies must be flexible and responsive to the primary cause of the conflict. Moore (2003) indicates that possible interventions to resolve structural conflicts include modifying the means of influence used by the parties and changing the time constraints. Adler (2000) indicates that the success of developing an administrative water code for the State of Hawaii in the United States focused on getting preliminary agreements on the scope of issues to be discussed, the procedural protocols, and the "table manners" that the parties will live by. Adler (2000) suggests that it is more important to get to "maybe" rather than getting to "yes" when discussing the upsides and downsides to all potential solutions, and once "maybe" is reached, then what follows is mutually focused thinking and productive talk where the participants begin "learning" from each other. One approach used in other environmental conflicts is a process referred to as collaborative learning.

According to Daniels and Walker (2001), collaborative learning approaches are well suited for natural resource, environmental and community decision-making situations that include (1) multiple parties, (2) multiple issues, (3) scientific and technical uncertainty, and (4) legal and jurisdictional constraints. The advantages of collaborative learning approaches to conflict management over water resources include the following:

- It is learning-based public participation;
- Stakeholders learn from one another;
- Agencies interact as stakeholders;
- Technical/scientific and traditional/local knowledge are respected;
- Public participation activities are accessible and inclusive.

Collaborative learning draws upon systems thinking, conflict management, and alternative dispute resolution (Daniels and Walker, 2001). The use of collaborative learning needs not only to focus on developing a good working relationship between the disputants, but also to begin the process of getting high quality data and information on the table. Adler (2000) indicates that it is impossible to meet this challenge for disputes over water without good process and good working relationships.

Conflicts over water underscore the importance of the evolving field of "mediator engineering scientific solutions to environmental disputes", or water professionals engaging in the practice of conflict resolution as opposed to simply serving as "experts" who are easily manipulated by legal practitioners and fuel the dispute as described by Wade (2004). If water professional are unwilling to accept a more integrated role in dispute resolution, they also risk the potential of distortions of their scientific knowledge and data during diplomatic political bargaining as described by Renevier and Henderson (2002). Adler (2000) suggests that "The real key to the next generation of programs and projects, however, will be a more avowedly self conscious philosophy of conflict resolution that is built on "mutual gains" problem solving, stronger analysis infused into the mediation process, and the ingraining of better information management in the face of contested science and scientific uncertainty. I believe the real "touchstone" for the future lies in a philosophy which seeks to enjoin science (which is all about truth-seeking) with politics (which is all about the constructive uses of power) in the service of better policymaking (which is all about the public "rules of the road").

In the book *Compass and Gyroscope*, Kai Lee (1993) calls this approach "civic science" and defines it as "...irreducibly public in the way responsibilities are exercised, intrinsically technical, and open to learning from errors and profiting from success." The outcomes of a true civic science should be environmental decisions that are at least as good, if not better than, what would happen otherwise in terms of their (1) conceptual soundness, (2) equity, (3) technical efficiency, and (4) practicability.

Tantamount to the increased role of water professionals in the process of water resources conflict resolution is patience with the process. The writer served as a volunteer within a team of experts in a community-based effort to define the boundaries of a groundwater protection plan for a fractured-rock aquifer located in the State of Wyoming located in the central United States. While nearly 2 years of meetings transpired to achieve consensus on the boundaries of the groundwater protection areas by the water professionals, the debate over the boundaries continues (City of Laramie, 2006). This example is consistent with the observations by Giordano et al. (2002) where they found as the geographic scale of the conflict drops, the likelihood and intensity of "violent" disagreement goes up leading to longer periods to mitigate the conflict.

19.6.2 Institutional Options for Conflict Management

Water management and governance are, by definition, conflict management. Management and governance of water are terms that are often used interchangeably in the literature focusing on the institutional capacity of entities in confronting the many difficult decisions that must be made regarding the use of water resources. Mukherji and Shah (2005) argue that surface water and groundwater "management" have traditionally focused on computer modeling of watersheds and aquifer systems by hydrologists and water managers to predict hydrologic responses to "stresses" imposed by the construction of dams, pipelines and irrigation systems or groundwater development to formulate and implement surface water and groundwater laws.

According to Conca (2006:8) governance of water involves "enduring, chronic, and sometimes raging controversies about local practices of resource management, conservation, and environmental protection in an increasingly transnational context". With such far-reaching issues, water governance has been defined as a holistic approach of inclusion, taking into account the concerns of water scientists and engineers, policy makers, and water users (Mukherji and Shah, 2005).

Where groundwater fits within the management of the commons, global environmental regimes, and international law remains problematic. While Kemper (2004) indicates that practical advances for groundwater management and governance are urgently needed, no "blueprint" for action exists. Finger et al.

(2006) argue that common property theory is better suited for water management at the subnational level. Many scholars indicate that laws regulating the international use of groundwater are in "embryonic" stages of development (McCaffrey, 2001), with transboundary management regimes in their infancy (Matsumoto, 2002) and often flouting the scientific principles of hydrology (Glennon, 2002).

Chasek et al. (2006) indicate that there must be "sufficient concern" within government and the public at large to develop effective global environmental regimes. The law of transboundary aquifers described in Chapter 3, coupled with the *US-Mexico Transboundary Aquifer Assessment Act* (Public Law No: 109-448) signed into law on December 22, 2006, provide indications that the threshold of sufficient concern has been reached. According to water law and policy professor and hydrogeologist Gabriel Eckstein (personal communication, 2007), the law, which had been pursued for many years, directs the United States to cooperate with the States of Arizona, California, New Mexico and Texas along the United States–Mexico border, and other appropriate entities, to systematically conduct a hydrogeological characterization, mapping, and modeling program for priority transboundary aquifers. The program is also intended to expand existing agreements between the United States Geological Survey, the four border States, relevant water resources research institutes, and appropriate United States and Mexican authorities to conduct joint scientific investigations and archive and share relevant data.

Table 19.1 provides a listing of the progressive levels of groundwater resources management at the national and international levels. With the exception of minimum legal controls over regulation of groundwater observed in countries such as India or China, the general trends in governance models either focus on an integration of groundwater as part of the surface water system as generally practiced in North America or Europe, or on the compartmentalization of groundwater as a unique hydrologic system as practiced in North Africa. It is important to note there is very little acknowledgement of shallow versus deep groundwater systems in the governance models.

The traditional approach to transboundary water management in sovereign states entered into mutually-agreed upon treaties and agreements sometimes referred to as international regimes as a means to maintain sovereignty over actions that may harm their respective environments or economies (Conca, 2006). And while this rule-based approach has a strong tradition in surface water resource agreements and treaties as described by Wolf and Giordano (2002), the success of this approach has been less than successful for groundwater resources for many reasons, including (1) the hidden nature of the resource, (2) the lack of monitoring and data collection, (3) the large uncertainty associated with the conceptual models of the groundwater resources, (4) scale mismatches, and (5) deeply rooted conflicts about authority, territory, and knowledge, all leading to a general lack of institutional capacity to accommodate groundwater management and governance.

Table 19.1 Levels of groundwater resource and governance (Adapted from Nanni et al., 2002; Jarvis, 2006)

Regulation Level	Implications	Limitations	Examples
Minimum Legal Control.	No control over groundwater abstractions.	Reduced natural discharge to ecosystems, pollution.	India, China.
Local Customary Rules.	Groundwater rights defined at local level; mechanisms for local conflict resolution.	Limited controls; no account of impacts to groundwater system, downstream users, water quality.	Pakistan, Iran.
Specific Groundwater Legislation.	Well construction and groundwater abstraction controlled.	Little consideration may be given to groundwater dependent ecosystems or water quality.	Philippines, United States.
Comprehensive Water Resources Legislation.	Surface water/groundwater subject to same regulation; both administered by same agency; water quality regulated under separate agency.	Pollution control may be deficient. Little to no recognition of shallow versus deep groundwater systems.	United States, Canada.
Fully-Integrated Water Resources Legislation.	Integrated catchment/groundwater body; emphasizes public awareness/participation; some transboundary issues recognized.	Best chance of implementing balanced and effective regulation policy. Deep aquifers identified if important to ecosystems or drinking water.	European Community.
International Agreement.	Water quality protection, allocation, recharge, extraction.	Surface water/groundwater interdependence vaguely recognized. Only one agreement in effect for groundwater.	French Prefect de Haute-Savoie & Swiss Canton of Geneva.
River Basin Organization.	Management and stakeholder involvement at river basin level.	Marginal recognition of groundwater rights in licensing arrangements.	Murray-Darling Basin, Australia.
Aquifer Management Organization.	Acknowledgement of limited interaction with surface water resources in arid areas.	Only one in effect for groundwater. No recognition of underlying aquifers.	Regional development of Nubian Sandstone in North Africa.
International Conventions.	Surface water/groundwater part of international watercourse.	Best chance of international participation. Not ratified. Deeper "confined" aquifers not covered.	Convention on the Law of the Non-navigational Uses of International Water Courses.
	Transboundary aquifers approach with integration of use and water quality protection.	ILC acknowledges importance of conceptual hydrogeologic model.	Convention on the La Transboundary Aquif

In comparison to surface water resources, the institutional capacity for groundwater is less robust with approximately 48 bilateral and multilateral treaties dating back to 1824 using the listings of treaties compiled by Teclaff and Utton (1981) and supplemented by Burchi and Mechlam (2005). Of the 109 freshwater treaties or agreements with provisions for groundwater inventoried by Matsumoto (2002), only the agreement between France and Switzerland regarding the Lake Geneva Basin groundwater is considered truly a unique and successful example of shared groundwater policy dating back to 1978 (Wohlwend, 2002; Eckstein and Eckstein, 2003; Hardberger, 2004).

Politics is about decision making, and geopolitics takes into account the geographical elements that can influence decisions (Bisson and Lehr, 2004). With river basins, states are most concerned about flows and what is dependent on flows whether it is for allocation, water quality, or ecosystems. For groundwater, Matsumoto (2002) found that the freshwater treaties and agreements focused more on the management of specific springs or wells in border areas, or within a particular geographic region such as the Southern Africa Development Community (SADC) as opposed to a specific aquifer system, how much water is stored in the aquifer systems and what "ecosystems", either natural or anthropogenic, are dependent on the change in storage within an aquifer system.

The findings by Wolf et al. (2003) that a lack of institutional capacity to absorb change within a river basin may increase the likelihood and intensity of disputes suggests that a large number of disputes may occur over groundwater resources and that there should be a way to predict "aquifers at risk" in a manner similar to the prediction of "basins at risk". Jarvis (2006) tested this hypothesis by comparing the treaties containing groundwater provisions with the transboundary aquifer system inventoried by Struckmeir et al. (2006). Of the 98 regionally important transboundary aquifer systems identified by Struckmeir et al. (2006), Jarvis (2006) determined that only 36 of these transboundary aquifers can be found within river basins that have treaties that contain provisions to groundwater. Acknowledging the importance of the different storage characteristics of the various groundwater systems identified by Struckmeir et al. (2006) as porous, fractured rock, and karst, Jarvis (2006) determined that none of the surveyed treaties recognized the differences between the types of aquifers.

Part of the problem focuses on the lack of a fundamental unit of analysis. Conca (2006:29) indicates that the building of regimes for international river basins is built around approaches used for traditional regimes including (1) a territorially-bounded construction of the problem, (2) strong presumptions of state authority, and (3) an "optimistic, universalizing, rationalist understanding of knowledge". Garduño et al. (2004) followed by Puri et al. (2005) suggest the differences between river systems and aquifers in terms of governance is that river systems are dominated by flow, whereas groundwater systems are dominated by storage. The implications are significant in that the time case of groundwater flow system is orders of magnitude slower, and for upstream–downstream considerations, neither predominates or are fixed in time and space. For the basins at risk described by Wolf et al. (2003), the confines of watershed boundaries served as the unit of analysis. During the course of developing the maps for WHYMAP, Struckmeir et al. (2004), followed by Foster et al. (2003), and Foster and Loucks (2006), recognized that (1) integrated water management either by nation or by river basin is not appropriate especially in arid areas where surface water catchments and deep aquifers are totally different, and (2) aquifers and groundwater systems are to be considered as relevant water management in regions where little to no recharge is available.

Another reason why river basins might not serve as a good metric for regime building focuses on the general lack of certainty associated with groundwater systems as opposed to surface water systems. Conca (2006:22) posits that regimes demand "definitive outcomes to the struggles over knowledge that are apparent to environmental politics" and that regimes tend not to form "when the understandings of a problem and its solution remain highly contested for an indefinite period". Similarly, Struckmeir et al. (2006) identified several transboundary aquifers that are located in coastal regions which fall outside of the metric of watersheds. Fisher (2005) reports significant quantities of freshwater underlying the oceans which may be targeted for future use. All of these observations imply that watershed boundaries are poorly suited for groundwater resource management and governance.

Likewise, the type of aquifer system in terms of permeability architecture is important given that fractured rocks and karst have poor storage characteristics. Analyses of the nuances of transboundary aquifers and political boundaries by Barberis (1991) and Eckstein and Eckstein (2003, 2005) typically address alluvial aquifers or aquifers with homogeneous hydraulic properties. Scholars addressing groundwater legislation and regulatory provisions or the dimensions of groundwater within river basin planning usually assume that the aquifers of concern are porous media and develop management and governance strategies under this limiting assumption (Nanni et al., 2002; Garduño et al., 2004). Yet given that all but unconsolidated subsurface materials are fractured to some degree, and that fractured rocks typically have unpredictable and poor storage characteristics, the modeling management of these types of aquifers as uniform or mildly nonuniform porous media is inappropriate (Moench, 2004).

Garduño et al. (2004) indicate that specific hydrogeological settings require different approaches to governance. They offer that groundwater systems of limited extent within a river basin catchment may require specific local management plans with recognition that groundwater recharge may be dependent on upstream flow, and downstream river flow may be dependent on aquifer discharge. Other special hydrogeologic settings include river basins underlain by extensive shallow groundwater systems that may be fully integrated into a water resource planning and management program. Extensive deep groundwater systems in arid regions where the groundwater system dominates and where there is little to no permanent surface water interaction makes establishing a River Basin Organization (RBO) meaningless. Groundwater systems characterized as having patchy distribution, shallow depths, and low production potential should be acknowledged as important to the socio-economic well being of a rural water system to justify attention to the optimum design of wells (Garduño et al., 2004).

Recognizing the limitation in the traditional arenas of water management, Moench et al. (2003) suggest rethinking groundwater management approaches as the intensity of development increases with increased need for food security. They recommend moving towards adaptive management strategies that acknowledge existing social trends and responses to a limited number of prioritized groundwater problem areas.

Of the many international water treaties, few have monitoring provisions, and nearly all have no enforcement mechanisms (Chalecki et al., 2002). Moench (2004) and Morris et al. (2003) indicate that more emphasis should be placed on regular and accountable monitoring of groundwater use, levels, and quality. Building upon the work of the above scholars, Jarvis (2006) also determined that an inadequate understanding or consensus of the socio-political and cultural values of groundwater leads to poor definition of the management or governance boundaries for the groundwater resources.

According to Wolf (1997) and Giordano et al. (2002), there is room for optimism, though, notably in the global community's record of resolving water-related disputes along international watercourses. For example, the record of acute conflict over international water resources is overwhelmed by the record of cooperation. Despite the tensions inherent in the international setting, riparians have shown tremendous creativity in approaching regional development, often through preventive diplomacy, and the creation of "baskets of benefits" which allow for positive-sum, integrative allocations of joint gains. Moreover, the most vehement enemies around the world either have negotiated water sharing agreements, or are in the process of doing so as of this writing, and once cooperative water regimes are established through treaty, they turn out to be impressively resilient over time, even between otherwise hostile riparians, and even as conflict is waged over other issues. Violence over water does not seem strategically rational, hydrographically effective, or economically viable. With all of the potential triggers for conflict, Llamas and Martinez-Santos (2005) report that there are no documented cases where intensive groundwater use in a medium or large-sized aquifer has caused social or economic disturbances or serious social conflicts. Shared interests along a water course seem to consistently outweigh water's conflict-inducing characteristics.

How can we learn from previous experiences in view of the continuing demand for groundwater? Despite the large storage of groundwater, especially in the deeper aquifers as inventoried by Shiklomanov and Rodda (2003), intensive use of transboundary aquifers can be expected to be inevitable. The time is ripe for water resources professionals to address this important issue. There are many contrasts between transboundary rivers and aquifers. Some of these are listed in Table 19.2 and these peculiarities need to be accounted for as existing regimes are modified or new agreements are enacted.

Table 19.2 Comparison between physical characteristics of transboundary surface water and groundwater and institutional issues (Adapted from Puri et al., 2005)

Transboundary Rivers	Transboundary Aquifers	Institutional Lessons for Aquifers
Catchment or Watersheds.	Bulk 3-dimensional systems.	"Boundaries matter". Good conceptual hydrogeologic models needed.
Use of resources generally limited to vicinity of the river channel.	Resources may be extracted from and used extensively over outcrop and subcrop.	Large diversity of users (industry, agriculture, mining, tourism, spiritual).
Replenishment always from upstream resources.	Recharge from up or downstream, or managed.	The resource planning mandate has to be wide.
Rapid and time-constrained gained from replenishment.	Recharge could be slow or none. Long-term storage.	Planning horizon must be related to the aquifer response time.
Abstraction has an immediate downstream impact.	Abstraction impact can be much slower—can be 10's to 100's of years.	Planning horizon must be related to the aquifer response time.
Allocation of flow.	Allocation of storage.	Acknowledge storage depletion.
Some data.	Few data.	Direct measurement vs. predictions needed.
Pollution transported rapidly downstream.	Slow movement of pollution.	Relate to the response time.
Little impact on upstream riparian sites.	Impacts on both upstream and downstream riparian sites.	A mandate for multinational linkage of institutions.
Pollutant transport invariably downstream, upstream source may be unaffected.	Pollutant transport controlled by local hydraulics; an operating well may induce "upstream" movement towards itself.	"Silent trade" vis-à-vis Basel Convention.

19.7 Conclusion

Conflicts over water resources are predicted to increase as the demand for freshwater increases with population growth. With the vast majority of the freshwater reserves found underground, groundwater is the world's most extracted raw material, with withdrawal rates approaching 600–700 km^3/year. The hidden nature, the lack of international law governing shared aquifers, and uncertainties associated with predicting the response of groundwater systems by intensive use aquifers invite misunderstandings leading to conflict. Likewise, access to water in drought prone regions has led to the first "War of the Well" in 2006.

Skills in conflict resolution are not part of the typical toolbox for water professionals, but the complexity of groundwater science and engineering, coupled with the complexity of the disputes over water resources dictate that groundwater professionals will have to acquire these skills as part of their professional practice. The reason is simple—groundwater professionals have much to learn about the resource beyond the mathematical, physical, chemical, and biological properties of the resource. In looking forward to avoiding future water management failures, Berndtsson et al. (2005) indicate that "...future water engineers will have to understand more of the complex interaction between water and society and how to include the human dimension in water projects in order to contribute in an active way to societal development and stability." History and values are staples of groundwater use which also dictate how boundaries are drawn to manage and govern the resource. While the existing institutional capacity for surface water resources provides a foundation for managing conflicts over groundwater resources, these tools alone are inadequate for groundwater because the physical characteristics and values of groundwater are different than surface water resources.

Acknowledgements The Institute for Water and Watersheds at Oregon State University provided the bulk of travel support to participate in the NATO Advanced Study Institute. The author acknowledges the National Science Foundation for the supplemental travel stipend. Dr. Aaron Wolf at Oregon State University is acknowledged for providing key research resources and guidance for this chapter. Professor Gabriel Eckstein of Texas Tech University is acknowledged for the information on the passage of the US-Mexico Transboundary Aquifer Assessment Act (Public Law No: 109-448).

References

Adams, W. M., Brockington, D., Dyson, J., and Vira, B., 2003, Managing tragedies: Understanding conflict over common pool resources, *Science* **302**:1915–1916.

Adler, P. S., 2000, *Water, Science, and the Search for Common Ground*, (August 10, 2007); http://www.mediate.com/articles.

Anderson, E. W., 1999, Geopolitics: International boundaries as fighting places, in: *Geopolitics, Geography, and Strategy*, C. S. Gray, and G. Sloan, eds., Frank Cass, Portland, OR.

Bakker, M., 2006, *Transboundary River Floods: Vulnerability of Continents, International River Basins and Countries*, Unpublished Ph.D. Dissertation, Oregon State University.

Barberis, J., 1991, The development of international law of transboundary groundwater, *Natural Resources Journal* **31**:167–186.

Berndtsson, R., Falkenmark, M., Lindh, G., Bahri, A., and Jinno, K., 2005, Educating the compassionate water engineer—A remedy to avoid future water management failures. *Hydrological Sciences* **50**(1):7–16.

Bisson, R. A., and Lehr, J. H., 2004, *Modern Groundwater Exploration: Discovering New Water Resources in Consolidated Rocks Using Innovative Hydrogeologic Concepts, Exploration, Aquifer Testing, and Management Methods*, Wiley Interscience, Hoboken, NJ.

Blomquist, W. A., and Ingram, H. M., 2003, Boundaries seen and unseen: Resolving transboundary groundwater problems, *Water International* **28**(2):162–169.

Blomquist, W. A., and Schlager, E., 2005, Political pitfalls of integrated watershed management, *Society and Natural Resources* **18**:101–117.

Buck, S. J., 1998, *The Global Commons: An Introduction*, Island Press, Washington, DC.

Burchi, S., and Mechlem, K., 2005, *Groundwater in International Law: Compilation of Treaties and Other Legal Instruments*, FAO Legislative Study 86, UNESCO and FAO, Rome, Italy.

Chalecki, E. L., Gleick, P. H., Larson, K. L., Pregenzer, A. L., and Wolf, A. T., 2002, Fire & water: An examination of the technologies, institutions, and social issues, in: *Arms Control and Transboundary Water-Resources Agreements*, Pacific Institute, Oakland, CA, (August 2, 2007); http://www.pacinst.org/ reports/fire_and_water/fire_and_water.pdf.

Chasek, P. S., Downie, D. L., and Welsh, J. W., 2006, *Global Environmental Politics,* Westview Press, Boulder, CO.

City of Laramie/Albany County Environmental Advisory Committee, 2006, *Laramie Regional Drinking Water Protection Plan: The Casper Aquifer Protection Plan,* (August 2, 2007); http://www.uwyo.edu/enr/enrschool/RiskAnalysis/Casper%20Aquifer%20Protection%20Plan%20-%20revised.pdf.

Conca, K., 2006, *Governing Water: Contentious Transnational Politics and Global Institution Building,* The MIT Press, Cambridge, MA.

Daniels, S. E., and Walker, G. B., 2001, *Working Through Environmental Conflict: The Collaborative Learning Approach,* Prager, Westport, CT.

de Marsily, G., 1994, Hydrogeological nature reserves? Future groundwater resources at risk. *IAHS Publication No. 22,* pp. 403–407.

Dietz, T., Dolsak, N., Ostrom, E., and Stern, P. C., 2002, The drama of the commons, in: *The Drama of the Commons,* E. Ostrom, T. Dietz, N. Dolsak, P. C. Stern, S. Stonich, and E. U. Weber, eds., National Academy Press, Washington, DC, pp. 3–35.

Eckstein, G. E., 2004, Protecting a hidden treasure: The U.N. International Law Commission and the International Law of transboundary ground water resources, *Sustainable Development Law & Policy* **V**(1):6–12.

Eckstein, G. E., and Eckstein, Y. E., 2003, A hydrogeological approach to transboundary ground water resources and international law, *American University of International Law Review* **19**:201–258.

Eckstein, Y., and Eckstein, G. E., 2005, Transboundary aquifers: Conceptual models for development of international law, *Journal of Groundwater* **43**(5):679–690.

Finger, M., Tamiotti, L., and Allouche, J., 2006, Introduction: conceptual elements, in: *The Multi-Governance of Water: Four Case Studies,* M. Finger, L. Tamiotti, and J. Allouche, eds., State University of New York Press, Albany, NY, pp. 1–42.

Fisher, A. T., 2005, Marine hydrogeology: Recent accomplishments and future opportunities, *Hydrogeology Journal* **13**(1):69–97.

Fitzhugh, J. H., and Dozier, D. P., 2001, *Finding the Common Good: Sugarbush Water Withdrawal,* (August 2, 2007); http://www.mediate.com/articles/dozier.cfm.

Foster, S., Nanni, M., Kemper, K., Garduño, H., and Tuinhof, A., 2003, *Utilization of Non-Renewable Groundwater,* Briefing Note Series No. 11, The World Bank, Washington, DC.

Foster, S., and D. P. Loucks, eds., 2006, *Non-Renewable Groundwater Resources: A Guidebook on Socially-Sustainable Management for Water Policy Makers,* IHP-VI, Series on Groundwater No. 10, United Nations Educational, Scientific and Cultural Organization, Paris.

Freeze, R. A., 2000, *The Environmental Pendulum,* University of California Press, Berkeley, CA.

Freeze, R. A., and Cherry, J. A., 1979, *Groundwater,* Prentice-Hall, Inc., Englewood Cliffs, NJ.

Giordano, M., 2003, The geography of the commons: Role of scale and space, *Annals of the Association of American Geographers* **93**(2):365–375.

Giordano, M., Giordano, M., and Wolf, A., 2002, The geography of water conflict and cooperation: Internal pressures and international manifestations, *Geographical Journal* **168**(4):293–312.

Garduño, H., Foster, S., Nanni, M., Kemper, K., Tuinhof, A., and Koudouori, P., 2004, *Groundwater Dimensions of National Water Resource and River Basin Planning. Sustainable Groundwater Management Concepts & Tools,* Briefing Note Series No. 10, The World Bank Washington, DC.

Gleick, P., 2006, Environment & security: Water conflict chronology, version 2006–2007, in: *The World's Water 2006–2007,* P. Gleick, ed., Island Press, Washington DC., pp. 189–218.

Glennon, R., 2002, *Water Follies: Groundwater Pumping and the Fate of America's Freshwaters,* Island Press, Washington, DC.

Hardberger, A., 2004, What lies beneath: Determining the necessity of international groundwater policy along the United States-Mexico border and a roadmap to an agreement, *Texas Tech Law Review* **35**(4):1212–1258.

Jarvis, W. T., 2006, *Transboundary Groundwater: Geopolitical Consequences, Commons Sense, and the Law of the Hidden Sea*, Unpublished Ph.D. Dissertation, Oregon State University.

Jurdi, M., 2002, Transboundary movement of hazardous wastes into Lebanon: Part 2. Environmental Impacts and the need for remedial actions, *Journal of Environmental Health* **64**:15–19.

Kemper, K. E., 2004, Groundwater-from development to management, *Hydrogeology Journal* **12**:3–5.

Lee, K. N., 1993, *Compass and Gyroscope: Integrating Science and Politics for the Environment*, Island Press, Washington DC.

Llamas, M. R., and Martinez-Santos, P., 2005, Intensive groundwater use: Silent revolution and potential source of social conflicts, *American Society of Civil Engineers Journal of Water Resources Planning and Management* **131**(4):337–341.

Matsumoto, K., 2002, *Transboundary Groundwater and International Law: Past Practices and Current Implications*, Unpublished Research Paper, Oregon State University.

McCaffrey, S. C., 1997, Water scarcity: Institutional and legal responses, in: *The Scarcity of Water: Emerging Legal and Policy Responses,* E. H. P. Brans et al. (eds.), London, The Hague, Boston, pp. 43–58.

McCaffrey, S. C., 2001, *The Law of International Watercourses, Non-Navigational Uses*, Oxford University Press, New York, NY.

Moench, M., Burke, J., and Moench, Y., 2003, *Rethinking the Approach to Ground water and Food Security*, FAO Water Reports 24, Food and Agriculture Organization of the United Nations, Rome, Italy.

Moench, M., 2004, Groundwater: the challenge of monitoring and management, in: *The World's Water 2004–2005*, P. Gleick, ed., Island Press, Washington DC, pp. 79–100.

Moore, C. W., 2003, *The Mediation Process*, Jossey-Bass, A Wiley Company, San Francisco, CA.

Morris, B. L., Lawrence, A. R. L., Chilton, P. J. C., Adams, B., Calow, R. C., and Klinck, B. A., 2003, *Groundwater and its Susceptibility to Degradation: A Global Assessment of the Problem and Options for Management*, Early Warning and Assessment Report Series, RS. 03-3, United Nations Environment Programme, Nairobi, Kenya.

Mukherji, A., and Shah, T., 2005, Groundwater socio-ecology and governance: A review of institutions and policies in selected countries, *Hydrogeology Journal* **13**(1):328–345.

Nanni, M., Foster, S., Dumars, C., Garduño, H., Kemper, K., and Tuinhof, A., 2002, *Groundwater Legislation and Regulatory Provision. Groundwater Dimensions of National Water Resource and River Basin Planning. Sustainable Groundwater Management Concepts & Tools*, Briefing Note Series No. 4, The World Bank, Washington, DC.

Postel, S. L., and Wolf, A. T., 2001, Dehydrating conflict, *Foreign Policy*, September/October issue.

Puri, S., Appelgren, B., Arnold, G., Aureli, A., Burchi, S., Burke, J., Margat, J., Pallas, P., and von Igel, W., 2001, *Internationally Shared (Transboundary) Aquifer Resources Management, Their Significance and Sustainable Management: A Framework Document*, IHP-VI, IHP International Hydrological Programme Non Serial Publications in Hydrology SC-2001/WS/40, UNESCO, Paris, France.

Puri, S., Gaines, L., Wolf, A., and Jarvis, T., 2005, Lessons from intensively used transboundary river basin agreements for transboundary aquifers, in: *Groundwater Intensive Use*, A. Sahuquillo, J. Capilla, L. Martinez-Cortina, and X. Sanchez-Vila, eds., International Association of Hydrogeologists Selected Papers No. 7, A.A. Balkema Publishers, Leiden, The Netherlands, pp. 137–145.

Renevier, L., and Henderson, M., 2002, Science and scientists in international environmental negotiations, in: *Transboundary Environmental Negotiation—New Approaches to Global Cooperation*, L. Susskind, W. Moomaw, and K. Gallagher, eds., Jossey-Bass, A Wiley Company, San Francisco, CA., pp. 107–129.

Robertson, K., 2004, *Design Considerations for an International Facility to Promote Cooperation Between States Sharing a Common Water Resource: A Feasibility Study on the International Water Cooperation Facility Initiative*, Unpublished M.S. thesis, UNESCO-IHE Institute for Water Education, The Netherlands.

Rothman, J., 1997, *Resolving Identity-Based Conflict in Nations, Organizations and Communities*, Jossey-Bass, A Wiley Company, San Francisco, CA.

Schlager, E., 2004, Common-pool resource theory, in: *Environmental Governance Reconsidered: Challenges, Choices, and Opportunities*, R. F. Durant, D. J. Fiorino, R. O'Leary, eds., MIT, Cambridge, MA, pp. 145–175.

Schlager, E., 2007, Community management of groundwater, in: *The Agricultural Groundwater Revolution Opportunities and Threats to Development*, M. Giordano, and K. Villhoth, eds., CABI, Wallingford, Oxon, UK.

Shiklomanov, I. A., and Rodda, J. C., eds., 2003, *World Water Resources at the Beginning of the 21st century*, UNESCO International Hydrology Series, Cambridge University Press, Cambridge, UK.

Struckmeir, W. F., Gilbrich, W. H., Richts, A., and Zaepke, M., 2004, *WHYMAP and the Groundwater Resources of the World Map*, Special Edition from World-wide Hydrogeological Mapping and Assessment Programme (WHYMAP) for the 32nd International Geological Congress, Scale 1:50,000,000, BGR/UNESCO, Paris, France and Hanover, Germany.

Struckmeir, W., Richts, A., Acworth, I., Arduino, G., Bocanegra, E., Cunningham, W., Droubi, A., de Franca, N., Gilbrich, W., Girman, J., van der Gun, J., Klingbeil, R., Margat, J., Poitrinal, D., Puri, S., Rivera, A., Safar-Zitoun, M., Tujchneider, O., Vasak, S., Vrba, J., Winter, P., Zaepke, M., Zaisheng, H., and Zekster, I., 2006, *WHYMAP and the World Map of Transboundary Aquifer Systems*, Special Edition from World-wide Hydrogeological Mapping and Assessment Programme (WHYMAP) for the 4th World Water Forum., Scale 1:50,000,000, BGR/UNESCO, Paris, France and Hanover, Germany.

Teclaff, L. A., and Utton, A. E., eds., 1981, *International Groundwater Law*, Oceana Publishers, Inc., New York, NY.

Thomasson, F., 2005, *Local Conflict and Water: Addressing Conflicts in Water Projects*, Swedish Water House, (August 2, 2007); http://www.swedishwaterhouse.se/images/partners/20051017114417Conflicts_Water_Projects_050823.pdf.

Trondalen, J. M., and M. Munasinghe, 2004, *Water and Ethics: Ethics and Water Resources Conflicts*, Series on Water and Ethics, Essay 12, United Nations Educational, Scientific and Cultural Organization, Paris.

Wade, J. H., 2004, Dueling experts in mediation and negotiation: How to respond when eager expensive entrenched expert egos escalate enmity, *Conflict Resolution Quarterly* **21**(4): 419–436.

Wallensteen, P., 2002, *Understanding Conflict Resolution: War, Peace and the Global System*, Sage, London.

Wax, E, 2006, Dying for water in Somalia's drought, *The Washington Post*, Washington, DC. April 14.

Wohlwend, B. J., 2002, *Workshop on Harmonization of Developing Interests in the Use of Shared Water Resources, an Overview of Groundwater in International Law, a Case Study: The Franco-Swiss Genovese Aquifer*; (August 2, 2007); http://www.bjwconsult.com?The%20Genevese%20Aquifer.pdf.

Wolf, A., 1998, Conflict and cooperation along international waterways, *Water Policy* **1**(2): 251–265.

Wolf, A. T., Natharius, J., Danielson, J., Ward, B. S., and Pender, J. K, 1999, International river basins of the World, *International Journal of Water Resources Development* **15**(4): 387–427.

Wolf, A. T., and Giordano, M. A., 2002, *Atlas of International Freshwater Agreements*, Early Warning and Assessment Report Series, RS. 02-4, United Nations Environment Programme, Nairobi, Kenya.

Wolf, A. T., Yoffe, S. B., and Giordano, M., 2003, International waters: Identifying basins at risk, *Water Policy* **5**:29–60.

Zeitoun, M., and Warner, J., 2006, Hydro-hegemony—a framework for analysis of trans-boundary water conflicts, *Water Policy* **8**:435–460

Zekster, I. S., and Everett, L. G., eds., 2004, *Groundwater resources of the world and their use*, IHP-VI, Series on Groundwater No. 6, UNESCO.

Chapter 20
Hydropolitics and Hydroeconomics of Shared Groundwater Resources: Experience in Arid and Semi-arid Regions

Manuel R. Llamas Madurga*, **Pedro Martínez-Santos**, **África de La Hera**

Abstract Interest on shared or transboundary aquifers management is relatively recent, but has already been useful in order to increase the awareness about the role of groundwater among politicians and high-level water decision-makers. However, the process to achieve some kind of international convention or agreement on this issue will probably take a long time. Implementing international agreements on shared aquifers in arid and semi-arid regions will be difficult if the current chaos in groundwater management is not previously corrected. In this regard, this chapter analyses the causes of this situation, and the concerns these raise in regard to sustainability. Later, the chapter considers the political and economics factors that affect groundwater development, as well as the social and legal aspects involved. Finally, the chapter reproduces the call for actions included in the Alicante Declaration, approved in January 2006 on occasion of the International Symposium on Groundwater Sustainability (Ragone et al. 2007).

Keywords Groundwater overexploitation, groundwater sustainable development, groundwater intensive use silent revolution, groundwater economics, groundwater management institutions, groundwater conflicts

Manuel R. Llamas Madurga
Royal Academy of Sciences, Valverde 22. 28034 Madrid, Spain

Pedro Martínez-Santos
Universidad Complutense de Madrid, Jose Antonio Novais s/n. 28040 Madrid, Spain

África de La Hera
Geological Survey of Spain, Ríos Rosas 23. 28003 Madrid, Spain

* To whom correspondence should be addressed. E-mail: mrllamas@geo.ucm.es

C.J.G. Darnault (ed.), *Overexploitation and Contamination of Shared Groundwater Resources*.

20.1 Introduction

In recent years, transboundary aquifer management has begun to receive much deserved attention, partly thanks to initiatives such as UNESCO's, which have been successful in raising awareness among high-level water decision-makers.

However, transboundary water management is still under discussion for many surface water bodies, let alone aquifers, which, with few exceptions have traditionally been the "Cinderella" of water resources management. Thus, the process to achieve an international convention on groundwater management is likely to still take a number of years. Provided, that is, that the current usual administrative and legal chaos in groundwater management in most arid and semi-arid regions can be corrected.

While inventories of transboundary aquifers already exist and thoughtful analyses of the implications of transboundary aquifer management have been carried out in the recent past (Puri and El Naser, 2003), the practical difficulties of groundwater management have perhaps received less attention. UNESCO expert Dr. A. Aureli who has been involved in this issue from its inception presents in this Advanced Study Institute (ASI) an updated outlook of the world situation on shared aquifers.

Groundwater is estimated to provide 50% of the world's drinking water supplies, and accounts for almost 100% in countries like Austria or Denmark. Nevertheless, it is in the field of irrigation where groundwater pumping has experienced a spectacular increase in the last decades. It cannot be forgotten that irrigation amounts to 70% of the world's water uses (reaching over 90% in many arid and semi-arid regions). Today, the volume pumped from the world's aquifers for this purpose probably amounts to over 1,000 km^3/year.

Particularly in the arid and semi-arid regions of the world, groundwater management is far from adequate. Uncontrolled development is widespread, thus raising concerns as to the viability of the currently prevailing groundwater management paradigms. This is largely a consequence of the "silent revolution" of intensive groundwater use (Llamas and Martínez-Santos, 2005): the widespread availability of drilling and pumping techniques, together with the reliability of groundwater supplies during dry periods and their availability on demand, remove a significant uncertainty from farmers minds. Thus, despite the fact that groundwater is usually more expensive than subsidized surface water, pumping costs are generally a very small fraction of the guaranteed crop value.

Therefore, it can be said that the spectacular growth of intensive groundwater use for irrigation is mostly driven by economic reasons, and has provided great benefits to millions of mostly modest farmers, facilitating a relatively rapid social transition from illiterate farming communities to a middle class societies.

However, some adverse effects have also occurred, such as groundwater quality degradation (Vrba, 2003) or ecological impacts on aquatic ecosystems. While the significance of these effects varies depending on the outlook, these have frequently been exaggerated, thus giving rise to the hydromyth of the general fragility or unsustainability of groundwater resources development.

In any case, farmers usually pay little attention to such ideas and continue to drill and pump, often with scarce control on the part of water authorities. These have traditionally suffered from what some authors term "hydroschizophrenia", that is, placing the emphasis of water resources management on surface water infrastructures while largely neglecting the role of groundwater. This is the main cause behind the current disarray in groundwater resources management. As long as this situation continues, potential international regulations in regard to transboundary aquifer management are unlikely to be enforced.

20.2 Scope and Aim

Given the above, hydropolitics and hydroeconomics of transboundary aquifers are heavily conditioned by the need for adequate management, which in turn tends to raise sustainability concerns. This chapter focuses on analyzing the relevance of the "silent revolution" of intensive groundwater use, a relatively recent paradigm shift in groundwater resources development that cannot be ignored, discussing its sustainability in view of the current situation.

This chapter refers almost exclusively to irrigation. As mentioned before, the reason is that despite the undeniable importance of urban supply, irrigation usually accounts for over 90% of all the consumptive uses in arid and semi-arid countries. As a consequence, potential conflicts about groundwater in shared, or not shared, aquifers are likely to be related to irrigation uses.

20.3 Hydroeconomics of Groundwater Resources in Arid and Semi-arid Regions: The "Silent Revolution" of Intensive Groundwater Use for Irrigation

Perhaps the single most relevant factor in regard to transboundary groundwater management is the economic value generated by the "silent revolution" of intensive groundwater use (Briscoe 2005; Fornés et al., 2005; Llamas and Martinez-Santos, 2005), which will be described in the following paragraphs. This new paradigm has given rise to management difficulties which need to be resolved in order to achieve adequate groundwater management.

Intensive groundwater use for irrigation has been carried out mostly by the personal initiative of millions of modest farmers in pursuit of the significant short-term benefits groundwater usually triggers. These farmers have frequently received incentives as soft loans or energy subsidies from governmental agriculture departments. In contrast, governmental water agencies have been mainly concerned with the planning, operation, maintenance and control of surface water irrigation systems, while frequently paying less attention to groundwater development. This attitude has been commonplace in India, Mexico, Spain, and many other arid and semi-arid regions worldwide. As a consequence, the current situation concerning groundwater development has been described as a "colossal anarchy" not only in South Asia (Shah et al., 2006, Deb Roy and Shah, 2003), but also in many other arid and semi-arid regions worldwide.

Figure 20.1 presents a qualitative overview of the different water policy stages induced by intensive groundwater use in arid and semi-arid countries. Each of the five stages is roughly equivalent to one generation (about 15–25 years). While the beginning of these can be traced back to the moment when intensive groundwater development begins, their end point might not be so easily identified. Thus, some overlapping between stages may occur: take for instance hydroschizophrenic attitudes, which may still persist in many countries.

Science and technology have played a key role in the "silent revolution", since the advances in hydrogeology and well-drilling techniques, and the popularization of the submersible pump, have significantly reduced abstraction costs over time. The total direct cost of groundwater abstraction today—not taking into account externalities—is, in most cases, only a small fraction of the economic value of the guaranteed crop. Thus, the "silent revolution" is largely a market driven phenomenon, although in very poor rural areas it is mainly driven by a subsistence livelihood effort (Polak, 2005a,b).

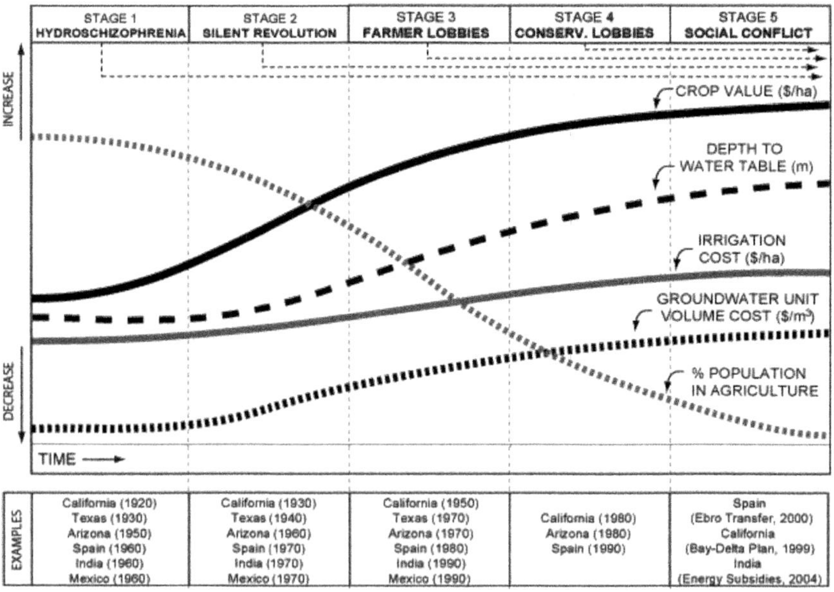

Fig. 20.1 General groundwater-related trends in arid and semi-arid regions (after Llamas and Martínez-Santos, 2005)

Surface water is heavily subsidized in most countries, and therefore its price for irrigation is generally cheaper than groundwater's. Yet, many farmers prefer groundwater. Several motives exist for this seemingly sub-optimal choice: one is that groundwater can be obtained individually, thus by-passing negotiations with other farmers and government officers, often an arduous task. A second and more important motive is the resilience of aquifers to dry periods. In this regard, most farmers resort to conjunctive use when possible, using subsidized surface water whenever available and groundwater whenever surface supplies fail. Many irrigated cash crops, which usually require large investments from farmers, depend today on groundwater, either totally or on conjunctive use with surface water. Garrido et al. (2006) show how Spanish irrigation is a typical case of this situation.

Another important and seldom mentioned benefit of the "silent revolution" is its positive effect on the social and economic transition of many farmers. Relatively low pumping costs, and the protection groundwater provides against drought, have allowed poor farmers to gradually progress into a middle class status, enabling them to provide a better education for their children. After one or two generations, those children have been trained as teachers, technicians and so forth, thus contributing to the overall progress of society. At the same time, those who choose to continue as farmers are in a position to use better agricultural technologies to grow cash crops that demand less water (Moench, 2003). During the twentieth century and almost in every country, significant changes in the labour force have occurred. Generally the population share in the agrarian sector has decreased dramatically as a consequence of technological advances. For example, during the last half century the proportion of the Spanish labour force in the agrarian sector has decreased from about 50% to less than 6%. Groundwater irrigation has been an excellent catalyst for such social and economic transition.

Perhaps one of the most significant aspects of the "silent revolution" is the manner in which farmers, as they become richer and more educated, move from low value crops to cash crops. This is mainly due to the intrinsic reliability of groundwater: encouraged by the expectation of enhanced revenues, farmers invest in better irrigation technology and, in turn, shift to higher value crops. As crop value is related to crop type, climatic and other natural and social variables of each site, and subject to trade constraints, it ranges widely: in Europe for instance, between US$ 500 per hectare (e.g. cereals) and more than US$ 60,000 per hectare for tomatoes, cucumber and other greenhouse crops. Frequently, the ratio between crop value and groundwater irrigation cost is greater than 20 (Llamas and Custodio, 2003). However, in aquifers with low permeability and storativity located in densely populated areas, this ratio can be substantially smaller (even if energy is heavily subsidized). This appears to be the case in some very poor regions of India, mainly located on hardrock (poor) aquifers.

Despite the illusory accuracy of global irrigation data and the variability of the existing estimates, rough calculations yield the following conclusion: groundwater-based irrigation seems to be twice as efficient as surface water irrigation in hydrological terms (m^3/ha), a ratio that increases to between three and ten from the social and economic points of view (US$/$m^3$ and jobs/m^3). Regional scale analyses carried out in Spain seem to confirm these figures (Hernandez-Mora et al., 2001; Vives, 2003). Thus, it appears relevant and urgent to assess the comparative hydrological and socio-economic efficiency of surface and groundwater irrigation at a world scale. Assessing the implications of this "silent revolution" should constitute a valuable contribution to the debate about global irrigation needs as perceived by many water experts. The "more crops and jobs per drop" motto has been considered crucial in order to avoid a "looming water crisis". This is because of the large share of irrigation in global water use, and irrigation's often low efficiency. However, few water experts or decision-makers are aware that the goal behind such motto is now often achieved by groundwater irrigation. Really, in arid and semi-arid regions in industrialized or rich countries the new motto is "more cash and nature per drop".

The required investment to assess the value and efficiency of groundwater versus surface water irrigation can be afforded by most governments, and the same holds for the promotion of hydrogeological education campaigns. For many countries, the cost of such undertakings would probably be in most cases only a small fraction of the amount invested every year in the construction of conventional hydraulic infrastructures. In many cases, the issue might be more of an ethical nature, related to the lack of political willingness to fight ignorance, arrogance, institutional inertia or corruption (Delli Priscoli and Llamas, 2001). In this regard, it is pertinent to note that groundwater development is less prone to corruption that traditional surface water irrigation systems, due to the smaller investment and shorter timeframe required for the implementation of groundwater supplies. Perhaps in some regions, this lesser susceptibility to corruption, and not the lack of economic means, may explain the lack of political willingness to truly assess the value of groundwater development (Valencia Statement, 2005).

20.3.1 Potential Problems in Intensive Groundwater Use

Intensive groundwater use is not a panacea that will necessarily solve the world's water problems (Mukherji, 2006). In fact, should the prevailing anarchy continue, serious problems may appear in the mid or long-term (two to three generations). Some are already well documented, although at a lesser scale, and are usually related to water table depletion, groundwater quality degradation, land subsidence or ecological impacts on aquatic ecosystems.

Intensive groundwater use frequently depletes the water table. Drawdowns in the order of 0.5 m/year are frequent, although rates up to 5–10 m/year have been reported (Llamas and Custodio, 2003). Farmers are seldom concerned by this issue, except in the case of shallow aquifers: the increase in pumping costs is usually a small problem in comparison with potential groundwater-quality degradation or equity issues such as the drying up of shallow wells owned by the less resourceful farmers. The opposite phenolmenon (rise of the water table due to surface water over-irrigation) is also a problem for example in Punjab, India, and Pakistan or in San Joaquin valley in California. Rising of the water table often results in significant social and economic troubles due to soil water logging and/or salinization.

Groundwater unit volume cost (Figure 20.1) increases with depth to groundwater, as more energy is required for pumping and deeper wells might be needed. These costs usually range between US$ 0.02 and US$ 0.20/m^3 (Figure 20.2) depending on the country and the aquifer. Groundwater irrigation cost per hectare also increases with time, albeit at a lower rate. This is because farmers begin to use a more efficient technology and switch (if soil and climate allow) to less water consuming crops: from maize or rice to grapes or olive trees, for instance. It is estimated that groundwater irrigation cost generally ranges between US$ 20 and US$ 1,000 per hectare and year.

Documented cases where intensive groundwater use in a medium-sized or large "good" aquifers (those with a surface larger than 500 km^2 and medium to high transmisivity and storage capacity values) has already caused social or economic disturbances are practically unknown; at least not in the degree of magnitude of those caused by soil water-logging and salinization (India, Pakistan or California) or the serious social conflicts in relation to people displaced or ousted by the construction of large dams (Briscoe, 2005; Shah et al., 2006).

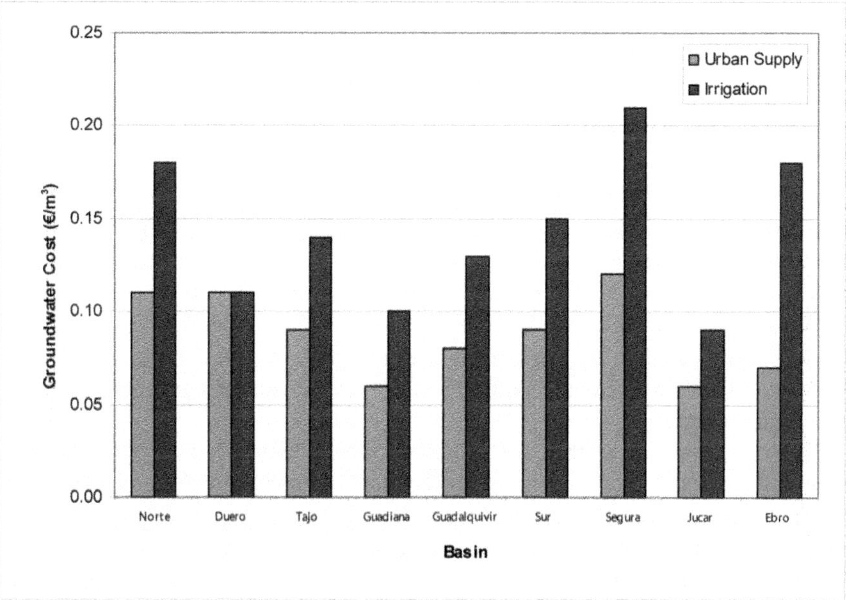

Fig. 20.2 Unit volume cost of groundwater in Spain's administrative water basins (after Llamas and Martínez-Santos, 2005)

Most aquifers present a large storage volume in relation to their renewable resources (often two or three orders of magnitude). A practical consequence is that the potential problems mentioned above do not usually become serious in the short term (within one or two generations). Besides, the social transition triggered by groundwater together with the implementation of more efficient irrigation technologies can often result in a sustainable use in the mid-term. The general reduction of global poverty and the transfer of population from agricultural sector to other more productive sectors (Sachs 2005, Sala-Martin, 2006) also are contribution to this transition. However, adequate groundwater management and governance remains an important challenge to ensure long-term sustainability. To this effect, the education of stakeholders and widespread presence of groundwater user associations is crucial for an adequate participatory bottom–up management approach.

The reality is that even some poor aquifers, such as the Indian "hard rock aquifers", have played a key role in increasing food production. In India groundwater irrigated surface has increased in more than 40 million hectares during the last decades (Shah et al, 2006; Deb Roy and Shah, 2003). As a consequence, India, despite a 100% increase of its population in the last 50 years, has not only achieved food security in practice, but also become an important grain exporter.

However, uncontrolled aquifer development in arid and semi-arid regions worldwide raises sustainability concerns, particularly whenever the natural rate of recharge is low.

20.4 Groundwater Sustainability Versus Aquifer Overexploitation

The concept "sustainable development" was first coined in the 1980s, and has been expressed in a variety of ways over the years. Rogers (2006), for instance, quotes the existence of 50 widely used definitions. Perhaps the better-known (and widely contested) meaning of sustainability was given by the United Nation's Commission on Sustainable Development in 1987: "*to satisfy current needs without compromising the needs of future generations*". In a more recent book, Rogers et al. (2006) present a thorough study on the general concepts of sustainable development.

Thus, it seems clear that sustainability means different things to different people. A reason for this is the multi-dimensional nature of the concept. There may be as many as ten different aspects to be considered in assessing whether a given development can be labelled sustainable (Shamir, 2000). However, even if all these are taken into account, it may not be so easy to reach a univocal conclusion, that is, the different dimensions of sustainable development may at times clash.

Let us take a look at an example. At a given aquifer, pumping rates for irrigation may prove "sustainable" from the hydrological viewpoint (provided that storage and/or average recharge are large enough). However, water table drawdowns may induce degradation of valuable groundwater-dependent ecosystems such as wetlands, which may be considered unsustainable from the ecological point of view. Would a restrain from pumping be the most "sustainable" course of action?

The answer to this question is difficult. If farmer livelihoods rely heavily on groundwater resources, a ruthless push towards wetland restoration may not be the most sensible solution to the problem. In that case, like in many real life situations, the social and economic aspects of sustainability come into play, and may eventually offset environmental conscience.

Llamas et al. (2006a) provide a succinct overview of nine different aspects of groundwater sustainability: hydrological, ecological, economic, social, legal, institutional, inter and intra-generational and political. Throughout that text, a distinction is often made between developed and developing regions. This is because perceptions as to what is sustainable vary across geographical boundaries, and are often rooted on cultural, political aspects and socio-economic situations. In this regard, the *Hydrogeology Journal* theme issue of March 2006 presents the socio-economic analyses of a dozen of case studies from all over the world (Llamas et al., 2006b).

Whenever adverse effects of groundwater development begin to be felt, it is common to hear about overexploitation, a term usually equated to pumping in excess of the recharge. While this practice is often dismissed as "unsustainable", the concept of overexploitation is conceptually complex. This is the reason why a significant number of authors consider it simplistic and potentially misleading (Selborne, 2001; Delli Priscoli and Llamas, 2001; Delli Priscoli et al., 2004, Abderrahman, 2003; Llamas, 2004). Probably the most complete analysis is the one by Custodio (2002). As a consequence, more and more authors are changing to the expression "intensive use of groundwater" instead of using "groundwater overexploitation".

Intensive groundwater use is that which induces significant changes on natural aquifer dynamics (Llamas and Custodio, 2003, Custodio et al., 2005). In contrast with "aquifer overexploitation", "intensive groundwater use" does not convey a positive or negative connotation. It merely refers to a change in flow patterns, groundwater quality or interrelations with surface water bodies.

As explained before, such changes may be perceived as beneficial or detrimental, and this perception may also vary with time. For instance, until the mid twentieth century, wetlands were considered barren land and a potential source of disease. Many a decision was made to desiccate groundwater-dependent wetlands by depleting the water table for, back then, that was perceived as a service to society. With the advent of the environmental movement and the advances in medical sciences, wetlands ceased to be wastelands to become ecological sanctuaries, to the point that nowadays advocates of wetland desiccation are generally frowned upon.

20.5 Social, Legal and Hydropolitical Issues in Intensively Exploited Aquifers

In the preface of the last *Hydrogeology Journal* theme issue, Llamas et al. (2006b) warn about the scarcity of analyses on the social sciences aspects about groundwater role in the general water resource policy. As a matter of fact, that theme issue tries to set the pace to increase that type of studies.

The following sections are mainly taken from the paper by Llamas et al. (2006a) presented in the recent International Symposium on Groundwater Sustainability (Alicante, Spain, 23–27 January, 2006).

20.5.1 Social Sustainability

As mentioned before, groundwater irrigation has proven an excellent catalyst for the positive social transition of farmers in arid and semi-arid regions worldwide (Moench, 2003, Steenberger and Shah, 2003). This is largely a consequence of groundwater's resilience against drought. Secured access to water during dry periods removes a significant perception from farmers' minds. These are thus encouraged to invest in new technologies, both from the agricultural (selective seeds, agrochemicals) and the technical point of view (drip irrigation). Increased revenues result, and allow for a greater degree of social welfare. In addition, farmers become able to provide a better education for their children, who may either move on to other economic sectors or return to agriculture with a more productive outlook.

20.5.2 Legal Issues

From the legal viewpoint, transboundary aquifers present two main issues of concern. The first one relates to whether groundwater resources should be public or private property. The second refers to the way groundwater rights should be inventoried and to whether the possibility should be allowed to trade with them. This second aspect, usually equated with "water banks" is perhaps subordinated to the first in terms of importance, even if significant informal markets already exist in some places (Mukherji, 2006).

In relation to property rights, groundwater is usually public and can be accessed by means of governmental permits (sometimes called "concessions"). This is the case in Israel, a number of states of the United States, Mexico and many other countries. In other places, like California, Chile, India or Texas, groundwater is under private ownership.

Spain presents a particularly interesting example of a mixed system. Wells drilled after January 1, 1986 require governmental permission, while those operational before 1986 remain private. Private groundwater may remain so for 50 years (provided that the well-owner reaches an agreement with the government in exchange for "administrative protection") or perpetually (if the owner wishes to preserve his/her rights under the 1879 Water Act).

In any case, the Spanish situation is far more complex due to the lack of a reliable registry of groundwater rights. While the government is currently carrying out a series of remedial initiatives, these are insufficient in the eyes of some authors. Fornes et al. (2005) for instance point out that these ignore a significant share of existing wells, and that the registry or inventory is therefore incomplete.

While some voices seem to disagree (Blomquist, 1992), the current situation may be considered unsustainable in the long run, particularly if a strong political willingness to apply the laws is lacking. It seems clear that a reliable inventory of groundwater rights is desirable in order to ensure adequate management, whether it be transboundary or not.

Given the "polycentric" nature of groundwater development, a bottom–up approach seems the best way to achieve adequate management. Surface water irrigation communities constitute a good example. Seven thousand of such communities (some of them centuries old) currently exist in Spain (Murcia, Valencia and Alicante being the better known ones) (Mass and Andersen, 1978).

However, there is an essential difference between surface and groundwater. A gatekeeper may ultimately control surface water, while groundwater is usually subject to the individual decisions of hundreds (perhaps thousands) of independent users with direct access to the resource. Thus, top–down control has proven insufficient in most places due to this intrinsic complexity of groundwater governance. This is the reason why user communities are often advocated as the most plausible solution to ensure adequate groundwater resources management.

Groundwater user associations are still fairly scarce. Under Spain's 1985 Water Act, an attempt was made to impose these communities in "overexploited" aquifers, although this initiative has been far from successful in most places (Hernandez-Mora and Llamas, 2000, Lopez-Gunn, 2003; Lopez-Gunn and Martinez-Cortina, 2006). Water agencies in Texas and California are currently trying to organize these communities, albeit by means of economic incentives rather than by compulsion (Kretsinger and Narasimhan, 2006).

In any case, since groundwater user associations are a relatively new feature, their ultimate implications on groundwater management are yet to be seen (Schlager and Lopez-Gunn, 2006, Lopez-Gunn and Martinez-Cortina, 2006).

20.5.3 Hydropolitical Issues

Politics has at times been defined as "the art of that which is possible" (rather than "that which is reasonable"). Although in modern democratic societies decision-making is ultimately restricted to politicians, these are often influenced by more or less powerful lobbies. These usually defend the interests of large corporations or different sectors of the population (unions, NGOs and others). An excellent summary of the main factors in water governance can be read in Rogers (2006).

The motivations behind political decisions are so difficult to take into account that they are generally overlooked. In addition, they depend very heavily on social and cultural constrains, which are very different from country to country. Therefore this section is restricted to three brief examples as to how politics may come into play in regard to groundwater sustainability.

The first example refers to the 2005 events of the Upper Guadiana basin. The Guadiana water authority (dependent on Spain's central administration) issued an order to shut down a series of wells. While law-in-hand this seemed an appropriate course of action, a social uproar ensued, fuelled mostly by farmer unions. This led the regional government to oppose Madrid's orders. Up to date, the central water authority has been unable to shut down the wells.

A second case is described by Mukherji (2006). In 2004, the ruling political party of Andhra Pradesh (central India) stated that they would gradually stop electricity subsidies for pumping. This led to a significant resistance on the part of farmers. Seemingly as a result, the opposition won the 2004 election largely on the strength of opposing this measure. Electricity remains mostly free to this day.

Finally, the third example refers to California. In 2002, and after a long and arduous work, Prof. Sacks (University of California, Berkeley) developed a law to replace the old water act. This effort was motivated by the fact that the old law was conceptually obsolete since, among other erroneous assumptions, it practically ignored the unity of the hydrological cycle and equated groundwater to underground rivers. However, frontal opposition on the part of farmers and urban supply companies eventually caused the project to be rejected and the obsolete code to remain. However, as Kretsinger and Narasimhan (2006) describe, the Government of California has abandoned the "command and control" approach while implementing a policy of education and economic incentives whose results seem encouraging.

These three examples show how political constrains (namely voters) may lead to potentially unsustainable situations. Education of the general public is perhaps the only means to avoid these kinds of occurrences in the future. In the case of transboundary aquifers, this is may be a particularly relevant issue, since integrated political actions are required on both sides of the border.

20.6 Conclusion

Transboundary aquifer management initiatives are still recent, and likely to take a number of years before being fully operational. In arid and semi-arid regions, the current situation of groundwater management is generally far from adequate, which raises concerns as to the viability of transboundary management schemes in the long run.

While political motivations are highly volatile, the economic importance of groundwater resources, particularly for irrigation, suggest the need for urgent action to ensure these are managed first within each country's borders. Though not exempt of practical difficulties, a participative effort based around groundwater user associations appears as the most plausible way of attaining adequate management in the future. The Alicante Declaration (Ragone and Llamas, 2006) is included as an annex to this chapter, and constitutes call for action in regard to these practical issues.

References

Abderrahman, W., 2003, Should intensive use of no-renewable groundwater always be rejected? in: *Intensive Use of Groundwater*, R. Llamas and E. Custodio, eds., Balkema Publishers, The Netherlands, pp. 191–203.

Blomquist, W., 1992, *Dividing the Waters. Governing Groundwater in Southern California*, ICS Press, San Francisco, USA.

Briscoe, J., 2005, *India's Water Economy: Bracing for a Turbulent Future*, The World Bank, November 28, 2005, 95 pp.

Custodio, E., 2002, Aquifer overexploitation: What does it mean? *Hydrogeology Journal* **10**(2):254–277.

Custodio, E., Kretsinger, V., and Llamas, M. R., 2005. Intensive development of groundwater: Concept, facts and suggestions, *Water Policy* **7**:151–162.

Deb Roy, A., and Shah, T., 2003, Socio-ecology of groundwater irrigations in India, in: *Intensive Use of Groundwater*, R. Llamas and E. Custodio, eds., Balkema Publishers, The Netherlands, pp. 307–336.

Delli Priscoli, J., and Llamas, M. R., 2001, International perspective in ethical dilemmas in the water industry, in: *Navigating Rough Waters*, C. K. Davis and R. E. McGinn, eds., American Water Works Association, Denver, Colorado, pp. 41–64.

Delli Priscoli, J., Dooge, J., and Llamas, M. R., 2004, Overview, *Series on Water and Ethics*, Essay 1, UNESCO, Paris.

Fornés, J. M., Hera, A. de la., Llamas, M. R., 2005, The silent revolution in groundwater intensive use and its influence in Spain, *Water Policy* 7(3):253–268.

Garrido, A., Martínez-Santos, P., and Llamas, M. R., 2006, Groundwater irrigation and its implications for water policy in semiarid countries: The Spanish experience, *Hydrogeology Journal* **14**(3):340–349.

Hernández-Mora, N., and Llamas, M. R., 2000, The role of user groups in Spain: Participation and conflict in groundwater management, *CD-ROM of the X World Water Congress (Melbourne, 12–16 march 2000)*, International Association of Water Resources.

Hernández-Mora, N., Llamas, M. R., and Martínez-Cortina, L., 2001, Misconceptions in aquifer over-exploitation. Implications for water policy in southern Europe, in: *Agricultural Use of Groundwater. Towards Integration between Agricultural Policy and Water Resources Management,* C. Dosi, ed., Kluwer Academic Publishers, pp. 107–125.

Kretsinger, V., and Narasimhan, T. N., 2006, California's evolution toward integrated regional water management: A long-term view, *Hydrogeology Journal* **14**(3):407–423.

Llamas, M. R., 2004, Use of groundwater, Series on *Water and Ethics*, Essay 7, UNESCO, Paris, pp. 1–33.

Llamas, M. R., and Custodio, E., eds., 2003, *Intensive Use of Groundwater: Challenges and Opportunities*, Balkema Publishers, Dordrecht.

Llamas, M. R., and Martínez-Santos, P., 2005, Intensive groundwater use: Silent revolution and potential source of social conflicts, *Journal of Water Resources Planning and Management,* American Society of Civil Engineers, September/October, 337–341.

Llamas, M. R., Martinez-Santos, P., and Hera, A. de la., 2006a, Dimensions of sustainability in regard to groundwater resources development: An overview, *Proceedings of the International Symposium on Groundwater Sustainability* (Alicante, Spain 24–27,January, 2006), CD-ROM.

Llamas, M. R., Shah, T., and Mukherji, A., 2006b, Preface, *Hydrogeology Journal* **14**(3):269–274.

López-Gunn, E., 2003, The role of collective action in water governance: A comparative stay of groundwater uses association in La Mancha aquifer in Spain, *Water International* **28**(3):367–378.

López-Gunn, E., and Martínez-Cortina, L., 2006, Is self-regulation a myth case study on Spanish groundwater association and the role of higher-level authorities, *Hydrogeology Journal* **14**(3):361–379.

Mass, A., and Andersen, R. L., 1978, *And the Desert Shall Rejoice: Conflict, Groundwater and Justice in Arid Environments*, M.I.T. Press, Cambridge, MA, U.S.A.

Moench, M., 2003, Groundwater and poverty: Exploring the connections, in: *Intensive Use of Groundwater*, R. Llamas and E. Custodio, eds.,. Balkema Publishers, The Netherlands, pp. 441–455.

Mukherji, A., 2006, Is intensive use of groundwater a solution to World's water crisis?, in: *Water Crisis: Myth or Reality?*, Rogers et al., eds., Balkema Publishers, The Netherlands, pp. 181–195.

Polak, P., 2005a, Water and the other four revolutions needed to end rural poverty, *Water Science and Technology* **8**:133–146.

Polak, P., 2005b, The big potential of small farms, *Scientific American*, September 2005, 84–91.

Puri, S., and El Naser, H., 2003, Intensive use of groundwater in transboundary aquifers, in: *Intensive Use of Groundwater*, R. Llamas and E. Custodio, eds., Balkema Publishers, The Netherlands, pp. 415–440.

Ragone, S., and Llamas, M. R., 2006, Reader's forum, *Ground Water* **44**(4):500–503.

Ragone, S., De La Hera, A., Hernández-Mora, N., Plumbey, T., Bergkamp, G., Mckay, J., 2007, The global importance of groundwater in the 21st century, in: *Proceedings of the International Symposium on Groundwater Sustainability*, January, 24–27, 2006, Alicante, Spain, National Ground Water Association Press, Westerville, U.S.A., pp. 1–382.

Rogers, P., 2006, Water governance, water security and water sustainable, in: *Water Crisis: Myth or Reality?* Rogers et al., eds., Balkema Publishers, The Netherlands, pp. 43–58.

Rogers, P., Jalal, K. F., and Boyd, J. A., 2006, *An Introduction to Sustainable Development*, Harvard University Press, Cambridge, MA, U.S.A., 404 pp.

Sachs, J. F., 2005, Can extreme poverty be eliminated? *Scientific American*, September 2005, 56–65.

Sala-Martin, X., 2006, The world distribution of income: falling poverty and convergence, *Quarterly Journal of Economics* **121**(2):351–397.

Schlager, A., and López-Gunn, E., 2006, Collective systems for water management: Is the tragedy of common a myth, in: *Water Crisis: Myth or Reality?*, Rogers et al., eds., Balkema Publishers, The Netherlands, pp. 43–58.

Selborne, J., 2001, *The Ethics of Freshwater Use: A Survey*, COMEST-UNESCO, París, 58 pp.

Shah, T., Singh, O. P., and Mukherji, A., 2006, Some aspects of South Asia's groundwater irrigation economy analyses from a survey in India, Pakistan, Nepal Terai and Bangladesh, *Hydrogeology Journal* **14**(3):286–309.

Shamir, U., 2000, Sustainable management of water resources, *Transitive Towards Sustainability*, Interacademy Panel, Tokyo conference, May 2000, pp. 62–67.

Steenberger, F. van, and Shah, T., 2003, Rules rather then rights: Self-regulations in intensively used groundwater systems, in: *Intensive Use of Groundwater*, R. Llamas and E. Custodio, eds., Balkema Publishers, The Netherlands, pp. 241–256.

Valencia Statement, 2005, in: *Proceedings of the International Symposium on Groundwater Intensive Use*, Valencia, Spain 7–11 December 2002. Selected Papers in Hydrogeology. No. 7, Taylor and Francis, Ondon, pp. 385–386.

Vives, R., 2003, Economic and social posibility of water use for irrigation in Andalucía (Spain), *Water International* **28**(3):320–333.

Vrba, J., 2003, The impact of intensive groundwater use on groundwater quality, in: *Intensive Use of groundwater*, R. Llamas, and E. Custodio, eds., Balkema Publishers, The Netherlands, pp. 113–132.

ANNEX. The Alicante Declaration: The Global Importance of Ground Water and a Call for Action for its Responsible Use, Management and Governance (Ragone et al., 2007)

Water is essential for life. Groundwater—that part of all water resources that lie underneath land surface—constitutes more than 95% of the global, unfrozen freshwater reserves. Given its vast reserves and broad geographical distribution, its general good quality, and its resilience to seasonal fluctuations and contamination, groundwater holds the promise to ensure current and future world communities an affordable and safe water supply. Groundwater is predominantly a renewable resource which, when managed properly, ensures a long-term supply that can help meet the increasing demands and mitigate the impacts of anticipated climate change. Generally, groundwater development requires a smaller capital investment that surface water development and can be implemented in a shorter timeframe.

Groundwater has provided great benefits for many societies in recent decades through its direct use as a drinking water source, for irrigated agriculture and industrial development, and indirectly, through ecosystem and streamflow maintenance. The development of groundwater often provides an affordable and rapid way to alleviate poverty and ensure food security. Further, by understanding the complementary nature of ground and surface waters, thoroughly integrated water resources management strategies can serve to foster their efficient use and enhance the longevity of supply.

Instances of poorly managed groundwater development and inadvertent impact of inadequate land use practices have produced adverse effects such as water quality degradation, impairment of aquatic ecosystems, lowered groundwater levels, and consequently, land subsidence and drying of wetlands. As it is less costly and more effective to protect groundwater resources from degradation than to restore them, improved water management will diminish such problems and save money.

Call For Action

To make groundwater's promise a reality requires the responsible use, management and governance of groundwater. In particular, actions need to be taken by water users, who sustain their well-being through groundwater abstraction; decision makers, both elected and non-elected; civil society groups and associations; and scientists who must advocate for the use of sound science in support of better management. To this end, the undersigners recommend the following actions:

- *Develop more comprehensive water management, land use and energy development strategies that fully recognize groundwater's important role in the hydrologic cycle;*

This requires better characterization of groundwater basins, their interconnection with surface water and ecosystems, and a better understanding of the response of the entire hydrologic system to natural and human-induced stresses. More attention should be given to non-renewable and saline groundwater resources when such waters are the only resource available for use.

- *Develop comprehensive understanding of groundwater rights, regulations, policy and uses;*

Such information, including social forces and incentives that drive present-day water management practices, will help in the formulation of policies and incentives to stimulate socially and environmentally sound groundwater management practices. This is particularly relevant in those situations where aquifers cross cultural, political or national boundaries.

- *Make the maintenance and restoration of hydrologic balance a long-term goal of regional water-management strategies;*

This requires that water managers identify options to: minimize net losses of water from the hydrologic system; encourage effective and efficient water use, and ensure the fair allocation of water for human use as well as ecological needs, taking into account long-term sustainability. Hydrological, ecological, economic and socio-economic assessments should be an integral part of any water management strategy.

- *Improve scientific, engineering and applied technological expertise in developing countries;*

This requires encouraging science based decision-making as well as "north–south" and "south–south" cooperation. Further, it is important that funds be allocated for programs that encourage the design and mass dissemination of affordable and low energy consuming water harnessing devices for household and irrigation.

- *Establish ongoing coordinated surface water and groundwater monitoring programs;*

This requires that data collection become an integral part of water management strategies so that such strategies can be adapted to address changing socio-economic, environmental and climatic conditions. The corresponding data sets should be available to all the stakeholders in a transparent and easy way.

- *Develop local institutions to improve sustainable groundwater management;*

This requires that higher-level authorities become receptive to the needs of local groups and encourage the development and support of strong institutional networks with water users and civic society.

- *Ensure that citizens recognize groundwater's essential role in their community and the importance of its responsible use.*

This requires that science and applied technology serve to enhance education and outreach programs in order to broaden citizen understanding of the entire hydrologic system and its global importance to current and future generations.

Chapter 21
Hydrodiplomacy and Environmental Security

Christophe J. G. Darnault*

Abstract This chapter presents the concept of hydrodiplomacy and environmental security. Integrated shared water resources management aims at sharing and optimizing water resources in a sustainable and secured environment. Its implementation to transboundary systems is challenged by continuous changes in hydrology and hydraulic systems, society, climate and the environment. These changes create a milieu of complex management, turbulent society and vulnerable resources. The depletion of water resources, overexploitation, droughts, pollution and socio-economic development has caused both conflicts and cooperation. Sharing water is crucial to meet the goals of equitable allocation and efficient management of water resources as well as environmental integrity and security. Hydrodiplomacy and water sharing principles are presented with a focus on conflicts prevention, management and solution approaches. Sustainable sharing of water resources encompasses environmental security because of regional and global environmental concerns which may lead to conflicts.

Keywords Hydrodiplomacy, environmental security, transboundary water resources, conflict, cooperation, vulnerability, sustainability, water sharing, water scarcity, global environment change

21.1 Introduction

The geopolitical nature of water leads to different social and ecological adaptations, as well as power to control and exploit resources (Vlachos, 1998). The depletion of water resources, overexploitation, droughts and socio-economic development has resulted not only in conflicts but also in cooperation. At the beginning of the twenty-first century, the traditional reactive approaches to crises have been replaced by

Christophe J. G. Darnault
University of Illinois at Chicago, Department of Civil and Materials Engineering, 842 W. Taylor St., ERF# 2071, Chicago, IL 60607, U.S.A.

* To whom correspondence should be addressed. E-mail: darnault@uic.edu

C.J.G. Darnault (ed.), *Overexploitation and Contamination of Shared Groundwater Resources.*

risk assessment and proactive strategies including anticipatory action and multi-stakeholders involvement to avoid conflict. The problems of sustainable development and integrated water resources management are increased through the interdependence and vulnerability of the transboundary water resources systems as well as a water-scarce and water-stressed environment (UN, 2005; UNESCO, 2006). Therefore, the development of intergovernmental integrations on hydrological, political, transboundary and exogenous interdependencies is essential.

Because of water resources use interdependencies, the spatial and temporal variations, the surface water and groundwater interactions, as well as the upstream and downstream differentiations, mechanisms for sustainable water sharing need to be created through hydrodiplomacy. Sustainable sharing of water resources should also integrate environmental security due to regional and global environmental concerns which may evolve in conflicts. The development of indicators to prevent and mitigate water conflicts should be associated with indicators assessing water sharing and cooperation.

21.2 Hydrodiplomacy

Hydrodiplomacy is mostly related to water sharing principles and doctrines which are present in legal and management approaches governing the exploitation of water resources in a complex milieu and transboundary environment.

21.2.1 Water Sharing Principles

The use of terms reasonable, equitable and sustainable in management and policy demonstrates the emphasis on water sharing as a public good. The right to water is recognized in legal and political instruments in which water access is guaranteed without discrimination, in a permanent and sustainable approach, and at a socially and economically acceptable cost. The issues of subsidiarity, solidarity and co-operation are also addressed as well as the interests of disadvantaged populations and the importance of decision-making at local levels (UNESCO, 2006). Values, norms and practices that provide a policy, legal and administrative framework for sharing water are established through water institutions. Water sharing is supported by a general principle of conduct in international law, treaties, binding acts and judgments of international courts that shape the regulations and procedures of shared and transboundary water resources systems. The main legal principles that shape hydrodiplomacy, including intra-state practices from the drafts of the Institute of International Law (IIL), International Law Association (ILA) and UN International Law Commission (UNILC) are (Vlachos, 1998; UNESCO, 2006):

- Principle of international water and international watercourse concept;
- Principle of reasonable and equitable utilization;
- Obligation not to cause significant harm and to exercise due diligence in the utilization of an international watercourse;
- Principle of negotiations;
- Duty to cooperate, including data and information exchange.

The draft articles prepared by UNILC on the use of transboundary aquifers have been developed in a similar concept than groundwater resources principles that aim the states to focus on the functions and integrity of the aquifer systems (Chapter 3). Yet, the EU Water Framework Directive (WFD) established a process for community action in water policy to strengthen the normative aspects (e.g. social preferences, goals and established practices) of valuing and sharing water with an emphasis on public participation (EC, 2000). Water resources laws are now being further elaborated. Declarations, international law organizations' drafts and the creation of the World Water Council aim to expand the legislation and to focus on integrated approaches for water sharing from national to international scales and from regional to global scales.

21.2.2 Water Sharing Indicators

Quantitative and qualitative indicators that measure the performance of shared water resources systems, monitor the process of equitable sharing and provide the mechanisms for monitoring both the current state and changes in transboundary water systems need to be developed (MEA, 2003). Specific criteria to select these indicators include reliability of analytical tools, measurability through accurate data and information and utility to users for simple interpretation showing trends over time, response to change, as well as comparison of threshold or reference values. UNESCO (2006) has categorized the indicators in different dimensions as below.

- Operational/administrative: it is used for water sharing interdependencies. Indicators include number of international basins and transboundary aquifers; dependency on inflow from other river basins; impacts on upstream water diversions and impoundments, impacts on groundwater ecosystems; upstream and downstream integrative mechanisms; systematic considerations of water user and use interdependencies; high water stress, water scarcity, poverty conditions; basin-wide operational water planning and management; surface and groundwater conjunctive use; number of treaties, cooperative events.
- Cooperation/conflict: it focuses on institutional mechanisms and conflict resolution efforts. Indicators include existing conflict accommodation and resolution mechanisms; significant number of water treaties or conventions; economic, scientific or industrial agreements; unilateral projects, highly

centralized water projects; existence of laws and regulations for equitable water allocation; stakeholders involvement and participation mechanisms; publication of joint inventories of transboundary resources; effectiveness of community-based management.

- Vulnerability/fragility: it emphasizes the volatility and turbulence of the environment and society, including environmental security, risks and inability to adapt to threats and disasters. Indicators include rivalries and disputes within and between countries and regions; ratio of water demand to water supply; environmental and social fragility; decreasing water quality, degraded groundwater dependent ecosystems; poverty and lack of sanitary conditions; extreme hydrological events and periodic water disasters; demand changes and distribution; dependence on hydroelectricity. Fragility, volatility and carrying capacity have also become indicators of conflict or cooperation in shared water resources.
- Sustainability/development: it characterizes not only the cleavages in expectations and achievements, but also the current preoccupation with balances between environment, economy and society by addressing the issues of growth and carrying capacity as well as survival and fulfillment. Indicators include expressed and implemented water conservation measures; competence for dealing with and managing water conflicts; desire for and implementation of balanced environmental policies; capacity to recover water projects costs; integrated Water Resources Management (IWRM) (Chapter 15); and importance of virtual water in food trade. Virtual water is also an indicator of water sharing and therefore of water security. The concept of virtual water was first defined by Allan (2003) as the water embedded in commodities. In terms of globalization, not only it raises awareness about water interdependencies, but also it serves as a mean for improving water efficiency.
- Socio-economic: indicators of income distribution, environmental damages, effects on cultural heritage, freedom of action, resilience, adaptability, environmental integrity and security, and integrated water resources management (IWRM) should also be considered.

Specific indicators of groundwater resources include total resource, recharge rates, total abstraction, depletion rates, risks and composite measurements of conjunctive water use.

21.2.3 Water Vulnerability

Water resources are linked to vulnerability which is a function of economic and political conditions, water availability, population growth, climate variability and the extent to which a source of water supply is shared (UNESCO, 2006). Extreme hydrological events, institutional problems and growing populations are exacerbating problems in water-scarce regions. These regions are characterized by three particular vulnerabilities (UNESCO, 2006):

- Ecological vulnerability: arid regions and regions of limited resources;
- Economic vulnerability: concerned with past practices of traditional exploitation and state economics;
- Social vulnerability: overexploitation of resources, complex social economic and ecological forces affecting a region natural equilibrium.

21.2.4 Doctrines in Transboundary Water Resources

The doctrines present in legal and management approaches that govern the exploitation by States of shared and transboundary water resources systems are (Buck, et al., 1993; Vlachos, 1998; Wolf, 1999):

21.2.4.1 Absolute Territorial Sovereignty

Absolute territorial sovereignty is one of the most extreme doctrine of water rights. The doctrine of absolute territorial sovereignty relies on hydrography and implies unilateral control over waters within its territory (Giordano and Wolf, 2001). This doctrine is usually the first claim by upstream riparians during a treaty negotiation, and has rarely been used in actual treaties and has never been invoked in an international law (Wolf, 1999).

21.2.4.2 Absolute Territorial Integrity

The doctrine of absolute territorial integrity is also one of the most extreme doctrines of water rights and is the first bargaining position for downstream riparians who will emphasize the importance of historical use and the right to water flowing through their territories (Giordano and Wolf, 2001). The doctrine of absolute sovereign integrity has rarely been applied in treaties and international law (Wolf, 1999).

21.2.4.3 Limited Territorial Sovereignty

The doctrine of limited territorial sovereignty is a moderate doctrine in terms of water rights. It recognizes the right to reasonable and equitable use of international waters while inflicting no significant harm to other riparian states (Giordano and Wolf, 2001). It also ensures downstream states with a reasonable share of water in reasonable conditions (Vlachos, 1998). Upstream riparians are mostly concerned with reasonable and equitable use while downstream riparians are primary concerned with the no significant harm clause (Wolf, 1999).

21.2.4.4 Community Co-riparian States

The doctrine of community co-riparian states is a moderate water rights doctrine. The doctrine of community co-riparian states refers to a community of interests in the watershed development across state boundaries (Giordano and Wolf, 2001).

21.2.5 Etiology of a Complex Milieu

The UN publication on the Comprehensive Management of the Freshwater Resources of the World (UN CSD, 1997) established the need for people to change not only their thoughts on water, but also their approaches to water resources management. It results that new comprehensive, participatory and environmentally sound policies need to be developed. States and regions are focusing on streamlining existing administrative mechanisms and implementing innovative institutional measures with regard to both quantitative and qualitative aspects of their water resources. Hydropolitics is therefore essential not only because of water scarcity in highly populated regions, but also because of transboundary water resources systems. In this complex milieu, the etiological categories of concern are according to Vlachos (1998):

- Ecosystemic changes: land degradation, natural resources, ecosystemic hazards, global warming and climate change, soil exhaustion, desertification or sea level rise;
- Human-induced disasters: industrial disasters and the effects of projects such as dams, environmental destruction, resource exploitation and degradation, deterioration of major watersheds and deforestation;
- Eco-political disruption: shifting, fading and disputed frontiers, confrontations, revolutions and wars, collapse of regimes, nationalism;
- Socio-economic dislocations: collapse of expectations, shifts in international economy, wrong development strategies, water scarcity, access to resources and social turmoil.

Based on this multifaceted milieu, a new governance of water resources should reflect new judicial norms, flexible institutions, demand-driven water policies, water concepts—blue and green water, or virtual water—as well as sustainability, transparency and public participation (UNESCO, 2006). Conflict prevention and interdependence concepts are essential to sustainable water resources sharing. The development of international agreements is difficult due to cooperation obstacles such as the increasing split between the North and the South, the persistence of national sovereignty and the lack of sufficient incentives for nations to negotiate. However, legal approaches can be strengthened by various flexible mechanisms such as (Vlachos, 1998; UNESCO, 2006):

- Hydrodiplomacy or environmental diplomacy;
- Alternative Dispute Resolution (ADR);
- Technical/professional or independent panels of experts;
- Public awareness, participation and mobilization.

The operational terms for international law and management mechanisms of transboundary water resources systems are complementarity in their implementation (Vlachos, 1998, UNESCO, 2006). However, the problems encountered by multilateral agencies include historical and cultural inertia of past differences and practices, calculation of all costs involved in shared water resources development and incorporation of social and environmental concerns related to effective and efficient water sharing. These problems accentuate the need for negotiation, third party expertise and constructive dialogue based on information and data exchange.

The agreements between parties resulting from hydrodiplomacy integrate comprehensive national planning, watershed management, multi-purpose development and water quality control. Flexible and durable regimes which are capable of not only enhancing ecosystem protection, but also serving the interests of all parties involved through reduction of uncertainty, stabilization of expectations and promotion of conflict resolution or conflict management, need to be created (Vlachos, 1998).

21.2.6 Hydrodiplomacy: To Prevent, Manage and Resolve Water Shared Conflicts

Water resources are unequally distributed and water scarcity and abundance are affected by policy, exploitation, management and climate. The water distribution and availability result in demographic changes and uneven development, which are sources of socio-economic differentiations. The scarcity of water is replacing that of oil as a flashpoint for conflict between high-risk countries (Gleditsch, 1997). Like water, ecological degradation and political instability can lead to conflicts or cooperation. Competition for water demands exists also between its uses: urban versus rural, present uses versus future demands, competing regions, water quantity versus water quality and water concerns versus social priorities. The various types of conflicts encountered were characterized by UNESCO (2006) as:

- Direct conflicts originated from competing and conflicting demands;
- Indirect conflicts resulting from migration, environmental or seasonal demands;

- Structural conflicts due to a socio-economic context with limited institutional and social capacity, fragmented authority, transboundary interdependencies and insufficient public participation.

21.2.6.1 Conflicts

Water conflicts encompass issues ranging from access to needed water supply to deliberate attacks on water resources. As population increases, urban and economic development requires more water for agricultural, municipal and industrial uses, and the risk of water conflicts increases. Simultaneously, water scarcity and water availability are becoming water barriers below which constraints to development arise (Falkenmark, 1999). Water scarcity and climate change are factors that will affect water resources availability and therefore intensify regional water conflicts. UNESCO (2006) reported that sources of potential water conflicts are:

- Permanent and temporary water scarcity;
- Different goals and objectives;
- Complex social and historical factors;
- Misunderstandings or ignorance of circumstances and data;
- Asymmetric power between localities, regions or nations;
- Significant data gaps or questions of validity and reliability;
- Specific hydropolitical issues at stake;
- Non-cooperative settings and value conflicts (e.g. water mythology, culture and water symbolism).

Water conflicts are constituted of three phases (UNESCO, 2006): creation, management and resolution. The conflict creation promotes diagnosis, anticipation and prevention, as well as joint search for information and data. The conflict management represents a trust-building stage through mediation, arbitration and neutral expert search for information and data. The conflict resolution includes consensus-building and depolarization of conflicting interests through public search for information and data or adjudication.

21.2.6.2 Alternative Dispute Resolution

Alternative dispute resolution is an essential element to manage and resolve conflicts (UNESCO, 2006). Alternatives to legal institutions for arbitration have been developed not only because of the saturation of legal mandates, but also because of the increasing litigation and confrontation cases. Mediation, a compromised discussion between disputants managed by a third party, is a viable alternative to adversarial processes. Adjudication, arbitration, conciliation and principled negotiation constitute also valuable alternative processes of dispute resolution. Public participation and negotiation strengthen agreements through the desirability and interests of particular outcomes. The outcome of conflicts for water regimes depend on several parameters such as, number of actors involved, external factors, preventive cooperation and information and data availability to all parties concerned, administrative and historical animosities. These parameters enhance the importance of demographic, social, economical and political pressures with resource scarcity and environmental degradation in water resources management and conflict resolution.

21.2.6.3 Capacity-building for Cooperation and Crisis Avoidance

Water is a catalyst for cooperation (UNESCO, 2006). Cooperative agreements for shared water resources, institution-building, comprehensive management and alternative dispute resolution efforts aim at dealing with future sources of conflict. According to UNESCO (2006), the future sources of conflict are likely to combine internal and external elements and to be linked to natural resources degradation and depletion, political confrontations, land-use and climate-based environmental changes.

The combination of the competitive demands and stakeholders interests as well as the evolving need for political compromise and the proactive movement to avoid conflicts have resulted in a shift from confrontation to cooperation, from monologue to dialogue and from dissent to consensus (UNESCO, 2006). Therefore, mechanisms for cooperation and crisis avoidance include collaboration and public involvement (UNESCO, 2006). Collaboration addressing environmental disputes involves three phases: problem setting (problem architecture), direction setting (problem negotiations) and implementation (systematic management of inter-organizational relations and monitoring of agreements). A proactive public can lead to conflict management and increased consensus, and in international cases, a reinforcement of the notions of transnational commons. Public awareness is a one way information to alert community on sensitive issues, while public involvement is a two way communication engaging community members in information and data exchange as well as dialogue (UNESCO, 2006). Public participation is the most intense form of interaction between authorities, experts and citizens, and implies shared leadership, joint planning and power delegation. Also, the broader the base of public participation is, the more is potential influence on managing, avoiding and resolving conflict there is.

21.3 Environmental Security

Integrated shared water resources management aims at sharing and optimizing water resources in a sustainable and secured environment (Chapter 15). The implementation of integrated shared water resources management to transboundary systems is challenged by continuous changes in hydrology and hydraulic systems, society, climate and the environment. These changes create a milieu of complex management, turbulent society and vulnerable resources. Sharing water is crucial to meet the goals of equitable allocation and efficient management of water resources as well as environmental integrity and security.

21.3.1 Security Concept

Security was introduced by Cicero (106 BC–43 BC) and Lucretius (ca. 99 BC–ca. 55 BC) in reference to a philosophical and psychological state of mind. Security as a political value is related to individual or societal value systems (Brauch 2003). The perception of security threats, challenges, vulnerabilities and risks depends on world-views and views of policy-makers (Brauch 2003, 2005). Three concepts have been distinguished by the English school (Bull 1977, Wight 1991): Hobbesian pessimism where power is the key category; Kantian optimism where international law and human rights are crucial and Grotian pragmatism where cooperation is vital (Brauch 2003, 2004). From an American perspective, Snyder (2004) distinguished three rival theories: realism, liberalism and idealism.

It is important to note that not only the scope of security has changed, but also the referent object has shifted from a national to a human security concept, both within the UN system and in the academic community (UNDP 1994; Waever 1997; UNESCO 1997, 1998; UNU 2002, Brauch, 2005). Security studies focus not only on military security, environmental change, but also on the concepts of environmental and human security and their linkages (Brauch, 2005). From a realist Hobbesian approach, the environmental and human security challenges are not perceived as threats and are often non-existing. From a pragmatic Grotian perspective, the environmental security challenges expose the societal vulnerability leading to a survival dilemma for those with a high degree of societal vulnerability which may be most seriously affected by natural or man-made environmental hazards (Brauch 2002, 2004). From a Kantian view, the liberal perspective of environmental treaties and regimes pose obligations for governments and people. Since 1990, gradually a fundamental reconceptualization of security has emerged (Ullman, 1983; Mathews, 1989; Myers, 1989, 1994; Buzan, et al., 1998; Abdus Sabur, 2003; Brauch, 2005; Brauch et al., 2006). Moller (2001, 2003) distinguishes a national and three expanded concepts of societal, human and environmental security. Oswald (2001) added gender security and introduced a human and gender security concept. Bogardi (2004) and Brauch (2003) suggested focusing the human security concept on the environmental dimension, especially on the interactions between the individual or humankind as the cause and victim of factors of global environmental change, both in anthropogenic and natural variability contexts. The UN system (position V) may be described as Grotian pragmatism in terms of security and as equity oriented pragmatic in terms of environmental perspectives where cooperation is important and is required to solve problems (Ohlsson, 1999, Brauch, 2005).

21.3.2 Security in a Changing Environment

Global environmental change encompasses changes in nature and society affecting humankind. Its impacts will depend upon human responses ranging from the actions of individuals to the creation of multilateral environmental agreements and the reactions of the global civil society. It will affect human who are both cause and victim of this change (UN, 2005; UNESCO, 2006). However, those who have caused global environmental change and those who are the most vulnerable to and affected by it are not always the same. Global environmental change affects and combines the ecosphere and anthroposphere. The interaction between processes in the ecosphere and anthroposphere can be represented in a survival hexagon of three resource challenges (Brauch 2002, 2003, 2005): air (climate change), land (soil, ecosystem degradation) and water (scarcity, degradation, floods) and three social challenges: human population (growth, changes of its value systems), urban systems (services, industries, pollution, health) and rural systems (securing food and fiber). These six factors interact differently and contribute to environmental scarcity of soil, water and food, intensifying environmental degradation and leading to environmental stress under specific socio-economic conditions both in national and international contexts (Brauch, 2005).

The nature and human induced factors of Global Environmental Change (GEC) may contribute, trigger or intensify ethnic, religious or political conflicts and lead to violence or need for peacemaking. Four different socio-economic scenarios of the complex interplay of the structural causes have occurred (UNESCO, 2006):

- Domestic societal conflicts;
- Resource and border conflicts (Klare 2001);
- Regional violence with different security perceptions in South and North;
- Militarization of non-military causes of conflicts.

The political process on the intra and international level has contributed to increase human mobility within the South and migration from the South to the North which result in tensions and internal or regional crises (UNESCO, 2006). These may lead to either successful crises resolution through cooperation, or to crises evolving into conflicts at the national or international level.

21.4 Conclusion

Water sharing is challenged by the quest for hydrodiplomacy and environmental security. The development of cooperative water agreements integrating strategic issues through hydrodiplomacy for the prevention, management and solution of water conflicts is crucial. Previously, military conflicts were the main concern, now environmental disasters needs to be included. Security incorporates contested water resources that, according to the UN, are issues of environmental and human security involving human life and dignity. Water conflicts and hydrodiplomacy, not only address strategic socio-economic and political issues, but also environmental interventions that affect claims on water resources and water dependent ecosystems. Reducing the risk of water resources conflicts and promoting equitable water sharing require regional and international hydrodiplomacy. As new paradigms, hydrodiplomacy and environmental security will be the critical issues of the twenty-first century for human security.

References

Abdus Sabur, A. K. M., 2003, Evolving a theoretical perspective on human security: The south Asian context, in: *Human Security in South Asia: Energy, Gender, Migration and Globalization*, P. R. Chari, and S. Gupta, eds., Social Science Press, New Delhi, pp. 35–51.

Allan, J. A., 2003, Virtual water—the water, food, and trade nexus: Useful concept or misleading metaphor? *Water International* **28**(1):4–11.

Bogardi, J., 2004, Hazards, risks and vulnerabilities in a changing environment: The unexpected onslaught on human security? *Global Environmental Change* **14**:361–365.

Brauch, H. G., 2002, A survival pact for the Mediterranean: Linking "virtual water" and "virtual sun", in: *Johannesburg and Beyond. Towards Concrete Action,* R. K. Pachauri, and G. Vasudeva, eds., Teri, New Delhi, pp. 151–190.

Brauch, H. G., 2003, Security and environment linkages in the Mediterranean: Three phases of research on human and environmental security and peace, in: *Security and Environment in the Mediterranean. Conceptualising Security and Environmental Conflicts*, H. G. Brauch, P. H. Liotta, A. Marquina, P. Rogers, and M. E. Selim, eds., Springer, Berlin-Heidelberg, pp. 35–143.

Brauch, H. G., 2004, From a Hobbesian security to a Grotian survival dilemma, in: *Peace and Conflict in a Time of Globalization*, Proceedings of the 40th Anniversary Conference of IPRA, Sopron, Hungary 5–9 July, 2004, (June 21, 2007); http://www.afes-press.de/pdf/Sopron_Survival%20Dilemma.pdf.

Brauch, H. G., 2005, *Environment and Human Security, Towards Freedom from Hazard Impacts*, United Nations University, Institute for Environment and Human Security, Interdisciplinary Security Connections, Publication Series, **No.2**, 60 pp.

Brauch, H. G., Grin, J., Mesjasz, C., Dunay, P., Behera, N. C., Chourou, B., Oswald Spring, U., Liotta, P. H., and Kameri-Mbote, P., eds., 2006, *Globalization and Environmental Challenges: Reconceptualizing Security in the 21st Century*, Springer-Verlag, Berlin.

Buck, S. J., Gleason, G. W., and Jofuku, M. S., 1993, The institutional imperative: Resolving transboundary water conflict in arid agricultural regions of the United States and the Commonwealth of independeant states, *Natural Resources Journal.* **33**:595–628.

Bull, H., 1977, *The Anarchical Society. A Study of Order in World Politics*, Columbia University Press, New York; Macmillan, London.

Buzan, B., Waever, O., and Wilde, J., 1998, *Security. A New Framework for Analysis*, Boulder-London: Lynne Rienner.

European Commission, (EC), 2000, Directive 2000/60/EC of the European parliament and of the council of 23 October 2000 establishing a framework for community action in the field of water policy. *Official Journal of the European Communities*, Brussels, EC.
Falkenmark, M., 1999. Forward to the future: A conceptual framework for water dependence, *Ambio* **28**(4):356–361.
Giordano, M. A., and Wolf, A. T., 2001, Incorporating equity into international water agreements, *Social Justice Research* **14**(4):349–366.
Gleditsch, N. P., ed., 1997, *Conflict and the Environment*, Kluwer Academic, Dordrecht.
Klare, M. T., 2001, *Resource Wars: The New Landscape of Global Conflict*, Henry Holt and Metropolitan Books, New York, U.S.A.
Mathews, J. T., 1989, Redefining Security, *Foreign Affairs* **68**(2):162–177.
Millennium Ecosystem Assessment, (MEA), 2003, *Ecosystems and Human Well-Being*, Island Press and World Resources Institute, Washington D.C., U.S.A.
Moller, B., 2001, National, societal and human security: General discussion with a case study from the Balkans, in: *What Agenda for Human Security in the Twenty-first Century,* UNESCO, ed., First International Meeting of Directors of Peace Research and Training Institutions, UNESCO, Paris, France, pp. 41–62.
Moller, B., 2003, National, societal and human security: Discussion—A case study of the Israeli-Palestine conflict, in: *Security and Environment in the Mediterranean. Conceptualizing Security and Environmental Conflicts*, H. G. Brauch, P. H. Liotta, A. Marquina, P. Rogers, and M. E. Selim, eds., Springer, Berlin-Heidelberg, Germany, pp. 277–288.
Myers, N., 1989, Environment and security, *Foreign Policy* **74**:23–41.
Myers, N., 1994, *Ultimate Security. The Environmental Basis of Political Stability*, W. W. Norton, New York, U.S.A.
Ohlsson, L., 1999, *Environment, Scarcity, and Conflict—A Study of Malthusian Concerns*, Department of Peace and Development Research, Gothenburg University.
Oswald Spring, U., 2001, Sustainable development with peace building and human security, in: *Our Fragile World. Challenges and Opportunities for Sustainable Development*, M. K. Tolba, ed., Forerunner to the Encyclopedia of Life Support Systems, Eolss Publishers, Oxford, U.K., **1**:873–916.
Snyder, J., 2004, One world, rival theories, *Foreign Policy* **145**:53–62.
Ullman, R., 1983, Redefining security, *International Security* **8**(1):129–153.
United Nations, (UN), 2005, *Third Report on Shared Natural Resources*, UN Doc A/CN.4/551
United Nations Commission on Sustainable Development, (UN CSD), 1997, *Comprehensive Assessment of the Freshwater Resources of the World*, Economic and Social Council, Fifth session, 5–25 April, E/CN. 171997/9.
UNDP, 1994, *Human Development Report*, Oxford University Press, New York, U.S.A.
United Nations Educational Scientific and Cultural Organization, (UNESCO), 1997, *Insecurity—Culture of Peace, International Symposium, From Partial Insecurity to Global Security*, in: Proceedings, UNESCO Headquarters, 12–14 June 1996, UNESCO, Paris, France.
United Nations Educational Scientific and Cultural Organization, (UNESCO), 1998, *What Kind of Security?* UNESCO, Paris, France.
United Nations Educational Scientific and Cultural Organization, (UNESCO), 2006, *Water A Shared Responsibility*, The United Nations World Water Development Report 2, UNESCO, ed., UNESCO, Paris, France and Berghahn, New York, U.S.A., 601 pp.
United Nations University, (UNU), 2002, *Advancing Knowledge for Human Security and Development*, The UNU Strategic Plan 2002, UNU, Tokyo, Japan.
Vlachos, E., 1998, Practicing hydrodiplomacy in the 21st century, *Water Resources Update* **111**:76–82.

Waever, O., 1997, Self-referential concepts of security as an instrument for reconstruction of an open-ended realism in IR, in: Concepts of Security, O. Waever, ed., Copenhagen: Institute of Political Science, University of Copenhagen, pp. 347–373.

Wight, M., 1991, *International Theory. The Three Traditions*, G. Wight and B. Porter, eds., Leicester University Press, Leicester-London, U.K.

Wolf, A. T., 1999, Criteria for equitable allocations: The heart of international water conflict, *Natural Resources Forum* **23**:3–30.

INDEX

GPSR Compliance

The European Union's (EU) General Product Safety Regulation (GPSR) is a set of rules that requires consumer products to be safe and our obligations to ensure this.

If you have any concerns about our products, you can contact us on ProductSafety@springernature.com

In case Publisher is established outside the EU, the EU authorized representative is:

Springer Nature Customer Service Center GmbH
Europaplatz 3
69115 Heidelberg, Germany

Batch number: 10755284

Printed by Printforce, the Netherlands